A TEXTBOOK OF
VECTOR CALCULUS

A TEXTBOOK OF
VECTOR CALCULUS

(With Applications)

For the Students of B.Sc. and B.A. (Pass & Honours) Courses

REVISED EDITION

SHANTI NARAYAN

Former Dean of Colleges
and Principal, Hans Raj College
University of Delhi
Delhi

Revised by

P.K. MITTAL

M.Sc., Ph.D.

Head of Mathematics Department
Government Post Graduate College, Rishikesh
Uttarakhand

S. CHAND
PUBLISHING

S Chand And Company Limited

(ISO 9001 Certified Company)

S. CHAND
PUBLISHING

S Chand And Company Limited

(ISO 9001 Certified Company)

Head Office: D-92, Sector–2, Noida – 201301, U.P. (India), Ph. 91-120-4682700
Registered Office: A-27, 2nd Floor, Mohan Co-operative Industrial Estate, New Delhi – 110 044 | Phone: 011-49731800
www.**schandpublishing.com**; e-mail: **info@schandpublishing.com**

Marketing Offices:

Chennai : Ph: 23632120; chennai@schandpublishing.com
Guwahati : Ph: 2738811, 2735640; guwahati@schandpublishing.com
Hyderabad : Ph: 40186018; hyderabad@schandpublishing.com
Jalandhar : Ph: 4645630; jalandhar@schandpublishing.com
Kolkata : Ph: 23357458, 23353914; kolkata@schandpublishing.com
Lucknow : Ph: 4003633; lucknow@schandpublishing.com
Mumbai : Ph: 25000297; mumbai@schandpublishing.com
Patna : Ph: 2260011; patna@schandpublishing.com

First Edition 1955
Subsequent Editions and Reprints 1958, 62, 66, 69, 72, 75, 76, 82, 83, 85, 86, 91, 94, 95, 96, 2003, 2006, 2009, 2012, 2014 (Twice), 2016, 2017 (Twice), 2019, 2020,2021, 2022

Reprint 2024

ISBN: 978-81-219-0161-1 **Product Code:** H6VCL61MATH10ENZX0XO

PRINTED IN INDIA

By Vikas Publishing House Private Limited, Plot 20/4, Site-IV, Industrial Area Sahibabad, Ghaziabad – 201 010 and Published by S Chand And Company Limited, A-27, 2nd Floor, Mohan Co-operative Industrial Estate, New Delhi – 110 044.

PREFACE TO THE REVISED EDITION

The book originally written about 45 years ago, has during the intervening period been revised and re-printed several times. Due to the changing trends of University studies and the demand of the students a thorough revision of the book was overdue. I very humbly took the challenge of revising a perfect work of Dr. Shanti Narayan and tried to meet the demands, of those who always loved and liked the books of Dr. Shanti Narayan.

To meet out the changing trends in the course structure of different Indian Universities, the book of thorough revised and a large number of solved examples have been given in each topic to meet the need of the students.

The book, in the present form, is an humble effort to make it more useful to the students and teachers. I owe my special gratitude to Sri Ravindra Kumar Gupta, Managing Director, S. Chand & Company Ltd., for giving me opportunity to revise the books of late Dr. Shanti Narayan, an eminent Indian Mathematician.

I also acknowledge my sincere thanks to Sri Navin Joshi, General Manager (Sales & Marketing) and Shri Dharmendra Jha, editor, M/s. S. Chand & Company Ltd., for providing necessary assistance in this revision work.

As the need for improvement is never ending, I will look forward to receive valuable and useful suggestions from our users for further improvements in the book.

Rishikesh **P.K. Mittal**

PREFACE

The present book is a companion volume to *Vector Algebra* by one of us and deals with the theory and applications of Vector Calculus of Differential Geometry and various branches of Applied Mathematics.

Chapter I of the book deals with vector functions of one and more scalar variables and introduces in a systematic and rigorous manner the notions of limit, continuity and derivability in relation to the same. A section on Indefinite integration is also given. Chapters II and III deal with Differential geometry and consider curves as well as surfaces. Cartesian Formulation of the various results arrived at is also given.

Chapter IV and V on Mechanics special attention has been paid to *Moving axes* and the *Kinematics* of a rigid body involving the existence of Angular velocity of a rigid body moving about a fixed point and reduction to *Screw Motion* in the general case have been given a systematic treatment. Euler's equations of motion have been obtained. The various problems in relation to Dynamics of a particle have also been dealt with.

Chapter VI deals with gradient, divergence and curl of point functions in terms of cartesian rectangular as well as curvilinear co-ordinate axes.

Chapter VII deals with different types of line, surface and volume integrals and the two integral transformation theorems of Gauss and Stokes.

In Chapter VIII on Hydrodynamics the notion of pressure in relation to perfect fluids as a scalar point function has been arrived at and the equation of continuity and Euler's equation of motion have been obtained. The equations for the steady irrotational motion of incompressible fluids have been deduced.

The final chapter on Electromagnetic Theory covers most of the topics in Electrostatics, Magnetism and Electromagnetism included in the post-graduate courses. The two final sections deal with Maxwell's equations and electromagnetic waves.

The text has been throughout profusely illustrated by means of a large number of solved examples. A large number of exercises has also been given.

In all applications it has been our constant endeavour to see that the treatment is systematic and as much self-contained as possible.

It is earnestly hoped that as a result of the study of the book a serious student will be enabled to acquire facility and confidence in the use of vector calculus so that he could employ the same in an effective manner to various applications of a suitable nature.

The exercises needing answers are such that they require only direct application of formulas and accordingly, taking into account the possible standard of the readers of the book, the answers have not been given.

We are grateful to our esteemed friend Dr. P.L. Bhatnagar, D. Sc., who kindly went through a part of the manuscript and made a number of useful suggestions.

The authors would gratefully welcome notices of mistakes and suggestions for improvements.

Delhi **AUTHOR**

A TEXT BOOK OF VECTOR CALCULUS (with applications)

First publication : 1958 — *Latest edition:*

The book has been written jointly with Prof. J.N. Kapur, Department of Mathematics, Indian Institute of Technology, Kanpur.

It is a companion volume to 'Vector Algebra' and deals with the theory and applications of vector calculus and differential geometry and various branches of Applied Mathematics. They text has been throughout profusely illustrated by means of a large number of solved examples. A large number of exercises has also been given.

CONTENTS

1

Vector Valued Functions of Scalar Variables

Introduction. In this chapter, we shall introduce the concepts of Vector Valued Function of a single and of several scalar variables and those of their limits, continuity and derivability. Basic properties of derivatives will also be obtained. In addition to these, the notion of *'Indefinite integral'* of a vector valued function of a single scalar variable will also be introduced.

The notion of *'Derivatives'* is a development of the concept of *'Instantaneous Rate of Change'*.

It will be seen that the treatment of fundamentals in this chapter is a good deal analogous to that of Scalar Calculus and in fact a number of theorems of this latter subject will be used to prove the corresponding theorems in Vector Calculus.

It should be noted that the symbols **i, j, k** denote a fixed triad of mutually orthogonal unit vectors.

1.1. VECTOR VALUED FUNCTIONS OF A SINGLE SCALAR VARIABLE

If to each value of a scalar variable t in some interval $[a, b]$, there corresponds, by any law whatsoever, a value of a variable vector **r** then we say that **r** is a vector function of the scalar variable 't' defined in the interval $[a, b]$ and write

$$\mathbf{r} = \mathbf{f}\,(t),$$

where **f** denotes the law of correspondence, in question. Also as usual, $\mathbf{f}\,(c)$ denotes the particular vector which the law associates to the value 'c' of t.

In the following, we shall not, in general, mention the intervals of definitions of vector functions and it will always be understood that, for any given function, the values of the scalar variable which we consider are included in the interval of definition of the function.

Illustrations.

(*i*) If a particle moves in any manner so that at any given time t, it is at a point P whose position vector relative to an origin O is $\mathbf{r}$ then the vector $\mathbf{r}$ is a function of the scalar variable t. The velocity and acceleration of a moving particle are also vector functions of the time t which is scalar.

(*ii*) Analytically a curve can be given by expressing the position vector $\mathbf{r}$ of any point on the curve as a function of some scalar variable t known as parameter. In particular, we may take the arc length c of the point from some fixed point on the curve as the parameter.

Thus,

$$\mathbf{r} = a \cos t\, \mathbf{i} + b \sin t\, \mathbf{j} + 0\mathbf{k},$$

$$\mathbf{r} = at^2\, \mathbf{i} + 2at\, \mathbf{j} + 0\mathbf{k},$$

are the vector equations of an ellipse and a parabola respectively; a, b being constant scalars.

1.1.2. Decomposition of a Vector Functions

Decomposing the vector $\mathbf{f}\,(t)$ as a linear combination of **i, j, k,** we may write

$$\mathbf{f}\,(t) = f_1(t)\,\mathbf{i} + f_2(t)\,\mathbf{j} + f_3(t)\,\mathbf{k},$$

where $f_1(t)$, $f_2(t)$, $f_3(t)$ are three scalar functions of t.

1.2. LIMIT OF A VECTOR FUNCTION

$$\lim_{t \to c} \mathbf{f}(t) = \mathbf{1}.$$

Def. *A vector function* $\mathbf{f}\,(t)$ *is said to tend, as its limit, to a vector* $\mathbf{1}$ *when* t *tends to* c, *if, to any preassigned positive number* ε *however small, there corresponds a positive number* δ *such that*

$$|\,\mathbf{f}\,(t) - \mathbf{1}\,| < \varepsilon \text{ when } 0 < |\,t - c\,| \leq \delta.$$

If $\mathbf{f}\,(t)$ tends to a limit when t tends to c, we say that

$$\lim_{t \to c} \mathbf{f}(t)$$

exists.

Note. In the investigation concerning limits, we shall need the following results :

(*i*) If **a, b, c** be three mutually orthogonal vectors, then

$$|\,\mathbf{a}\,| < |\,\mathbf{a} + \mathbf{b} + \mathbf{c}\,|, \quad |\,\mathbf{b}\,| < |\,\mathbf{a} + \mathbf{b} + \mathbf{c}\,|, \quad |\,\mathbf{c}\,| < |\,\mathbf{a} + \mathbf{b} + \mathbf{c}\,|.$$

This follows from the fact that the length of a diagonal of a *rect-angular* parallelopiped exceeds that of any edge of the same.

(*ii*) For *any* three vectors **a, b, c,**

$$|\,\mathbf{a} + \mathbf{b} + \mathbf{c}\,| \leq |\,\mathbf{a}\,| + |\,\mathbf{b}\,| + |\,\mathbf{c}\,|.$$

This follows from the fact that the length of any diagonal of a parallelopiped is less than or equal to the sum of the lengths of its three coterminous edges.

(*iii*) $$|\mathbf{a}\,.\,\mathbf{b}| = |\mathbf{a}|\;|\mathbf{b}|\cos\theta\,| \le |\mathbf{a}|\;|\mathbf{b}|.$$

$$|\mathbf{a}\times\mathbf{b}| = |\mathbf{a}|\;|\mathbf{b}|\sin\theta\,| \le |\mathbf{a}|\;|\mathbf{b}|.$$

1.2.1. Theorem

If

(*i*) $lim\ \mathbf{f}(t) = \mathbf{1}$ *when* $t \to c$,

(*ii*) $\mathbf{f}(t) = f_1(t)\,\mathbf{i} + f_2(t)\,\mathbf{j} + f_3(t)\,\mathbf{k}$,

and (*iii*) $\mathbf{l} = l_1\mathbf{i} + l_2\mathbf{j} + l_3\mathbf{k}$,

then, when $t \to c$,

$$lim\ f_1(t) = l_1,\ \ lim\ f_2(t) = l_2,\ \ lim\ f_3(t) = l_3,$$

and conversely.

Let ε be any preassigned positive number.

Since

$$\lim_{t\to c} \mathbf{f}(t) = \mathbf{1},$$

there exists a positive number δ such that

$$|\mathbf{f}(t) - \mathbf{1}| < \varepsilon \text{ when } 0 < |t - c| \le \delta.$$

Thus, for

$0 < |t - c| \le \delta$, we have

$$\left|\overline{(f_1(t)\,\mathbf{i} + f_2(t)\,\mathbf{j} + f_3(t)\,\mathbf{k})} - \overline{(l_1\mathbf{i} + l_2\mathbf{j} + l_3\mathbf{k})}\right| < \varepsilon$$

or

$$|[f_1(t) - l_1]\,\mathbf{i} + [f_2(t) - l_2]\,\mathbf{j} + [f_3(t) - l_3]\,\mathbf{k} < \varepsilon.$$

Therefore, by (*i*) of the note after § 1.2, we have for $0 < |t - c| \le \delta$,

$$|f_1(t) - l_1| < \varepsilon,\ \ |f_2(t) - l_2| < \varepsilon,\ \ |f_3(t) - l_3| < \varepsilon.$$

Hence, when $t \to c$,

$$\lim\ f_1(t) = l_1,\ \ \lim\ f_2(t) = l_2,\ \ \lim\ f_3(t) = l_3.$$

To prove the converse, we suppose that, when $t \to c$,

$$\lim\ f_1(t) = l_1,\ \ \lim\ f_2(t) = l_2,\ \ \lim\ f_3(t) = l_3,$$

so that if ε be any preassigned positive number, there exists positive number δ_1, δ_2, δ_3 such that

$$|f_1(t) - l_1| < \frac{1}{3}\varepsilon \text{ when } 0 < |t - c| \le \delta_1,$$

$$|f_2(t) - l_2| < \frac{1}{3}\varepsilon \text{ when } 0 < |t - c| \le \delta_2,$$

$$|f_3(t) - l_3| < \frac{1}{3}\varepsilon \text{ when } 0 < |t - c| \le \delta_3.$$

Let

$$\delta = \min(\delta_1, \delta_2, \delta_3).$$

Then each of the three inequalities above holds for $0 < |t - c| \leq \delta$. Thus, $0 < |t - c| \leq \delta$,

$$|\mathbf{f}(t) - \mathbf{l}| = |[f_1(t) - l_1]\,\mathbf{i} + [f_2(t) - l_2]\,\mathbf{j} + [f_3(t) - l_3]\,\mathbf{k}|$$

$$\leq |f_1(t) - l_1| + |f_2(t) - l_2| + |f_3(t) - l_3|,$$

[Refer (*ii*) of note after § 1.1.2]

since

$$|\mathbf{i}| = |\mathbf{j}| = |\mathbf{k}| = 1.$$

Hence,

$$|\mathbf{f}(t) - \mathbf{l}| < \varepsilon \text{ when } 0 < |t - c| \leq \delta.$$

Thus,

$$\lim \mathbf{f}(t) = \mathbf{l} \text{ when } t \to c.$$

1.2.2. The Limiting Operation in Relation to Algebraic Operations

We shall now obtain the fundamental theorems on limits which correlate the limiting operation with algebraic operations.

Theorems. *If* $\mathbf{f}(t)$ *and* $\mathbf{F}(t)$ *are two vector functions of a scalar 't' such that, when* $t \to c$,

$$\lim \mathbf{f}(t) = \mathbf{l}, \text{ limit } \mathbf{F}(t) = \mathbf{m};$$

then

(*i*) $\lim [\mathbf{f}(t) \pm \mathbf{F}(t)] = \mathbf{l} \pm \mathbf{m}$,

(*ii*) $\lim [\mathbf{f}(t) \cdot \mathbf{F}(t)] = \mathbf{l} \cdot \mathbf{m}$,

(*iii*) $\lim [\mathbf{f}(t) \times \mathbf{F}(t)] = \mathbf{l} \times \mathbf{m}$.

Also if $\phi(t)$ *is any scalar function of 't' such that, when* $t \to c$,

$$\lim \phi(t) = p,$$

then

(*iv*) $\lim [\phi(t)\,\mathbf{f}(t)] = p\mathbf{l}$.

Finally,

(*v*) $\lim |\mathbf{f}(t)| = |\mathbf{l}|$.

The proofs as based on the corresponding theorems in Scalar Calculus and the theorem of § 1.2.1 above will be given here and independent proofs are given in the appendix to this chapter. The theorem in the Scalar Calculus which will be used is that the limit of the sum, difference and product of two functions is equal to the sum, difference and product of the corresponding limits of the functions.

Let

$$\mathbf{f}(t) = f_1(t)\,\mathbf{i} + f_2(t)\,\mathbf{j} + f_3(t)\,\mathbf{k},\quad \mathbf{F}(t) = F_1(t)\,\mathbf{i} + F_2(t)\,\mathbf{j} + F_3(t)\,\mathbf{k};$$

$$\mathbf{l} = l_1\mathbf{i} + l_2\,\mathbf{j} + l_3\mathbf{k},\quad \mathbf{m} = m_1\mathbf{i} + m_2\,\mathbf{j} + m_3\mathbf{k}.$$

As $\lim \mathbf{f}(t) = \mathbf{l}$, $\therefore$ $\lim f_1(t) = l_1$, $\lim f_2(t) = l_2$, $\lim f_3(t) = l_3$.

As $\lim \mathbf{F}(t) = \mathbf{m}$, $\therefore$ $\lim F_1(t) = m_1$, $\lim F_2(t) = m_2$, $\lim F_3(t) = m_3$.

$$\therefore\ \lim(f_1 \pm F_1) = l_1 \pm m_1,\ \lim(f_2 \pm F_2) = l_2 \pm m_2,$$
$$\lim(f_3 \pm F_3) = l_3 \pm m_3,$$

omitting t, here and subsequently.

$$\lim(f_1F_1) = l_1m_1,\quad \lim(f_2F_2) = l_2m_2,\quad \lim(f_3F_3) = l_3m_3$$

$\therefore$ by § 1.2.1,

(*i*) $$\lim[\mathbf{f}(t) \pm \mathbf{F}(t)] = (l_1 \pm m_1)\,\mathbf{i} + (l_2 \pm m_2)\,\mathbf{j} + (l_3 \pm m_3)\,\mathbf{k} = \mathbf{l} \pm \mathbf{m}.$$

(*ii*) $$\begin{aligned}\lim[\mathbf{f}(t) \cdot \mathbf{F}(t)] &= \lim(f_1F_1 + f_2F_2 + f_3F_3)\\ &= l_1m_1 + l_2m_2 + l_3m_3\\ &= (l_1\mathbf{i} + l_2\,\mathbf{j} + l_3\mathbf{k}) \cdot (m_1\mathbf{i} + m_2\,\mathbf{j} + m_3\mathbf{k})\\ &= \mathbf{l} \cdot \mathbf{m}.\end{aligned}$$

(*iii*) $$\begin{aligned}\lim[\mathbf{f}(t) \times \mathbf{F}(t)] &= \lim[(f_2F_3 - f_3F_2)\,\mathbf{i} + (f_3F_1 - f_1F_3)\,\mathbf{j} + (f_1F_2 - f_2F_1)\,\mathbf{k}]\\ &= (l_2m_3 - l_3m_2)\,\mathbf{i} + (l_3m_1 - l_1m_3)\,\mathbf{j} + (l_1m_2 - l_2m_1)\,\mathbf{k}\\ &= \mathbf{l} \times \mathbf{m}.\end{aligned}$$

(*iv*) $$\begin{aligned}\lim[\phi(t)\,\mathbf{f}(t)] &= \lim(\phi f_1\mathbf{i} + \phi f_2\,\mathbf{j} + \phi f_3\mathbf{k})\\ &= pl_1\mathbf{i} + pl_2\,\mathbf{j} + pl_3\mathbf{k}\\ &= p\mathbf{l}.\end{aligned}$$

(*v*) $$\begin{aligned}\lim|\mathbf{f}(t)| &= \lim\sqrt{f_1^2 + f_2^2 + f_3^2)}\\ &= \sqrt{(l_1^2 + l_2^2 + l_3^2)}\\ &= |\mathbf{l}|.\end{aligned}$$

1.3. CONTINUITY

Def. *A vector function* $\mathbf{f}(t)$ *is said to be continuous for a value c, if*

$$\lim_{t \to c} \mathbf{f}(t) = \mathbf{f}(c),$$

Also $\mathbf{f}(t)$ *is said to be a continuous function, if it is continuous for every value of the interval of definition of the function.*

With the help of the theorem of § 1.2.2, we can easily show that *the sum, difference, scalar product and vector product of two continuous functions are also continuous and the product of a continuous vector function with a continuous scalar function is also continuous.*

Thus, for example, let us take the case of vector product.

The functions $\mathbf{f}(t)$ and $\mathbf{F}(t)$ being continuous for t = c, we have

$$\begin{aligned}\lim_{t\to c}[\mathbf{f}(t)\times\mathbf{F}(t)] &= \lim_{t\to c}\mathbf{f}(t)\times\lim_{t\to c}\mathbf{F}(t)\\ &= \mathbf{f}(c)\times\mathbf{F}(c)\\ &= \text{value of } \mathbf{f}(t)\times\mathbf{F}(t) \text{ for } t = \text{c}.\end{aligned}$$

Hence, $\mathbf{f}(t) \times \mathbf{F}(t)$ is continuous for t = c.

With the help of the theorem of § 1.2.1, we at once see that if

$$\mathbf{f}(t) = f_1(t)\,\mathbf{i} + f_2(t)\,\mathbf{j} + f_3(t)\,\mathbf{k}$$

is continuous, then $f_1(t)$, $f_2(t)$ and $f_3(t)$ are also continuous scalar functions and conversely.

1.4. DERIVATIVE

Def. *A vector function* $\mathbf{f}(t)$ *of a scalar variable 't' is said to be derivable for a value c if*

$$\lim_{h\to 0}\frac{\mathbf{f}(c+h)-\mathbf{f}(c)}{h}$$

exists and the limit, called the derivative or *differential coefficient* of $\mathbf{f}(t)$ at c, is denoted by $\mathbf{f}'(t)$. The process is called derivation or differentiation.

We can also write

$$\mathbf{f}'(c) = \lim_{t\to c}\frac{\mathbf{f}(t)-\mathbf{f}(c)}{t-c},$$

A function is said to be derivable if it is derivable for every value of the interval of definition of the function.

The derivative of a derivable function $\mathbf{f}(t)$ which is itself a vector function is denoted by the symbol

$$\mathbf{f}'(t) \text{ or } \frac{d\mathbf{f}}{dt},$$

For a derivable function $\mathbf{f}(t)$, we have

$$\frac{d\mathbf{f}}{dt} = \mathbf{f}'(t) = \lim_{h\to 0}\frac{\mathbf{f}(t+h)-\mathbf{f}(t)}{h},$$

the equality holding for every value of 't' belonging to the interval of definition of the function.

If as is usual in Scalar Calculus, we denote the changes in t and $\mathbf{f}(t)$ by δt and $\delta\mathbf{f}$ respectively, we have

$$\mathbf{f}'(t) = \frac{d\mathbf{f}}{dt} = \lim \frac{\delta \mathbf{f}}{\delta t}.$$

1.4.1. Differentials

The relation

$$d\mathbf{f} = \mathbf{f}'(t)\, dt.$$

We can write

$$\frac{\delta \mathbf{f}}{\delta t} = \frac{\mathbf{f}\,(t+\delta t) - \mathbf{f}(t)}{\delta t} = \mathbf{f}'(t) + \mathbf{h},$$

where $\mathbf{h}$ is a vector which $\to 0$ as $\delta t \to 0$.

Thus, we have

$$\delta \mathbf{f} = \delta t \mathbf{f}'(t) + \mathbf{h}\delta t.$$

As in the scalar calculus the change δt in the independent variable is called the differential of t denoted by dt and principal part of the change in $\delta \mathbf{f}$. *viz.*, $\delta t \mathbf{f}'(t)$ is called the differential of $\mathbf{f}$ denoted by $d\mathbf{f}$. Thus, in terms of our definitions of differentials,

$$d\mathbf{f} = \mathbf{f}'(t)\, dt.$$

1.4.2. Derivability in Relation to Continuity

Theorem. *Every derivable function is continuous,*

For any value 'c' of t, we have

$$\mathbf{f}(t) - \mathbf{f}(c) = (t-c)\frac{\mathbf{f}(t) - \mathbf{f}(c)}{t-c},$$

so that

$$\lim_{t \to c}\, [\mathbf{f}(t) - \mathbf{f}(c)] = \lim\, (t-c)\frac{\mathbf{f}(t) - \mathbf{f}(c)}{t-c}$$

$$= 0\mathbf{f}'\,(c) = 0.$$

Hence,

$$\lim_{t \to c}\, \mathbf{f}(t) = \mathbf{f}(c),$$

so that $\mathbf{f}\,(t)$ is continuous at c. At c may be any value, we see that *every derivable function is continuous.*

Note. The converse of the above theorem is not true. Thus

$$\mathbf{f}\,(t) = |\, t \,|\, \mathbf{i},$$

is continuous for every value of t but not derivable for $t = 0$. In fact, we have

$$|\, \mathbf{f}\,(t) - \mathbf{f}\,(0) \,| = |\, |\, t \,|\, \mathbf{i} - 0 \,| = |\, t \,|$$

so that

$$\lim_{t \to 0} \mathbf{f}(t) = 0 = \mathbf{f}(0).$$

Also

$$\frac{\mathbf{f}(t) - \mathbf{f}(0)}{t - 0} = \frac{|t|\,\mathbf{i}}{t}$$

so that the limit is **i** or – **i** according as t tends to zero through positive or through negative values. Hence **f**′(0) does not exist.

1.4.3. Theorem

If

$$\mathbf{f}(t) = f_1(t)\,\mathbf{i} + f_2(t)\,\mathbf{j} + f_3(t)\,\mathbf{k},$$

is a derivable function, then $f_1(t)$, $f_2(t)$, $f_3(t)$ *are also derivable functions and*

$$\frac{d\mathbf{f}}{dt} = \frac{df_1}{dt}\mathbf{i} + \frac{df_2}{dt}\mathbf{j} + \frac{df_3}{dt}\mathbf{k}.$$

Also conversely.

We have

$$\frac{\mathbf{f}(t+h) - \mathbf{f}(t)}{h} = \frac{f_1(t+h) - f_1(t)}{h}\mathbf{i} + \frac{f_2(t+h) - f_2(t)}{h}\mathbf{j} + \frac{f_3(t+h) - f_3(t)}{h}\mathbf{k}. \quad ...(1)$$

so that by the theorem of § 1.2.1,

$$\lim_{h \to 0} \frac{\mathbf{f}(t+h) - \mathbf{f}(t)}{h},$$

exists, if and only if, the three limits

$$\lim_{h \to 0} \frac{f_1(t+h) - f_1(t)}{h}, \ \lim_{h \to 0} \frac{f_2(t+h) - f_2(t)}{h}, \ \lim_{h \to 0} \frac{f_3(t+h) - f_3(t)}{h},$$

all exist. Also if these limits exist, we have, from (1),

$$\frac{d\mathbf{f}}{dt} = \frac{df_1}{dt}\mathbf{i} + \frac{df_2}{dt}\mathbf{j} + \frac{df_3}{dt}\mathbf{k}. \quad (\S\ 1.2.1)$$

1.5. DERIVABILITY IN RELATION TO ALGEBRAIC OPERATIONS

Theorem. *If* **f** (t) *and* **F** (t) *are two derivable functions of t then*

$$\mathbf{f}(t) \pm \mathbf{F}(t), \ \mathbf{f}(t) \,.\, \mathbf{F}(t), \ \mathbf{f}(t) \times \mathbf{F}(t).$$

are also derivable and

(*i*) $$\frac{d(\mathbf{f} \pm \mathbf{F})}{dt} = \frac{d\mathbf{f}}{dt} \pm \frac{d\mathbf{F}}{dt}.$$

(*ii*) $$\frac{d(\mathbf{f} \cdot \mathbf{F})}{dt} = \mathbf{f} \cdot \frac{d\mathbf{F}}{dt} + \mathbf{F} \cdot \frac{d\mathbf{f}}{dt}.$$ *(Kolkata, 97)* (*iii*) $$\frac{d\,(\mathbf{f} \times \mathbf{F})}{dt} = \mathbf{f} \times \frac{d\mathbf{F}}{dt} + \frac{d\mathbf{f}}{dt} \times \mathbf{F}.$$ (*Kolkata, 97*)

Also if $\phi(t)$ *is a derivable scalar function, then* $\phi(t)\,\mathbf{f}(t)$ *is also derivable and*

(*iv*) $$\frac{d\,(\phi\mathbf{f})}{dt} = \mathbf{f}\,\frac{d\mathbf{F}}{dt} + \frac{d\phi}{dt}\,\mathbf{f}.$$

The order of factors in the case of the derivative of vector product is *not* to be changed.

(*i*) Let

$$\mathbf{G}(t) = \mathbf{f}(t) \pm \mathbf{F}(t).$$

$$\therefore \quad \frac{\mathbf{G}(t+h) - \mathbf{G}(t)}{h} = \frac{\mathbf{f}(t+h) - \mathbf{f}(t)}{h} \pm \frac{\mathbf{F}(t+h) - \mathbf{F}(t)}{h}.$$

Taking limits, when $h \to 0$, we have, by § 1.2.1

$$\frac{d\,(\mathbf{f} \pm \mathbf{F})}{dt} = \frac{d\mathbf{G}}{dt} = \frac{d\mathbf{f}}{dt} \pm \frac{d\mathbf{F}}{dt}.$$

(*ii*) Let

$$g(t) = \mathbf{f}(t) \cdot \mathbf{F}(t).$$

$$\therefore \quad \frac{g(t+h) - g(t)}{h} = \frac{\mathbf{f}(t+h) \cdot \mathbf{F}(t+h) - \mathbf{f}(t) \cdot \mathbf{F}(t)}{h}$$

$$= \mathbf{f}(t+h) \cdot \frac{\mathbf{F}(t+h) - \mathbf{F}(t)}{h}$$

$$+ \mathbf{F}(t) \cdot \frac{\mathbf{f}(t+h) - \mathbf{f}(t)}{h}$$

Taking limits, (§ 1.2.1), we obtain

$$\frac{d\,(\mathbf{f} \cdot \mathbf{F})}{dt} = \frac{dg}{dt} = \mathbf{f} \cdot \frac{d\mathbf{F}}{dt} + \mathbf{F} \cdot \frac{d\mathbf{f}}{dt}.$$

Here

$$\lim_{h \to 0} \mathbf{f}(t+h) = \mathbf{f}(t),$$

for, being derivable, the function $\mathbf{f}(t)$ is necessarily continuous.

(*iii*) Let

$$\mathbf{G}(t) = \mathbf{f}(t) \times \mathbf{F}(t).$$

$$\therefore \quad \frac{\mathbf{G}(t+h) - \mathbf{G}(t)}{h} = \frac{\mathbf{f}(t+h) \times \mathbf{F}(t+h) - \mathbf{f}(t) \times \mathbf{F}(t)}{h}$$

$$= \mathbf{f}(t+h) \times \frac{\mathbf{F}(t+h) - \mathbf{F}(t)}{h} + \frac{\mathbf{f}(t+h) - \mathbf{f}(t)}{h} \times \mathbf{F}(t)$$

Taking limits, (§ 9.2.1), we obtain

$$\frac{d(\mathbf{f} \times \mathbf{F})}{dt} = \frac{d\mathbf{G}}{dt} = \mathbf{f} \times \frac{d\mathbf{F}}{dt} + \frac{d\mathbf{f}}{dt} \times \mathbf{F}.$$

(*iv*) Let

$$\mathbf{G}(t) = \phi(t)\, \mathbf{f}(t).$$

$$\therefore \quad \frac{\mathbf{G}(t+h) - \mathbf{G}(t)}{t} = \frac{\phi(t+h)\, \mathbf{f}(t+h) - \phi(t)\, \mathbf{f}(t)}{h}$$

$$= \phi(t+h) \frac{\mathbf{f}(t+h) - \mathbf{f}(t)}{h} + \mathbf{f}(t) \frac{\phi(t+h) - \phi(t)}{h}$$

Taking limits, (§ 9.2.1), we obtain

$$\frac{d(\phi\mathbf{f})}{dt} = \phi \frac{d\mathbf{f}}{dt} + \mathbf{f} \frac{d\phi}{dt}.$$

Cor.

$$\mathbf{f} \cdot \frac{d\mathbf{f}}{dt} = |\mathbf{f}| \frac{d|\mathbf{f}|}{dt}$$

We have

$$\mathbf{f} \,.\, \mathbf{f} = |\mathbf{f}|^2,$$

so that by (*ii*) above

$$\mathbf{f} \cdot \frac{d\mathbf{f}}{dt} + \frac{d\mathbf{f}}{dt} \cdot \mathbf{f} = 2|\mathbf{f}| \frac{d|\mathbf{f}|}{dt}.$$

or

$$\mathbf{f} \cdot \frac{d\mathbf{f}}{dt} = |\mathbf{f}| \frac{d|\mathbf{f}|}{dt}.$$

1.5.1. Derivation of Triple Products

If **f**, **g**, **h** *be three derivable functions of t, then*

(*i*) $$\frac{d[\mathbf{f}\ \mathbf{g}\ \mathbf{h}]}{dt} = \left[\frac{d\mathbf{f}}{dt}\mathbf{g}\ \mathbf{h}\right] + \left[\mathbf{f}\ \frac{d\mathbf{g}}{dt}\mathbf{h}\right] + \left[\mathbf{f}\ \mathbf{g} \frac{d\mathbf{h}}{dt}\right].$$

(*Awadh 99, Kolkata 97*)

(*ii*) $$\frac{d[\mathbf{f} \times (\mathbf{g} \times \mathbf{h})]}{dt} = \frac{d\mathbf{f}}{dt} \times (\mathbf{g} \times \mathbf{h}) + \mathbf{f} \times \left(\frac{d\mathbf{g}}{dt} \times h\right) + \mathbf{f} \times \left(\mathbf{g} \times \frac{d\mathbf{h}}{dt}\right).$$

(*i*) We have

$$\frac{d[\mathbf{f}\ \mathbf{g}\ \mathbf{h}]}{dt} = \frac{\mathbf{d}(\mathbf{f} \times \mathbf{g} \,.\, \mathbf{h})}{dt}$$

$$= \frac{d(\mathbf{f} \times \mathbf{g})}{dt} \cdot \mathbf{h} + (\mathbf{f} \times \mathbf{g}) \cdot \frac{d\mathbf{h}}{dt} \qquad \text{[By (ii) of § 1.5]}$$

$$= \left(\frac{d\mathbf{f}}{dt} \times \mathbf{g} + \mathbf{f} \times \frac{d\mathbf{g}}{dt}\right) \cdot \mathbf{h} + (\mathbf{f} \times \mathbf{g}) \cdot \frac{d\mathbf{h}}{dt} \qquad \text{[By } (iii) \text{ of § 1.5]}$$

$$= \frac{d\mathbf{f}}{dt} \times \mathbf{g} \cdot \mathbf{h} + \mathbf{f} \times \frac{d\mathbf{g}}{dt} \cdot \mathbf{h} + (\mathbf{f} \times \mathbf{g}) \cdot \frac{d\mathbf{h}}{dt}$$

$$= \left[\frac{d\mathbf{f}}{dt} \mathbf{g}\ \mathbf{h}\right] + \left[\mathbf{f} \frac{d\mathbf{g}}{dt} \mathbf{h}\right] + \left[\mathbf{f}\ \mathbf{g} \frac{d\mathbf{h}}{dt}\right].$$

(*ii*) It can be established as above by repeated application of the result (*iii*) of § 1.1 above for the derivative of vector product.

1.5.2. Derivability of Function of a Function

Theorem. *If f (t) is a derivable vector function of t and t is a derivable scalar function* $\phi(u)$ *of u then* **f** *is a derivable function of u and*

$$\frac{d\mathbf{f}}{du} = \frac{d\mathbf{f}}{dt}\frac{dt}{du} = \frac{d\mathbf{f}}{dt}\frac{d\phi}{du}.$$

Consider corresponding values of u, t and let δu, δ', $\delta\mathbf{f}$ be the corresponding changes in u, t and **f**.

As $t = \phi(u)$ is a derivable function of u, we have

$$\delta t = \delta u\, \phi'(u) + k\, \delta u,$$

where $k \to 0$ as $\delta u \to 0$.

Again, as $\mathbf{f}(t)$ is a derivable vector function of t, we have

$$\delta\mathbf{f} = \delta t \mathbf{f}'(t) + \mathbf{h}\delta t, \text{[§ 1.4.1]} \qquad ...(2)$$

where $\mathbf{h} \to 0$ as $\delta t \to 0$.

From (1) and (2),

$$\delta\mathbf{f} = \delta u \mathbf{f}'(t)\, \phi'(u) + \mathbf{h}\delta u \phi'(u) + k\delta u \mathbf{f}'(t) + k\mathbf{h}\delta u$$

or
$$\frac{d\mathbf{f}}{du} = \mathbf{f}'(t)\phi'(u) + \mathbf{h}\phi'(u) + k\mathbf{f}'(t) + k\mathbf{h}.$$

Thus,
$$\lim \frac{\delta\mathbf{f}}{\delta u} = \mathbf{f}'(t)\phi'(u),$$

i.e.,
$$\frac{d\mathbf{f}}{du} = \frac{d\mathbf{f}}{dt}\frac{dt}{du}.$$

1.6. CONSTANT VECTORS

Vectors with constant magnitude or with constant direction. A vector will change if either its magnitude changes or direction changes or both direction and magnitude change. We will now find criteria for a vector function to remain constant in magnitude or in direction or in both.

1.6.1. Constant Vectors

Theorem. *The necessary and sufficient condition for* $\mathbf{f}(t)$ *to be constant is*

$$\frac{d\mathbf{f}}{dt} = \mathbf{0}.$$

If $\mathbf{f}$ is constant then

$$\mathbf{f}(t+h) - \mathbf{f}(t) = \mathbf{0},$$

for every h and accordingly

$$\frac{d\mathbf{f}}{dt} = \lim_{h \to 0} \frac{\mathbf{f}(t+h) - \mathbf{f}(t)}{h} = \mathbf{0}.$$

Suppose now that the derivative is zero. Let

$$\mathbf{f}(t) = f_1(t)\,\mathbf{i} + f_2(t)\,\mathbf{j} + f_3(t)\,\mathbf{k},$$

so that

$$\mathbf{0} = \frac{d\mathbf{f}}{dt} = \frac{df_1}{dt}\mathbf{i} + \frac{df_2}{dt}\mathbf{j} + \frac{df_3}{dt}\mathbf{k},$$

$$\therefore \qquad \frac{df_1}{dt} = 0, \quad \frac{df_2}{dt} = 0, \quad \frac{df_3}{dt} = 0,$$

so that from Scalar Calculus, $f_1(t)$, $f_2(t)$ and $f_3(t)$ are constant scalars.

Hence, $\mathbf{f}(t)$ is constant.

1.6.2. Vectors with Constant Magnitude.

Theorem. *The necessary and sufficient condition for* $\mathbf{f}(t)$ *to have constant magnitude is*

$$\mathbf{f} \cdot \frac{d\mathbf{f}}{dt} = 0. \qquad \textit{(Kumaon, 99)}$$

We have

$$\mathbf{f}(t) \,.\, \mathbf{f}(t) = |\,\mathbf{f}(t)\,|^2,$$

so that

$$\mathbf{f} \cdot \frac{d\mathbf{f}}{dt} = |\,\mathbf{f}\,| \frac{d\,|\,\mathbf{f}\,|}{dt}.$$

Thus

$$\mathbf{f} \cdot \frac{d\mathbf{f}}{dt} = 0,$$

if and only if, $\dfrac{d\,|\,\mathbf{f}\,|}{dt}$ is zero, *i.e.*, $|\,\mathbf{f}\,|$ is constant for, without any loss of generality we may suppose that $\mathbf{f}$ is not constantly zero.

1.6.3. Vectors with Direction Constant

Theorem. *The necessary and sufficient condition for* $\mathbf{f}(t)$ *to have constant direction is*

$$\mathbf{f} \times \frac{d\mathbf{f}}{dt} = \mathbf{0}. \qquad \textit{(Lucknow, 99)}$$

Let

$$\mathbf{f}(t) = \phi(t)\,\mathbf{F}(t),$$

where $\phi(t)$ denotes the magnitude of $\mathbf{f}(t)$ and accordingly $\mathbf{F}(t)$ is a vector function with modulus unity for every value of t.

$$\therefore \qquad \frac{d\mathbf{f}}{dt} = \phi\frac{d\mathbf{F}}{dt} + \frac{d\phi}{dt}\mathbf{F},$$

$$\mathbf{f} \times \frac{d\mathbf{f}}{dt} = \phi\mathbf{F} \times \left(\phi\frac{d\mathbf{F}}{dt} + \frac{d\phi}{dt}\mathbf{F}\right) = \phi^2\mathbf{F} \times \frac{d\mathbf{F}}{dt}. \qquad \text{...(1)}$$

Firstly suppose that the direction of **f** remains constant. Then **F** is constant and accordingly by § 1.6.1.

$$\frac{d\mathbf{F}}{dt} = \mathbf{0}.$$

Thus, in this case

$$\mathbf{f} \times \frac{d\mathbf{f}}{dt} = \mathbf{0}.$$

Suppose now that $\mathbf{f} \times \dfrac{d\mathbf{f}}{dt} = \mathbf{0}.$

$\therefore$ From (1),

$$\phi^2\mathbf{F} \times \frac{d\mathbf{F}}{dt} = \mathbf{0}.$$

Without any loss of generality we may suppose that ϕ is not constantly zero. Thus,

$$\mathbf{F} \times \frac{d\mathbf{F}}{dt} = \mathbf{0}. \qquad \text{...(2)}$$

Also **F** being of constant length we have, by § 1.6.2,

$$\mathbf{F} \cdot \frac{d\mathbf{F}}{dt} = \mathbf{0}.$$

From (2) and (3), we have $\dfrac{d\mathbf{F}}{dt} = 0. \qquad \text{...(3)}$

$\therefore$ by § 1.6.1,

F is constant,

i.e., the direction of **f** remains the same.

Note. The student would do well to compare the results obtained in the preceding three sub-sections.

1.7. DERIVATIVES OF SECOND AND HIGHER ORDERS

We suppose that the derivative $d\mathbf{f}/dt$ of the derivable function $\mathbf{f}(t)$ is itself a derivable function. Then the derivative

$$\frac{d}{dt}\left(\frac{d\mathbf{f}}{dt}\right),$$

denoted by

$$\frac{d^2\mathbf{f}}{dt^2},$$

is called the second order derivative of **f**.

Higher order derivatives

$$\frac{d^3\mathbf{f}}{dt^3}, \frac{d^4\mathbf{f}}{dt^4}, \text{ etc.}$$

may also be defined as in the Scalar Calculus.

1.8. INDEFINITE INTEGRATION

If $\mathbf{F}(t)$ be the derivative of a function $\mathbf{f}(t)$, we say that $\mathbf{f}(t)$ is an indefinite integral or simply integral of $\mathbf{F}(t)$ with respect to t and write

$$\int \mathbf{F}(t)\, dt = \mathbf{f}(t).$$

As in Scalar Calculus, indefinite integral of a function is not unique. In fact, *if* $\mathbf{f}(t)$ *be any indefinite integral of* $\mathbf{F}(t)$, *then* $\mathbf{f}(t) + \mathbf{c}$, *where* $\mathbf{c}$ *is any constant vector, is the general indefinite integral of* $\mathbf{F}(t)$.

Firstly we see that

$$\frac{d\,[\mathbf{f}(t)+\mathbf{c}]}{dt} = \frac{d\mathbf{f}(t)}{dt},$$

for the derivative of the constant **c** is zero.

Secondly, if $\mathbf{g}(t)$ be any other indefinite integral of $\mathbf{F}(t)$, besides $\mathbf{f}(t)$ so that

$$\frac{d\,[\mathbf{f}(t)+\mathbf{f}(t)]}{dt} = \frac{d\mathbf{g}(t)}{dt} - \frac{d\mathbf{f}(t)}{dt} = \mathbf{F}(t) - \mathbf{F}(t) = 0,$$

then, by § 1.6.1,

$$\mathbf{g}(t) - \mathbf{f}(t),$$

is a constant say, **c**. Thus,

$$\mathbf{g}(t) - \mathbf{f}(t) = \mathbf{c} \quad \text{or} \quad \mathbf{g}(t) = \mathbf{f}(t) + \mathbf{c}.$$

Hence, $\mathbf{g}(t)$ is of the form $\mathbf{f}(t) + \mathbf{c}$.

Note. The reader would be advised to bear in mind the following two useful results concerning indefinite integrals :

$$2\int \mathbf{f}\cdot\frac{d\mathbf{f}}{dt}\,dt = \mathbf{f}\cdot\mathbf{f} + c = \mathbf{f}^2 + c.$$

$$\int \mathbf{f}\times\frac{d^2\mathbf{f}}{dt^2}\,dt = \mathbf{f}\times\frac{d\mathbf{f}}{dt} + \mathbf{c}.$$

1.8.1. Definite Integral

If $\int \mathbf{f}(t)\,dt = \mathbf{F}(t)$, then $\mathbf{F}(b) - \mathbf{F}(a)$ is called the definite integral of $\mathbf{f}(t)$ between the limits $t = a$ and $t = b$ and is denoted as

$$\int_a^b \mathbf{f}(t)\,dt.$$

$$\therefore \quad \int_a^b \mathbf{f}(t)\,dt = \left[\mathbf{F}(t)\right]_a^b = \mathbf{F}(b) - \mathbf{F}(a).$$

EXAMPLES

Example 1. *A particle moves along the curve $x = 2t^2$, $y = t^2 - 4t$, $z = 3t - 5$, where t is the time. Find the component of its velocity and acceleration at time $t = 1$ in the direction* $\mathbf{i} - 3\mathbf{j} + 2\mathbf{k}$.

Solution. $\mathbf{r} = 2t^2\mathbf{i} + (t^2 - 4t)\,\mathbf{j} + (3t - 5)\,\mathbf{k}$

$$\therefore \quad \text{Velocity} = \frac{d\mathbf{r}}{dt} = 4t\,\mathbf{i} + (2t - 4)\,\mathbf{j} + 3\mathbf{k}$$

$$= 4\mathbf{i} - 2\mathbf{j} + 3\mathbf{k} \text{ at } t = 1.$$

we know that $\dfrac{\mathbf{A}\cdot\mathbf{B}}{|\mathbf{A}|}$ is the projection of $\mathbf{B}$ in the direction of $\mathbf{A}$.

Hence, component of velocity vector in the direction of $\mathbf{i} - 3\mathbf{j} + 2\mathbf{k}$

$$= \frac{(\mathbf{i} - 3\mathbf{j} + 2\mathbf{k})}{\sqrt{(1 + 9 + 4)}}\cdot(4\mathbf{i} - 2\mathbf{j} + 3\mathbf{k}) = \frac{16}{\sqrt{14}}.$$

Again acceleration $\dfrac{d^2\mathbf{r}}{dt^2} = 4\mathbf{i} + 2\mathbf{j}$

Hence, component of acceleration in the given direction is

$$\frac{\mathbf{i} - 3\mathbf{j} + 2\mathbf{k}}{\sqrt{14}}\cdot(4\mathbf{i} + 2\mathbf{j}) = -\frac{2}{\sqrt{14}}$$

Example 2. *If* $\mathbf{e}$ *is the unit vector making an angle θ with x-axis, show that $\dfrac{de}{d\theta}$ is a unit vector obtained by rotating* $\mathbf{e}$ *through a right angle in θ increasing direction.*

Solution. Since $\mathbf{e}$ is a unit vector making an angle θ with x-axis.

$$\mathbf{e} = \cos\theta\,\mathbf{i} + \sin\theta\,\mathbf{j} \qquad ...(1)$$

Replacing θ by $90° + \theta$,

$$\mathbf{f} = -\sin\theta\,\mathbf{i} + \cos\theta\,\mathbf{j} \qquad ...(2)$$

Clearly **f** is also a unit vector. Again

$$\frac{d\mathbf{e}}{d\theta} = -\sin\theta\,\mathbf{i} + \cos\theta\,\mathbf{j} = \mathbf{f}.$$

Example 3. *Evaluate* $\frac{d^2}{dt^2} = \left[\mathbf{r}\,\frac{d\mathbf{r}}{dt}\,\frac{d^2\mathbf{r}}{dt^2}\right]$

(*Kolkata 97, Kumaon 98*)

Solution.
$$\frac{d}{dt} = \left[\mathbf{r}\,\frac{d\mathbf{r}}{dt}\,\frac{d^2\mathbf{r}}{dt^2}\right] = \left[\frac{d\mathbf{r}}{dt}\,\frac{d\mathbf{r}}{dt}\,\frac{d^2\mathbf{r}}{dt^2}\right] + \left[\mathbf{r}\,\frac{d^2\mathbf{r}}{dt^2}\,\frac{d^2\mathbf{r}}{dt^2}\right] + \left[\mathbf{r}\,\frac{d\mathbf{r}}{dt}\,\frac{d^3\mathbf{r}}{dt^3}\right]$$

$$= \left[\mathbf{r}\,\frac{d\mathbf{r}}{dt}\,\frac{d^3\mathbf{r}}{dt^3}\right] \quad \text{as first two scalar triple products vanish}$$

$$\therefore \quad = \frac{d^2}{dt}\left[\mathbf{r}\,\frac{d\mathbf{r}}{dt}\,\frac{d^2\mathbf{r}}{dt^2}\right] = \frac{d}{dt}\left[\mathbf{r}\,\frac{d\mathbf{r}}{dt}\,\frac{d^3\mathbf{r}}{dt^3}\right]$$

$$= \left[\frac{d\mathbf{r}}{dt}\,\frac{d\mathbf{r}}{dt}\,\frac{d^3\mathbf{r}}{dt^3}\right] + \left[\mathbf{r}\,\frac{d^2\mathbf{r}}{dt^2}\,\frac{d^3\mathbf{r}}{dt^3}\right] + \left[\mathbf{r}\,\frac{d\mathbf{r}}{dt}\,\frac{d^4\mathbf{r}}{dt^4}\right]$$

$$= \left[\mathbf{r}\,\frac{d^2\mathbf{r}}{dt^2}\,\frac{d^3\mathbf{r}}{dt^3}\right] + \left[\mathbf{r}\,\frac{d\mathbf{r}}{dt}\,\frac{d^4\mathbf{r}}{dt^4}\right].$$

Example 4. *If* **r** = **a** *cos* ω*t* + **b** *sin* ω*t, show that*

$$\mathbf{r} \times \frac{d\mathbf{r}}{dt} = \omega\,\mathbf{a} \times \mathbf{b} \text{ and } \frac{d^2\mathbf{r}}{dt^2} = -\omega^2\mathbf{r} \qquad (\textit{Awadh, 98})$$

where **a**, **b** *and* ω *are constants.*

Solution.
$$\frac{d\mathbf{r}}{dt} = -\mathbf{a}\,\omega\sin\omega t + \mathbf{b}\,\omega\cos\omega t$$

$$\frac{d^2\mathbf{r}}{dt^2} = -\mathbf{a}\,\omega^2\cos\omega t - \mathbf{b}\,\omega^2\sin\omega t$$

$$= -\omega^2(\mathbf{a}\cos\omega t + \mathbf{b}\sin\omega t) = -\omega^2\mathbf{r}$$

$$\mathbf{r} \times \frac{d\mathbf{r}}{dt} = (\mathbf{a}\cos\omega t + \mathbf{b}\sin\omega t) \times (-\mathbf{a}\,\omega\sin\omega t + \mathbf{b}\,\omega\cos\omega t)$$

$$= \omega\,[\mathbf{a} \times \mathbf{b}\cos^2\omega t - \mathbf{b} \times \mathbf{a}\sin^2\omega t]$$

$$= \omega\,(\mathbf{a} \times \mathbf{b})(\cos^2\omega t + \sin^2\omega t) = \omega\,(\mathbf{a} \times \mathbf{b})$$

Example 5. *If* **r** *is a vector function of a scalar t and module of* **r** *be r, and* **a** *is a constant vector, then differentiate the following w.r.t t.*

$$(i)\ \frac{\mathbf{r}\times\mathbf{a}}{\mathbf{r}\cdot\mathbf{a}},\qquad (ii)\ \frac{r\,\mathbf{a}}{\mathbf{a}\times\mathbf{r}},$$

Solution. (*i*) $\frac{d}{dt}\left(\frac{\mathbf{r}\times\mathbf{a}}{\mathbf{r}\cdot\mathbf{a}}\right)=\frac{d}{dt}(\mathbf{r}\times\mathbf{a})\frac{1}{\mathbf{r}\cdot\mathbf{a}}+(\mathbf{r}\times\mathbf{a})\frac{d}{dt}\left(\frac{1}{\mathbf{r}\cdot\mathbf{a}}\right)$

$$=(\mathbf{r}\times\mathbf{a})\cdot\frac{1}{\mathbf{r}\cdot\mathbf{a}}+(\mathbf{r}\times\mathbf{a})\left\{-\frac{1}{(\mathbf{r}\cdot\mathbf{a})^2}(\mathbf{r}\cdot\mathbf{a})\right\}$$

$$=\frac{\mathbf{r}\times\mathbf{a}}{\mathbf{r}\cdot\mathbf{a}}-\frac{(\mathbf{r}\times\mathbf{a})(\mathbf{r}\cdot\mathbf{a})}{(\mathbf{r}\cdot\mathbf{a})^2}$$

$$=\frac{(\mathbf{r}\times\mathbf{a})(\mathbf{r}\cdot\mathbf{a})-(\mathbf{r}\times\mathbf{a})(\mathbf{r}\cdot\mathbf{a})}{(\mathbf{r}\cdot\mathbf{a})^2}$$

(*ii*) $\frac{d}{dt}\left(\frac{r\,\mathbf{a}}{\mathbf{a}\times\mathbf{r}}\right)=\frac{d}{dt}(r\,\mathbf{a})\cdot\frac{1}{\mathbf{a}\times\mathbf{r}}+(r\,\mathbf{a})\frac{d}{dt}\left(\frac{1}{\mathbf{a}\times\mathbf{r}}\right)$

$$=(r\,\mathbf{a}+0)\cdot\frac{1}{\mathbf{a}\times\mathbf{r}}+(r\,\mathbf{a})\left[-\frac{1}{(\mathbf{a}\times\mathbf{r})^2}\{(\mathbf{a}\times\mathbf{r})+0\}\right]$$

$$=\frac{r\,\mathbf{a}}{\mathbf{a}\times\mathbf{r}}-\frac{(r\,\mathbf{a})\cdot(\mathbf{a}\times\mathbf{r})}{(\mathbf{a}\times\mathbf{r})^2}$$

Example 6. *If* $\hat{\mathbf{r}}$ *denotes a unit vector, prove that*

$$\hat{\mathbf{r}}\times d\hat{\mathbf{r}}=(\mathbf{r}\times d\mathbf{r})/r^2\quad\textit{where } r=r\hat{\mathbf{r}}.$$

Solution. $\hat{\mathbf{r}}=\frac{\mathbf{r}}{r}\quad\therefore\quad d\hat{\mathbf{r}}=\frac{1}{r}d\mathbf{r}+\mathbf{r}\left(-\frac{1}{r^2}\right)dr$

$$\therefore\quad \hat{\mathbf{r}}\times d\hat{\mathbf{r}}=\hat{\mathbf{r}}\times\left[\frac{1}{r}d\mathbf{r}-\frac{\mathbf{r}}{r^2}dr\right]$$

$$=\frac{\mathbf{r}}{r}\times\left[\frac{1}{r}d\mathbf{r}-\frac{\mathbf{r}}{r^2}dr\right]=\frac{\mathbf{r}\times d\mathbf{r}}{r^2}.$$

Example 7. *Find* **r** *from the equation* $\frac{d^2\mathbf{r}}{dt^2}=\mathbf{a}t+\mathbf{b}$ *given that both* **r** *and* $\frac{d\mathbf{r}}{dt}$ *vanish when t = 0.*

Solution. Integrating the given relation w.r.t. t

$$\frac{d\mathbf{r}}{dt}=\frac{1}{2}\mathbf{a}t^2+\mathbf{b}t+\mathbf{c},$$

where **c** is a constant vector.

when $t = 0$, $\quad \dfrac{d\mathbf{r}}{dt} = 0; \quad \therefore \quad \mathbf{c} = 0$

$$\therefore \quad \frac{d\mathbf{r}}{dt} = \frac{1}{2}\mathbf{a}t^2 + \mathbf{b}t$$

Integrating, $\quad \mathbf{r} = \dfrac{1}{6}\mathbf{a}t^3 + \dfrac{1}{2}\mathbf{b}t^2 + \mathbf{d}$

when $t = 0$, $\mathbf{r} = 0$; $\quad \therefore \mathbf{d} = 0$

$$\therefore \quad \mathbf{r} = \frac{1}{6}\mathbf{a}t^3 + \frac{1}{2}\mathbf{b}t^2.$$

Example 8. *Given that* $\mathbf{r}(t) = 2\mathbf{i} - \mathbf{j} + 2\mathbf{k}$, *when* $t = 2$

$= 4\mathbf{i} - 2\mathbf{j} + 3\mathbf{k}$, *when* $t = 3$

show that $\displaystyle\int_2^3 \mathbf{r} \cdot \frac{d\mathbf{r}}{dt}\, dt = 10.$ *(Rohilkhand 98, Kumaon 2000)*

Solution. $\displaystyle\int_2^3 \mathbf{r} \cdot \frac{d\mathbf{r}}{dt}\, dt = \frac{1}{2}\left[r^2\right]_2^3$

$$= \frac{1}{2}\left[\left(r^2 \text{ at } t = 3\right) - \left(r^2 \text{ at } t = 2\right)\right]$$

$\mathbf{r}^2$ at $t = 2$ is $(2\mathbf{i} - \mathbf{j} + 2\mathbf{k})^2 = 4 + 1 + 4 = 9$

$\mathbf{r}^2$ at $t = 3$ is $(4\mathbf{i} - 2\mathbf{j} + 3\mathbf{k})^2 = 16 + 4 + 9 = 29$

$\therefore$ given integral $\dfrac{1}{2}[29 - 9] = 10.$

Example 1. *Evaluate* $\displaystyle\int \mathbf{a} \cdot \left(\mathbf{r} \times \frac{d^2\mathbf{r}}{dt^2}\right) dt$

Solution. We have $\dfrac{d}{dt}\left(\mathbf{r} \times \dfrac{d\mathbf{r}}{dt}\right) = \dfrac{d\mathbf{r}}{dt} \times \dfrac{d\mathbf{r}}{dt} + \mathbf{r} \times \dfrac{d^2\mathbf{r}}{dt^2} = \mathbf{r} \times \dfrac{d^2\mathbf{r}}{dt^2}$

$$\therefore \quad \int \left(\mathbf{r} \times \frac{d^2\mathbf{r}}{dt^2}\right) dt = \mathbf{r} \times \frac{d\mathbf{r}}{dt} + \mathbf{r} \times \frac{d^2\mathbf{r}}{dt^2} + \mathbf{c}.$$

where $\mathbf{c}$ is an arbitrary constant vector.

Now, $\quad \displaystyle\int \mathbf{a} \cdot \left(\mathbf{r} \cdot \frac{d^2\mathbf{r}}{dt^2}\right) dt = \mathbf{a} \cdot \int \mathbf{r} \times \frac{d^2\mathbf{r}}{dt^2}\, dt$

$$= \mathbf{a} \cdot \left[\mathbf{r} \times \frac{d\mathbf{r}}{dt} + \mathbf{c}\right] = \mathbf{a} \cdot \mathbf{r} \times \frac{d\mathbf{r}}{dt} + d$$

where $d = \mathbf{a} \,.\, \mathbf{c}$ is an arbitrary constant scalar.

Example 10. *The equation of motion of a particle P of mass m is given by* $m\left(d^2\mathbf{r}/dt^2\right)=f(r)\,\hat{\mathbf{r}}$, *where* $\mathbf{r}$ *is the position vector of P measured from an origin O,* $\hat{\mathbf{r}}$ *is a unit vector in the direction of* $\mathbf{r}$ *and* $f(r)$ *is a function of the distance of P from O, show that* $\mathbf{r}\times\left(\dfrac{d\mathbf{r}}{dt}\right)=\mathbf{c}$, *where* $\mathbf{c}$ *is a constant vector.*

Solution. We have

$$m\left(\frac{d^2\mathbf{r}}{dt^2}\right)=f(r)\,\hat{\mathbf{r}}$$

$$\Rightarrow \qquad m\mathbf{r}\times\left(\frac{d^2\mathbf{r}}{dt^2}\right)=f(r)\,\mathbf{r}\times\hat{\mathbf{r}}$$

$$\Rightarrow \qquad m\mathbf{r}\times\left(\frac{d^2\mathbf{r}}{dt^2}\right)=0 \;\Rightarrow\; \mathbf{r}\times\left(\frac{d^2\mathbf{r}}{dt^2}\right)=0 \qquad ...(1)$$

$$\left[\because\; \mathbf{r}\times\hat{\mathbf{r}}=\mathbf{r}\times\frac{\mathbf{r}}{r}=0\right]$$

But, $\dfrac{d}{dt}\left(\mathbf{r}\dfrac{d\mathbf{r}}{dt}\right)=\dfrac{d\mathbf{r}}{dt}\times\dfrac{d\mathbf{r}}{dt}+\mathbf{r}\times\dfrac{d^2\mathbf{r}}{dt^2}=\mathbf{r}\times\dfrac{d^2\mathbf{r}}{dt^2}$

Hence, (1) becomes

$$\frac{d}{dt}\left(\mathbf{r}\times\frac{d\mathbf{r}}{dt}\right)=0$$

Integrating, $\mathbf{r}\times\dfrac{d\mathbf{r}}{dt}=\mathbf{c}.$

EXERCISES

1. If

$$\mathbf{r}=\mathbf{a}\cos\omega t+\mathbf{b}\sin\omega t,$$

show that

$$\mathbf{r}\times\frac{d\mathbf{r}}{dt}=\omega\mathbf{a}\times\mathbf{b}.$$

and

$$\frac{d^2\mathbf{r}}{dt^2}=-\omega^2\mathbf{r};$$

a, b, ω being constants.

2. If

$$\mathbf{r}=a\cos t\,\mathbf{i}+a\sin t\,\mathbf{j}+at\tan\alpha\mathbf{k},$$

find

$$\left|\frac{d\mathbf{r}}{dt}\times\frac{d^2\mathbf{r}}{dt^2}\right| \text{ and } \left[\frac{d\mathbf{r}}{dt}\;\frac{d^2\mathbf{r}}{dt^2}\;\frac{d^2\mathbf{r}}{dt^2}\right].$$

(*Rohilkhand, 2000*)

3. Given $r = 4a \sin^3 \theta \mathbf{i} + 4a \cos^3 \theta \mathbf{j} + 3b \cos 2\theta \mathbf{k}$, find

$$\left| \frac{d\mathbf{r}}{d\theta} \times \frac{d^2\mathbf{r}}{d\theta^2} \right| \text{ and } \left[\frac{d\mathbf{r}}{d\theta} \; \frac{d^2\mathbf{r}}{d\theta^2} \; \frac{d^2\mathbf{r}}{d\theta^2} \right].$$

4. Evaluate

(i) $\frac{d}{dt}\left[\mathbf{r} \frac{d\mathbf{r}}{dt} \times \frac{d^2\mathbf{r}}{dt^2}\right]$ (ii) $\frac{d}{dt}\left\{\left(\mathbf{r} \times \frac{d\mathbf{r}}{dt}\right) \times \frac{d^2\mathbf{r}}{dt^2}\right\}$

(iii) $\frac{d^2}{dt^2}\left\{\left(\mathbf{r} \times \frac{d\mathbf{r}}{dt}\right) \times \frac{d^2\mathbf{r}}{dt^2}\right\}$.

$\mathbf{r}$ being a vector function of t.

5. Find the derivatives of the following :

(i) $\frac{\mathbf{r}}{r}$. (ii) $\mathbf{r}^n\mathbf{r}$.

(iii) $r^2 + \frac{1}{r^2}$. (iv) $\frac{\mathbf{r}}{\mathbf{a} \cdot \mathbf{r}} - \frac{\mathbf{r} \times \mathbf{a}}{r}$.

[Here r is the magnitude $|\mathbf{r}|$ of $\mathbf{r}$.]

6. A particle moves along the curve $x = t^3 + 1$, $y = t^2$, $z = 2t + 5$, where t is time. Find the components of its velocity and acceleration at $t = 1$ in the direction $\mathbf{i} + \mathbf{j} + 3\mathbf{k}$.

7. If $\mathbf{a} = \sin\theta\mathbf{i} + \cos\theta\mathbf{j} + \theta\mathbf{k}$, $\mathbf{b} = \cos\theta\mathbf{i} - \sin\theta\mathbf{j} - 3\mathbf{k}$ and $\mathbf{c} = \mathbf{i} + 2\mathbf{j} - 3\mathbf{k}$ then find $\frac{d}{d\theta}\{\mathbf{a} \times (\mathbf{b} \times \mathbf{c})\}$ at $\theta = \pi/2$.

8. If $\mathbf{r} = \mathbf{a}\, e^{nt} + \mathbf{b}\, e^{-nt}$ where $\mathbf{a}$ and $\mathbf{b}$ are constant vectors then prove that $\frac{d^2\mathbf{r}}{dt^2} = n^2\mathbf{r}$. *(Kumaon, 2000)*

9. If $\frac{d\mathbf{a}}{dt} = \mathbf{c} \times \mathbf{a}$ and $\frac{d\mathbf{b}}{dt} = \mathbf{c} \times \mathbf{b}$, then prove that

$$\frac{d}{dt} = (\mathbf{a} \times \mathbf{b}) = \mathbf{c} \times (\mathbf{a} \times \mathbf{b}).$$

(Ajmer 98, Rohilkhand 98)

10. A particle P is moving on a circle of radius $\mathbf{r}$ with constant angular velocity $\omega = d\theta/dt$. Show that the acceleration is $-\omega^2\mathbf{r}$.

11. If $\mathbf{r}(t) = 5t^2\mathbf{i} + t\mathbf{j} - t^3\mathbf{k}$, prove that

$$\int_1^2 \mathbf{r} \times \frac{d^2\mathbf{r}}{dt^2}\, dt = -14\mathbf{i} + 75\mathbf{j} - 15\mathbf{k}.$$

12. The acceleration of a particle at any time t is

$$e^t\mathbf{i} + e^{2t}\mathbf{j} + \mathbf{k},$$

find $\mathbf{v}$ given that $\mathbf{v} = \mathbf{i} + \mathbf{j}$ at $t = 0$.

13. Integrate $\mathbf{a} \times \dfrac{d^2\mathbf{r}}{dt^2} = \mathbf{b}$, where **a** and **b** are constant vectors.

14. Evaluate $\displaystyle\int_1^2 \mathbf{a} \cdot (\mathbf{b} \times \mathbf{c})\, dt$, whose

$\mathbf{a} = t\mathbf{i} - 3\mathbf{j} + 2t\mathbf{k}$, $\mathbf{b} = \mathbf{i} - 2\mathbf{j} + 2\mathbf{k}$, $\mathbf{c} = 3\mathbf{i} + t\mathbf{j} - \mathbf{k}$.

15. A particle moves with constant velocity **v**, starting from a point with position vector $\mathbf{r}_0$ at time $t = 0$. Given the position vector **r** of the particle at time t. Show that the particle will be closer to the co-ordinate origin at time $t = -(\mathbf{r}_0 \cdot \mathbf{v}) / (\mathbf{v} \cdot \mathbf{v})$.

16. Solve

$$\frac{d^2\mathbf{r}}{dt^2} = \mathbf{a},$$

where **a** is a constant vector; given that $\mathbf{r} = 0$ and $d\mathbf{r}/dt = 0$ when $t = 0$. Is the solution unique ? *(Rohilkhand 2000)*

17. Solve

$$\frac{d^2\mathbf{r}}{dt^2} = \mathbf{a}t + \mathbf{b},$$

where **a** and **b** are constant vectors, given that $\mathbf{r} = 0$ and $d\mathbf{r}/dt = \mathbf{u}$ when $t = 0$. Show that the solution is unique.

18. Show that

$$\mathbf{r} = \mathbf{a}e^{kt} + \mathbf{b}e^{lt},$$

is the solution of the linear differential equation

$$\frac{d^2\mathbf{r}}{dt^2} + p\frac{d\mathbf{r}}{dt} + q\mathbf{r} = 0, \qquad (p,\ q \text{ constant scalars})$$

where k, l are the distinct roots of the equation

$$m^2 + pm + q = 0,$$

and **a**, **b** are arbitrary constant vectors.

Also show that if the two roots of $m^2 + pm + a = 0$ are both equal then

$$(\mathbf{a} + l\mathbf{b})\, e^{kt}$$

is the solution of the given differential equation; k being the root.

19. The vector **r** satisfies the equation

$$m\frac{d^2\mathbf{r}}{dt^2} = e\mathbf{E} + \frac{e}{c}\frac{d\mathbf{r}}{dt} \times \mathbf{H},$$

where $\mathbf{E} = E\mathbf{j}$, $\mathbf{H} = H\mathbf{k}$ and e, m, c, E, H are constants.

Find the solution that satisfies the conditions

$$\mathbf{r} = 0, \quad \frac{d\mathbf{r}}{dt} = 0 \text{ when } t = 0.$$

20. If **r** is a unit vector, show that

$$\left| \mathbf{r} \times \frac{d\mathbf{r}}{dt} \right| = \left| \frac{d\mathbf{r}}{dt} \right|.$$

(Awadh 2000, Gorakhpur 2000, Lucknow 98)

1.9. VECTOR VALUED FUNCTIONS OF SEVERAL SCALAR VARIABLES

Like vector valued functions of a single scalar variable, we can also consider vector functions **f** $(u, v, w, ...)$ of several scalar variables u, v, w, ... We shall, however, generally attend to the developments in relation to functions of two variables only and the extension to the case of more than two variables can be made without any further difficulty.

Thus, we shall consider functions

$$\mathbf{f}(u, v),$$

so that to each pair of values of the scalar variables, u, v belonging to a given aggregate of pairs of the same, there corresponds a vector **f** (u, v).

Illustrations.

(*i*) If

$$\mathbf{r} = \mathbf{f}(u, v),$$

be thought of as the position vector of a point P then the locus of P for different pairs of values of u, v, is a surface and we say that $\mathbf{r} = \mathbf{f}(u, v)$ is the parametric vectorial equation or simply the equation of the surface.

(*ii*) If we consider any fluid in motion, then the velocity **q** of a fluid particle at any point and at any time is a vector function of four scalar variables x, y, z, t, where x, y, z are the co-ordinates of the point and t is the time.

1.10. DECOMPOSITION

Decomposition the vector **f** (u, v), as a linear combination of **i, j, k,** we have

$$\mathbf{f}(u, v) = f_1(u, v)\,\mathbf{i} + f_2(u, v)\,\mathbf{j} + f_3(u, v)\,\mathbf{k},$$

where f_1, f_2, f_3, are three scalar functions of the two variables u, v.

1.11. LIMIT OF A VECTOR FUNCTION

$$\lim_{(u, v)\to(c, d)} \mathbf{f}(u, v) = \mathbf{l}.$$

Def. *A vector function* **f** (u, v) *is said to tend to a limit* **l** *when u tends to c and v tends to d, if to every preassigned positive number* ε, *there corresponds a positive number* δ *such that*

$$|\mathbf{f}(u, v) - \mathbf{l}| < \varepsilon \text{ when } 0 < |u - c| \le \delta,\ \ 0 < |v - d| \le \delta.$$

As in the case of vector functions of a single variable (§ 1.2.1), we can show that *if*

(*i*) $\lim \mathbf{f}(u, v) = \mathbf{l}$ *when* $(u, v) \to (c, d)$, (*ii*) $\mathbf{f}(u, v) = f_1(u, v)\,\mathbf{i} + f_2(u, v)\,\mathbf{j} + f_3(u, v)\,\mathbf{k}$,

and (*iii*) $\mathbf{l} = l_1\mathbf{i} + l_2\mathbf{j} + l_3\mathbf{k}$,

when $(u, v) \to (c, d)$,

$$\lim f_1(u, v) = l_1, \quad \lim f_2(u, v) = l_2, \quad \lim f_3(u, v) = l_3$$

and conversely.

Again we can also easily establish theorems for vector valued functions of several variables corresponding to those for functions of a single variable given in § 1.2.2.

1.12. CONTINUITY

Def. A function $\mathbf{f}(u, v)$ is said to be continuous for $u = c$, $v = d$, if

$$\lim_{(u,v)\to(c,d)} \mathbf{f} = (u, v) = \mathbf{f}(c, d).$$

Again $\mathbf{f}(u, v)$ is said to be continuous function, if it is continuous for every pair of values u, v belonging to the region of definition of the function.

It can be easily shown (Refer § 1.11) that

$$\mathbf{f}(u, v) = f_1(u, v)\,\mathbf{i} + f_2(u, v)\,\mathbf{j} + f_3(u, v)\,\mathbf{k},$$

is continuous if, and only if,

$$f_1(u, v), \quad f_2(u, v), \quad f_3(u, v)$$

are continuous.

Again we may show that if $\mathbf{f}(u, v)$, $\mathbf{F}(u, v)$ are two continuous functions, then the functions

$$\mathbf{f} \pm \mathbf{F}, \quad \mathbf{f} \cdot \mathbf{F}, \quad \mathbf{f} \times \mathbf{F},$$

are also continuous.

1.13. PARTIAL DERIVATIVES

First order partial derivatives. If $\mathbf{f}(u, v)$ is any function, then

$$\lim_{h\to 0} \frac{\mathbf{f}(c+h, d) - \mathbf{f}(c, d)}{h}$$

if it exists, is called the partial derivatives of $\mathbf{f}(u, v)$ with respect to u for $u = c$, $v = d$.

Similarly

$$\lim_{h\to 0} \frac{\mathbf{f}(c, d+k) - \mathbf{f}(c, d)}{k}$$

if it exists, is the partial derivative of $\mathbf{f}(u, v)$ with respect to v for $u = c$, $v = d$.

The two partial derivatives are denoted by the symbols

$$\mathbf{f}_u(c, d), \quad \mathbf{f}_v(c, d)$$

respectively.

If we suppose that $\mathbf{f}(u, v)$ possesses partial derivatives with respect to u and v for every pair of values u, v then we obtain two vector functions denoted by

$$\frac{\partial \mathbf{f}}{\partial u}, \frac{\partial \mathbf{f}}{\partial v}$$

of (u, v) such that the value of $\partial \mathbf{f}/\partial u$ for $u = c, v = d$ is the partial derivative $\mathbf{f}$ with respect to u for $u = c, v = d$ and similarly for $\partial \mathbf{f}/\partial v$.

We may show as in § 1.4.3, that if

$$\mathbf{f}(u, v) = f_1(u, v)\,\mathbf{i} + f_2(u, v)\,\mathbf{j} + f_3(u, v)\,\mathbf{k},$$

then,

$$\frac{\partial \mathbf{f}}{\partial u} = \frac{\partial f_1}{\partial u}\mathbf{i} + \frac{\partial f_2}{\partial u}\mathbf{j} + \frac{\partial f_3}{\partial u}\mathbf{k},$$

and

$$\frac{\partial \mathbf{f}}{\partial v} = \frac{\partial f_1}{\partial v}\mathbf{i} + \frac{\partial f_2}{\partial v}\mathbf{j} + \frac{\partial f_3}{\partial v}\mathbf{k}.$$

1.13.1. Partial Derivatives of Second and Higher Orders

We suppose that each of the two vector functions

$$\frac{\partial \mathbf{f}}{\partial u}, \frac{\partial \mathbf{f}}{\partial v}$$

possesses partial derivatives w.r. to u and v. Then these partial derivatives, *viz.*,

$$\frac{\partial}{\partial u}\left(\frac{\partial \mathbf{f}}{\partial u}\right), \frac{\partial}{\partial v}\left(\frac{\partial \mathbf{f}}{\partial u}\right); \frac{\partial}{\partial u}\left(\frac{\partial \mathbf{f}}{\partial v}\right), \frac{\partial}{\partial u}\left(\frac{\partial \mathbf{f}}{\partial v}\right)$$

are denoted by the symbols

$$\frac{\partial^2 \mathbf{f}}{\partial u^2}, \frac{\partial^2 \mathbf{f}}{\partial v \partial u}, \frac{\partial^2 \mathbf{f}}{\partial v \partial u}, \frac{\partial^2 \mathbf{f}}{\partial v^2}$$

respectively and are called second order partial derivatives of $\mathbf{f}$. The two partial derivatives

$$\frac{\partial^2 \mathbf{f}}{\partial v \partial u}, \frac{\partial^2 \mathbf{f}}{\partial u \partial v},$$

involving inverted orders of derivation, are generally equal. In this connection, we shall have a theorem in the following subsection.

The first order partial derivatives of the second order partial derivatives of $\mathbf{f}(u, v)$ and so on for higher order derivatives of $\mathbf{f}$.

1.13.2. Inversion of the Order of Derivation

Theorem. *The two second order partial derivatives*

$$\frac{\partial^2 \mathbf{f}}{\partial u \partial v}, \frac{\partial^2 \mathbf{f}}{\partial v \partial u}$$

are equal, if each of them is a continuous function.

The proof will be based on the corresponding theorem in the Scalar Calculus which states that if the two derivatives

$$\frac{\partial^2 \phi}{\partial u \partial v}, \frac{\partial^2 \phi}{\partial v \partial u}.$$

of scalar function $\phi(u, v)$ are both continuous, then they are equal.

Let

$$\mathbf{f}(u, v) = f_1(u, v)\,\mathbf{i} + f_2(u, v)\,\mathbf{j} + f_3(u, v)\,\mathbf{k}.$$

$$\therefore \quad \frac{\partial \mathbf{f}}{\partial u} = \frac{\partial f_1}{\partial u}\mathbf{i} + \frac{\partial f_2}{\partial u}\mathbf{j} + \frac{\partial f_3}{\partial u}\mathbf{k}, \quad ...(1)$$

$$\frac{\partial \mathbf{f}}{\partial v} = \frac{\partial f_1}{\partial v}\mathbf{i} + \frac{\partial f_2}{\partial v}\mathbf{j} + \frac{\partial f_3}{\partial v}\mathbf{k}. \quad ...(2)$$

Also

$$\frac{\partial^2 \mathbf{f}}{\partial v \partial u} = \frac{\partial}{\partial v}\left(\frac{\partial \mathbf{f}}{\partial u}\right) = \frac{\partial^2 f_1}{\partial v \partial u}\mathbf{i} + \frac{\partial^2 f_2}{\partial v \partial u}\mathbf{j} + \frac{\partial^2 f_3}{\partial v \partial u}\mathbf{k}, \quad ...(3)$$

and

$$\frac{\partial^2 \mathbf{f}}{\partial u \partial v} = \frac{\partial}{\partial u}\left(\frac{\partial \mathbf{f}}{\partial v}\right) = \frac{\partial^2 f_1}{\partial u \partial v}\mathbf{i} + \frac{\partial^2 f_2}{\partial u \partial v}\mathbf{j} + \frac{\partial^2 f_3}{\partial u \partial v}\mathbf{k} \quad ...(4)$$

As $\partial^2 \mathbf{f}/\partial u \partial v$ is a continuous function, if follows that

$$\frac{\partial^2 f_1}{\partial u \partial v}, \frac{\partial^2 f_2}{\partial u \partial v}, \frac{\partial^2 f_3}{\partial u \partial v},$$

are also continuous. Similarly

$$\frac{\partial^2 f_1}{\partial v \partial u}, \frac{\partial^2 f_2}{\partial v \partial u}, \frac{\partial^2 f_3}{\partial v \partial u},$$

are continuous functions.

Thus, by the theorem quoted,

$$\frac{\partial^2 f_1}{\partial u \partial v} = \frac{\partial^2 f_1}{\partial v \partial u}, \frac{\partial^2 f_2}{\partial u \partial v} = \frac{\partial^2 f_2}{\partial v \partial u}, \frac{\partial^2 f_3}{\partial u \partial v} = \frac{\partial^2 f_3}{\partial v \partial u}. \quad ...(5)$$

Hence, from (3), (4) and (5),

$$\frac{\partial^2 \mathbf{f}}{\partial u \partial v} = \frac{\partial^2 \mathbf{f}}{\partial v \partial u}.$$

1.13.3. Derivation of Composite Functions

Theorem. *If*

$$\mathbf{r} = \mathbf{f}\,(u, v), \quad u = \phi\,(t, s), \quad v = \psi\,(t, s)$$

are three functions possessing continuous first order partial derivatives with respect to the respective variables then r, is a function of t, s possessing continuous first order partial derivatives with respect to the same and

$$\frac{\partial \mathbf{r}}{\partial t} = \frac{\partial \mathbf{f}}{\partial u}\,\frac{\partial u}{\partial t} + \frac{\partial \mathbf{f}}{\partial v}\,\frac{\partial v}{\partial t},$$

$$\frac{\partial \mathbf{r}}{\partial s} = \frac{\partial \mathbf{f}}{\partial u}\,\frac{\partial u}{\partial s} + \frac{\partial \mathbf{f}}{\partial v}\,\frac{\partial v}{\partial s}.$$

Let

$$\mathbf{r} = \mathbf{f}\,(u, v) = f_1(u, v)\,\mathbf{i} + f_2(u, v)\,\mathbf{j} + f_3(u, v)\,\mathbf{k}.$$

As f_1, f_2, f_3 possess continuous first order partial derivatives with respect to u, v we see, by the corresponding theorem in Scalar Calculus, that these functions possess continuous first order partial derivatives with respect to t and s and

$$\frac{\partial f_1}{\partial t} = \frac{\partial f_1}{\partial u}\,\frac{\partial u}{\partial t} + \frac{\partial f_1}{\partial v}\,\frac{\partial v}{\partial t}, \quad \frac{\partial f_1}{\partial s} = \frac{\partial f_1}{\partial u}\,\frac{\partial u}{\partial s} + \frac{\partial f_1}{\partial v}\,\frac{\partial v}{\partial s},$$

with similar expressions in relation to the functions f_2 and f_3. Thus, $\partial \mathbf{r}/\partial t$ and $\partial \mathbf{r}/\partial s$ are continuous and

$$\frac{\partial \mathbf{r}}{\partial t} = \frac{\partial \mathbf{f}}{\partial t} = \frac{\partial f_1}{\partial t}\mathbf{i} + \frac{\partial f_2}{\partial t}\mathbf{j} + \frac{\partial f_3}{\partial t}\mathbf{k}$$

$$= \left(\frac{\partial f_1}{\partial u}\,\frac{\partial u}{\partial t} + \frac{\partial f_1}{\partial v}\,\frac{\partial v}{\partial t}\right)\mathbf{i} + \left(\frac{\partial f_2}{\partial u}\,\frac{\partial u}{\partial t} + \frac{\partial f_2}{\partial v}\,\frac{\partial v}{\partial t}\right)\mathbf{j}$$

$$+ \left(\frac{\partial f_3}{\partial u}\,\frac{\partial u}{\partial t} + \frac{\partial f_3}{\partial v}\,\frac{\partial u}{\partial t}\right)\mathbf{k}.$$

$$= \left(\frac{\partial f_1}{\partial u}\mathbf{i} + \frac{\partial f_2}{\partial u}\mathbf{j} + \frac{\partial f_3}{\partial u}\mathbf{k}\right)\frac{\partial u}{\partial t} + \left(\frac{\partial f_1}{\partial v}\mathbf{i} + \frac{\partial f_2}{\partial v}\mathbf{j} + \frac{\partial f_3}{\partial v}\mathbf{k}\right)\frac{\partial v}{\partial t},$$

$$= \frac{\partial \mathbf{f}}{\partial u}\,\frac{\partial u}{\partial t} + \frac{\partial \mathbf{f}}{\partial v}\,\frac{\partial v}{\partial t}.$$

Similarly for $\partial \mathbf{r}/\partial s$.

1.13.4. Total Differential

The relation

$$d\mathbf{f} = \frac{\partial \mathbf{f}}{\partial u}du + \frac{\partial \mathbf{f}}{\partial v}dv.$$

Let

$$\mathbf{f}\,(u,\,u) = f_1(u,\,v)\,\mathbf{i} + f_2(u,\,v)\,\mathbf{j} + f_3(u,\,v)\,\mathbf{k}.$$

We suppose that **f** possesses continuous first order partial derivatives so that f_1, f_2, f_3 also do the same.

Now, by a theorem in Scalar Calculus, there exists scalars

$$\varepsilon_1,\ \rho_1;\ \varepsilon_2,\ \rho_2;\ \varepsilon_3,\ \rho_3$$

which $\to 0$ as $\delta u \to 0$ and $\delta v \to 0$ and such that

$$\delta f_1 = \frac{\partial f_1}{\partial u}\delta u + \frac{\partial f_1}{\partial b}\delta v + \varepsilon_1 \delta u + \rho_1 \delta v,$$

$$\delta f_2 = \frac{\partial f_2}{\partial u}\delta u + \frac{\partial f_2}{\partial u}\delta v + \varepsilon_2 \delta u + \rho_2 \delta v,$$

$$\delta f_3 = \frac{\partial f_3}{\partial u}\delta u + \frac{\partial f_3}{\partial u}\delta v + \varepsilon_3 \delta u + \rho_3 \delta v.$$

$$\therefore\quad \delta\mathbf{f} = \mathbf{f}\,(u + \delta u,\ v + \delta v) - \mathbf{f}\,(u,\ v)$$

$$= \delta f_1 \mathbf{i} + \delta f_2\,\mathbf{j} + \delta f_3 \mathbf{k}$$

$$= \frac{\partial \mathbf{f}}{\partial u}\delta u + \frac{\partial \mathbf{f}}{\partial v}\delta v + (\varepsilon_1 \mathbf{i} + \varepsilon_2 \mathbf{j} + \varepsilon_3 \mathbf{k})\ \delta u\ (\rho_1 \mathbf{i} + \rho_2 \mathbf{j} + \rho_3 \mathbf{k})\ \delta v.$$

Here $\varepsilon_1 \mathbf{i} + \varepsilon_2\,\mathbf{j} + \varepsilon_3 \mathbf{k}$ and $\rho_1 \mathbf{i} + \rho_2\,\mathbf{j} + \rho_3 \mathbf{k}$ both tend to zero when δu and δv tend to 0.

The increments δu, δv of the independent variables are also called the differentials of u, v denoted by du, dv respectively. Also the principal part

$$\frac{\partial \mathbf{f}}{\partial u}\partial u + \frac{\partial \mathbf{f}}{\partial v}\partial v$$

of the increment $\delta\mathbf{f}$ in **f** is called the differential of **f**, denoted by $d\mathbf{f}$. Thus, we have

$$d\mathbf{f} = \frac{\partial \mathbf{f}}{\partial u}du + \frac{\partial \mathbf{f}}{\partial v}dv.$$

EXAMPLES

Example 1. *If* $\mathbf{A} = (2x^2y - x^4)\,\mathbf{i} + (e^{xy} - y\sin x)\,\mathbf{j} + x^2 \cos y\,\mathbf{k}$, verify that $\dfrac{\partial^2 \mathbf{A}}{\partial y \partial x} = \dfrac{\partial^2 \mathbf{A}}{\partial x \partial y}$.

Solution. $\dfrac{\partial \mathbf{A}}{\partial x} = \dfrac{\partial}{\partial x}(2x^2 y - x^4)\,\mathbf{i} + \dfrac{\partial}{\partial x}(e^{xy} - y \sin x)\,\mathbf{j} + \dfrac{\partial}{\partial x}(x^2 \cos y)\,\mathbf{k}$

$$= (4xy - 4x^3)\,\mathbf{i} + (ye^{xy} - y\cos x)\,\mathbf{j} + 2x\cos y\,\mathbf{k}$$

$$\frac{\partial^2 \mathbf{A}}{\partial y \partial x} = \frac{\partial}{\partial y}\left(\frac{\partial^2 \mathbf{A}}{\partial x}\right) = 4x\mathbf{i} + (xye^{xy} - \cos x)\,\mathbf{j} - 2x \sin y\,\mathbf{k} \quad ...(1)$$

Again,

$$\frac{\partial \mathbf{A}}{\partial y} = 2x^2\mathbf{i} + (xe^{xy} - \sin x)\,\mathbf{j} - x^2 \sin y\,\mathbf{k}$$

$$\frac{\partial^2 \mathbf{A}}{\partial x \partial y} = \frac{\partial}{\partial x}\left(\frac{\partial \mathbf{A}}{\partial y}\right) = 4x\mathbf{i} + (xye^{xy} - \cos x)\,\mathbf{j} - 2x \sin y\,\mathbf{k}$$

From (1) and (2), the result follows.

Example 2. *If* $\phi(x, y, z) = xy^2z$ *and* $\mathbf{A} = xz\,\mathbf{i} - xy\,\mathbf{j} + yz^2\,\mathbf{k}$ *find* $\dfrac{\partial^3(\phi\mathbf{A})}{\partial x^2 \partial z}$ *at* $(2, -1, 1)$.

Solution. $\phi\mathbf{A} = xy^2z\,(xz\mathbf{i} - xy\mathbf{j} + yz^2\mathbf{k})$

$$= x^2y^2z^2\,\mathbf{i} - x^2y^2z\,\mathbf{j} + xy^3z^3\,\mathbf{k}$$

Now

$$\frac{\partial}{\partial z}(\phi\mathbf{A}) = 2x^2y^2z\,\mathbf{i} - x^2y^3\mathbf{j} + 3xy^3z^2\mathbf{k}$$

$$\frac{\partial^2}{\partial x \partial z}(\phi\mathbf{A}) = 4xy^2z\,\mathbf{i} - 2xy^3\mathbf{j} + 3y^3z^2\mathbf{k}$$

and

$$\frac{\partial^3}{\partial x^2 \partial z}(\phi\mathbf{A}) = 4y^2z\,\mathbf{i} - 2y^3\mathbf{j}$$

$$= 4(-1)^2(1)\,\mathbf{i} - 2(-1)^3\,\mathbf{j}$$

$$= 4\mathbf{i} + 2\mathbf{j} \text{ at the point } (2, -1, 1).$$

EXERCISES

1. Find

$$\frac{\dfrac{\partial \mathbf{r}}{\partial u} \times \dfrac{\partial \mathbf{r}}{\partial v}}{\left|\dfrac{\partial \mathbf{r}}{\partial u} \times \dfrac{\partial \mathbf{r}}{\partial v}\right|},$$

when

(*i*) $\mathbf{r} = u \cos v\mathbf{i} + u \sin v\,\mathbf{j} + ae^{mv}\mathbf{k}$,

(*ii*) $\mathbf{r} = \frac{1}{2}a\,(u+v)\,\mathbf{i} + \frac{1}{2}b\,(u-v)\,\mathbf{j} + \frac{1}{2}uv\,\mathbf{k}$.

(*iii*) $\mathbf{r} = u \cos v\,\mathbf{i} + u \sin v\,\mathbf{j} + c \log\left[u + \sqrt{(u^2 - c^2)}\right]\mathbf{k}$.

2. Find

$$\left[\frac{\partial \mathbf{r}}{\partial u}\ \frac{\partial \mathbf{r}}{\partial v}\ \frac{\partial^2 \mathbf{r}}{\partial u^2}\right], \left[\frac{\partial \mathbf{r}}{\partial u}\ \frac{\partial \mathbf{r}}{\partial v}\ \frac{\partial^2 \mathbf{r}}{\partial u \partial v}\right], \left[\frac{\partial \mathbf{r}}{\partial u}\ \frac{\partial \mathbf{r}}{\partial v}\ \frac{\partial^2 \mathbf{r}}{\partial v^2}\right]$$

are each of the functions given in the Ex. 1, above.

APPENDIX I

The proofs of the theorems on limits given in § 1.2.2, which are independent of the corresponding theorems in scalar calculus will now be given.

(*i*) Let ε be any positive number. As

$$\lim_{t \to c} \mathbf{f}(t) = \mathbf{l}, \quad \lim_{t \to c} \mathbf{F}(t) = \mathbf{m},$$

there exists a positive number δ such that

$$|\mathbf{f}(t) - \mathbf{l}| < \frac{1}{2}\varepsilon, \; |\mathbf{F}(t) - \mathbf{m}| < \frac{1}{2}\varepsilon$$

when

$$0 < | t - c | \leq \delta.$$

Thus, for $0 < | t - c | \leq \delta$,

$$\begin{aligned} | [\mathbf{f}(t) + \mathbf{F}(t)] - [\mathbf{l} + \mathbf{m}] | &= | [\mathbf{f}(t) - \mathbf{l}] + [\mathbf{F}(t) - \mathbf{m}] | \\ &\leq | \mathbf{f}(t) - \mathbf{l} | + [\mathbf{F}(t) - \mathbf{m} | \\ &< \frac{1}{2}\varepsilon + \frac{1}{2}\varepsilon = \varepsilon. \end{aligned}$$

and

$$\begin{aligned} | [\mathbf{f}(t) - \mathbf{F}(t)] - [\mathbf{l} - \mathbf{m}] | &= | [\mathbf{f}(t) - \mathbf{l}] - [\mathbf{F}(t) - \mathbf{m}] | \\ &\leq | \mathbf{f}(t) - \mathbf{l} | + [\mathbf{F}(t) - \mathbf{m} | \\ &< \frac{1}{2}\varepsilon + \frac{1}{2}\varepsilon = \varepsilon. \end{aligned}$$

Hence,

$$\lim [\mathbf{f}(t) \pm \mathbf{F}(t)] = \mathbf{l} \pm \mathbf{m} \text{ when } t \to c.$$

(*ii*) We write

$$\mathbf{f}(t) \,.\, \mathbf{F}(t) - \mathbf{l} \,.\, \mathbf{m} = \mathbf{f}(t) \,.\, [\mathbf{F}(t) - \mathbf{m}] + \mathbf{m} \,.\, [\mathbf{f}(t) - \mathbf{l}],$$

so that for every value of t,

$$\begin{aligned} | \mathbf{f}(t) \,.\, \mathbf{F}(t) - \mathbf{l} \,.\, \mathbf{m} | &\leq | \mathbf{f}(t) \,.\, [\mathbf{F}(t) - \mathbf{m}] | + | \mathbf{m} \,.\, [\mathbf{f}(t) - \mathbf{l}] | \\ &\leq | \mathbf{f}(t) | \; | \mathbf{F}(t) - \mathbf{m} | + | \mathbf{m} | \; | \mathbf{f}(t) - \mathbf{l}] | \end{aligned}$$

As $\lim_{t \to c} \mathbf{f}(t) = \mathbf{l}$, there exists a positive number δ_1 such that

$$| \mathbf{f}(t) - \mathbf{l} | < 1 \text{ when } 0 < | t - c | \leq \delta_1.$$

But

$$| \mathbf{f}(t) - \mathbf{l} | \geq | \mathbf{f}(t) | - | \mathbf{l} |, \text{ for every } t.$$

Thus, $0 < | t - c | \leq \delta_1$, we have

$$| \mathbf{f}(t) | - | \mathbf{l} | \leq | \mathbf{f}(t) - \mathbf{l} | < 1.$$

or

$$| \mathbf{f}(t) | < | \mathbf{l} | + 1.$$

$\therefore$ for $0 < |t - c| \leq \delta_1$, we have

$$|\mathbf{f}(t) \cdot \mathbf{F}(t) - \mathbf{l} \cdot \mathbf{m}| < (|\mathbf{l}| + 1)\,|\mathbf{F}(t) - \mathbf{m}| + |\mathbf{m}|\;|\mathbf{f}(t) - \mathbf{l}|.$$

Let now ε be any positive number. Then there exist positive number δ_2 and δ_3 such that

$$|\mathbf{F}(t) - \mathbf{m}| < \frac{\varepsilon}{2(|\mathbf{l}|+1)} \text{ when } 0 < |t-c| \leq \delta_2,$$

and $$|\mathbf{f}(t) - \mathbf{l}| < \frac{\varepsilon}{2(|\mathbf{m}|+1)} \text{ when } 0 < |t-c| \leq \delta_3.$$

If $\delta = \min. (\delta_1, \delta_2, \delta_3)$, we have, for $0 < |t - c| \leq \delta$,

$$|\mathbf{f}(t) \cdot \mathbf{F}(t) - \mathbf{l} \cdot \mathbf{m}| < (|\mathbf{l}|+1) \frac{\varepsilon}{2(|\mathbf{l}|+1)} + |\mathbf{m}| \frac{2}{2(|\mathbf{m}|+1)} < \varepsilon,$$

i.e., $$|\mathbf{f}(t) \cdot \mathbf{F}(t) - \mathbf{l} \cdot \mathbf{m}| < \varepsilon.$$

Hence, $$\lim_{t \to c} [\mathbf{f}(t) \cdot \mathbf{F}(t)] = \mathbf{l} \cdot \mathbf{m}$$

(*iii*) We write

$$\mathbf{f}(t) \times \mathbf{F}(t) - \mathbf{l} \times \mathbf{m} = \mathbf{f}(t) \times [\mathbf{F}(t) - \mathbf{m}] + [\mathbf{f}(t) - \mathbf{l}] \times \mathbf{m},$$

so that for every value of t,

$$|\mathbf{f}(t) \times \mathbf{F}(t) - \mathbf{l} \times \mathbf{m}| \leq |\mathbf{f}(t)|\;|\mathbf{F}(t) - \mathbf{m}| + |\mathbf{f}(t) - \mathbf{l}|\;|\mathbf{m}|.$$

The further process is now the same as in the case of scalar product above.

(*iv*) We write

$$\phi(t)\,\mathbf{f}(\mathbf{l}) - p\mathbf{l} = \mathbf{f}(t)\,[\phi(t) - p] + p\,[\mathbf{f}(t) - \mathbf{l}],$$

so that for every value of t,

$$|\phi(t)\,\mathbf{f}(t) - p\mathbf{l}| \leq |\mathbf{f}(t)|\;|\phi(t) - p| + |p|\;|\mathbf{f}(t) - \mathbf{l}|.$$

Again the further processes is the same as in the two cases above.

(*v*) We have

$$|\mathbf{f}(t)| - |\mathbf{l}| \leq |\mathbf{f}(t) - \mathbf{l}|, \qquad ...(1)$$

for every value of t.

As $\lim \mathbf{f}(t) = \mathbf{l}$, when $t \to c$,

there exists a positive number δ corresponding to any preassigned positive ε such that

$$|\mathbf{f}(t) - \mathbf{l}| < \varepsilon \text{ when } 0 < |t - c| \leq \delta. \qquad ...(2)$$

From (1) and (2), we have

$$\big|\,|\mathbf{f}(t)| - |\mathbf{l}|\,\big| < \varepsilon \text{ when } 0 < |t - c| \leq \delta.$$

Hence,

$$\lim_{t \to c} |\mathbf{f}(t)| = |\mathbf{l}|.$$

OBJECTIVE QUESTIONS

For each of the following questions, four alternatives are given for the answer. Only one of them is correct. Choose the correct alternative.

1. Let **f** (t) and **g** (t) be two vector functions such that $\lim_{t \to a} \mathbf{f}(t) = \mathbf{L}$ and $\lim_{t \to a} \mathbf{g}(t) = \mathbf{M}$, then $\lim_{t \to 0} [\mathbf{f}(t) \times \mathbf{g}(t)]$

(a) **L** (b) **M** (c) **L × M** (d) **L . M**

2. If $\mathbf{r} = \mathbf{f}(s)$ be a continuous and single-valued function of the scalar variable s, then $d\mathbf{r}/ds$ is a unit vector along

(a) Normal (b) Radius vector

(c) Tangent (d) None of these

3. Derivative of a vector of constant magnitude of

(a) zero (b) parallel to the vector

(c) perpendicular to the vector (d) None of these

4. Necessary and sufficient condition for the vector **v** (t) to have a constant direction is

(a) $\mathbf{v} \cdot \dfrac{d\mathbf{v}}{dt} = 0$ (b) $\mathbf{v} \times \dfrac{d\mathbf{v}}{dt} = 0$

(c) $\mathbf{v} = 0$ (d) $\dfrac{d\mathbf{v}}{dt} = 0$

5. Necessary and sufficient condition for the vector **v** (t) to have a constant magnitude is

(a) $\dfrac{d\mathbf{v}}{dt} = 0$ (b) $\mathbf{v} \times \dfrac{d\mathbf{v}}{dt} = 0$

(c) $\mathbf{v} \cdot \dfrac{d\mathbf{v}}{dt} = 0$ (d) None of these

6. If $\mathbf{r} = \mathbf{a} \cos \omega t + \mathbf{b} \sin \omega t$, then value of $\dfrac{d^2\mathbf{r}}{dt^2}$ is

(a) $\omega (\mathbf{a} \times \mathbf{b})$ (b) $-\omega^2 \mathbf{r}$ (c) $\omega^2 \mathbf{r}$ (d) 0

7. If $\mathbf{r} \times d\mathbf{r} = 0$, then

(a) $\hat{\mathbf{r}}$ = constant (b) $\hat{\mathbf{r}} = \mathbf{0}$

(c) $\hat{\mathbf{r}}$ is indeterminate (d) None of these

8. A particle P is moving on a circle of radius r with constant angular velocity $\omega = d\theta/dt$. The acceleration is

(a) 0 (b) $\omega^2 \mathbf{r}$ (c) $-\omega^2 \mathbf{r}$ (d) $\omega \mathbf{r}$

9. If $\frac{d\mathbf{a}}{dt} = \lambda \mathbf{b}$ and $\frac{d\mathbf{b}}{dt} = \mu \mathbf{a}$, then $\frac{d}{dt} = (\mathbf{a} \times \mathbf{b})$ is

(*a*) $\lambda\mu\ (\mathbf{a} \times \mathbf{b})$ (*b*) $\lambda\mu\ (\mathbf{b} \times \mathbf{a})$

(*c*) $\frac{\lambda}{\mu}(\mathbf{a} \times \mathbf{b})$ (*d*) zero

10. If $\mathbf{r} \cdot d\mathbf{r} = 0$, then

(*a*) $\hat{\mathbf{r}} = \mathbf{0}$ (*b*) $\hat{\mathbf{r}}$ is constant

(*c*) $|\,\mathbf{r}\,| =$ constant (*d*) $|\,\mathbf{r}\,| = 0$

ANSWERS

1. (*c*)	**2.** (*c*)	**3.** (*c*)	**4.** (*b*)	**5.** (*c*)
6. (*b*)	**7.** (*a*)	**8.** (*c*)	**9.** (*d*)	**10.** (*c*)

2

Differential Geometrv I

CURVES

Introduction. In this chapter we shall consider the Elements of Differential Geometry of curves. This will include the basic consideration of the triad of orthonormal vectors associated with any point of a curve and the well-known *Frenet's* formulae giving relations between the arc-derivatives of the vectors of the triad and the curvature and torsion. The problem of the determination of the involutes and evolutes of a given curve will also be taken up. The differential Geometry of Surfaces and the problem of the distribution of curvatures of curves on a surface through a point leading to *Meunier's* theorem and the notion of principal normal sections will be given a detailed treatment in the following chapter.

2.1. FUNDAMENTAL TRIADS OF LINES AND PLANES ASSOCIATED WITH ANY POINT ON A CURVE.

Associated with each point on a curve, there is a set of three mutually perpendicular lines known as

Tangent, Principal Normal, Binormal

and three mutually perpendicular planes determined by these in pairs and known as

Osculating plane, Normal plane, Rectifying plane.

In this section we shall define these important lines and planes and obtain their vector equations.

It may be carefully noted that a curve is analytically representable by an equation of the form

$$\mathbf{r} = \mathbf{f}(t),$$

such that the vector, **r**, corresponding to any value of the scalar parameter, t, is the position vector of a point on the curve with respect to some origin O. The point P on the curve corresponding to the value 't' may be denoted as $P(t)$.

For the following, we shall suppose that $\mathbf{f}(t)$ possesses continuous derivatives of every order which appear in any investigation. In general, we shall not be required to consider derivatives of order higher than three.

2.1.1. Tangent at a point

Let $P(t)$ be any point on the curve

$$\mathbf{r} = \mathbf{f}(t).$$

Take any other point $Q(t + \delta t)$ on the curve.

The tangent at P is the limiting position of the chord PQ when $Q \to P$.

Let

$$\mathbf{r} + \delta\mathbf{r} = \mathbf{f}(t + \delta t)$$

$$\therefore \quad \overrightarrow{PQ} = \overrightarrow{OQ} - \overrightarrow{OP}$$

$$= \mathbf{f}(t + \delta t) - \mathbf{f}(t) = \delta\mathbf{r}.$$

As, δt, is scalar, the vector

$$\frac{\delta\mathbf{r}}{\delta t} = \frac{\mathbf{f}(t+\delta t) - \mathbf{f}(t)}{\delta t},$$

is parallel to $\overrightarrow{PQ}$. Thus, taking the limit, when, $Q \to P$ and consequently $\delta t \to 0$, we see that the vector

$$\frac{d\mathbf{r}}{dt},$$

is parallel to the tangent at P.

Hence

$$\mathbf{R} = \mathbf{r} + u\frac{d\mathbf{r}}{dt}, \quad i.e., \quad \mathbf{R} = \mathbf{f}(t) + u\mathbf{f}'(t),$$

is the vector equation of the tangent at P.

Here 'u' is the scalar parameter; $\mathbf{r}$, is the position vector of the point P and $\mathbf{R}$, the position vector of any point on the tangent.

Note. A curve $\mathbf{r} = \mathbf{f}(t)$ is said to be **smooth** if $\mathbf{f}(t)$ possesses a continuous derivative at every point. Also a point on a smooth curve is said to be **regular** if the derivative $d\mathbf{r}/dt \neq$ orthereat. Thus we see from above that a smooth curve possesses a tangent at every regular point thereof.

2.2. OSCULATING PLANE AT A POINT

Def. *The limiting position of the plane which passes through the tangent at P and is parallel to the tangent at Q, when $Q \to P$, is called the osculating plane at P.* *Agra 98, Rohilkhand 2000*

The tangents at $P(t)$ and $Q(t + \delta t)$ are parallel to the vectors

$$\mathbf{f}'(t), \ \mathbf{f}'(t + \delta t),$$

so that the plane through the tangent at P and parallel to the tangent at Q is perpendicular to the vector

$$\mathbf{f}'(t) \times \mathbf{f}'(t + \delta t) = \mathbf{f}'(t) \times [\mathbf{f}'(t + \delta t) - \mathbf{f}'(t)]$$

and hence also to the vector

$$\mathbf{f}'(t) \times \frac{\mathbf{f}'(t+\delta t)-\mathbf{f}'(t)}{\delta t}$$

Thus, taking the limit, we see that the osculating plane at P is perpendicular to the vector

$$\mathbf{f}'(t) \times \mathbf{f}''(t) = \frac{d\mathbf{r}}{dt} \times \frac{d^2\mathbf{r}}{dt^2}$$

Hence the equation of the osculating plane at $P(t)$ is

$$[\mathbf{R} - \mathbf{f}(t)] \times \mathbf{f}'(t) \times \mathbf{f}''(t) = 0.$$

which is the same as

$$(\mathbf{R} - \mathbf{r}) \cdot \frac{d\mathbf{r}}{dt} \times \frac{d^2\mathbf{r}}{dt^2} = 0,$$

i.e.,
$$\left[\mathbf{R}\ \frac{d\mathbf{r}}{dt}\ \frac{d^2\mathbf{r}}{dt^2}\right] = \left[\mathbf{r}\ \frac{d\mathbf{r}}{dt}\ \frac{d^2\mathbf{r}}{dt^2}\right];$$

R being the position vector of any point on the osculating plane.

2.3. NORMAL PLANE AT A POINT

Def. *The plane through a point P perpendicular to the tangent thereat is called the Normal plane at P.*

Thus the equation of the normal plane at a point P with position vector, **r**, is

$$(\mathbf{R} - \mathbf{r}) \cdot \frac{d\mathbf{r}}{dt} = 0;$$

R, denoting the position vector of any point on the plane.

The normal plane is perpendicular to the osculating plane at any point.

2.4. NORMAL LINES AT A POINT

Two special normals. Principal Normal and Binormal. Any line perpendicular to the tangent line at Q is called a normal at P. The normal plane at a point is the locus of the normal through the point. Of all the normals at a point, two normals, known as principal normal and binormal play a specially important part. We shall now define these.

Principal Normal. *The normal lying in the osculating plane at any point is called the principal normal at the point.*

Thus the intersection of the normal plane and the osculating plane is the principal normal. Also the plane containing the tangent and the principal normal is the osculating plane.

Binormal. *The normal perpendicular to the principal normal is called Binormal* so that the binormal at any point is the line perpendicular to the osculating plane at the point.

The plane containing the tangent and binormal is called *Rectifying plane.*

Thus we have defined tangent, principal normal and binormal at any point; these being mutually perpendicular. They determine three mutually perpendicular planes, *viz.*,

(*i*) Osculating plane containing the tangent and the principal normal.
(*ii*) Normal plane containing the principal normal and binormal and
(*iii*) Rectifying plane containing the binormal and the tangent.

2.5. DIRECTIONS OF PRINCIPAL NORMAL AND BINORMAL

From the equation of the osculating plane (§ 2.2), we see that the binormal, being perpendicular to the osculating plane, is parallel to the vector

$$\frac{d\mathbf{r}}{dt} \times \frac{d^2\mathbf{r}}{dt^2}.$$

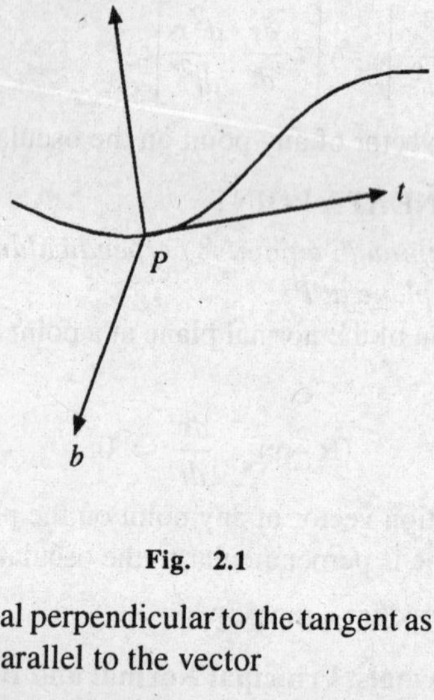

Fig. 2.1

The principal normal perpendicular to the tangent as well as the binormal and, accordingly, it is parallel to the vector

$$\left(\frac{d\mathbf{r}}{dt} \times \frac{d^2\mathbf{r}}{dt^2}\right) \times \frac{d\mathbf{r}}{dt} = \left(\frac{d\mathbf{r}}{dt}\right)^2 \frac{d^2\mathbf{r}}{dt^2} - \left(\frac{d\mathbf{r}}{dt} \cdot \frac{d^2\mathbf{r}}{dt^2}\right) \frac{d\mathbf{r}}{dt}.$$

Cor. We now take the arc length, s, as the parameter. We have

$$\frac{\delta \mathbf{r}}{\delta s} = \frac{\overrightarrow{PQ}}{\text{arc } PQ},$$

so that

$$\left|\frac{\delta \mathbf{r}}{\delta t}\right| = \frac{\text{chord } PQ}{\text{arc } PQ}.$$

Thus taking limit when $Q \to P$, we see that

$$\left|\frac{d\mathbf{r}}{dt}\right| = 1.$$

so that the vector $d\mathbf{r}/ds$ is of constant length unity.

Thus

$$\frac{d\mathbf{r}}{dt} \cdot \frac{d\mathbf{r}}{ds} = 1. \qquad ...(1)$$

Differentiating, we get

$$\frac{d\mathbf{r}}{ds} \cdot \frac{d^2\mathbf{r}}{ds^2} = 0. \qquad ...(2)$$

The principal normal is parallel to the vector

$$\left(\frac{d\mathbf{r}}{ds} \times \frac{d^2\mathbf{r}}{ds^2}\right) \times \frac{d\mathbf{r}}{ds} = \left(\frac{d\mathbf{r}}{ds} \cdot \frac{d\mathbf{r}}{ds}\right)\frac{d^2\mathbf{r}}{ds^2} - \left(\frac{d\mathbf{r}}{ds} \cdot \frac{d^2\mathbf{r}}{ds^2}\right)\frac{d\mathbf{r}}{ds}$$

$$= \frac{d^2\mathbf{r}}{ds^2}. \qquad \text{from (1) and (2)}$$

Thus we see that the tangent, principal normal and binormal are parallel to the vectors

$$\frac{d\mathbf{r}}{ds}, \frac{d^2\mathbf{r}}{ds^2}, \frac{d\mathbf{r}}{ds} \times \frac{d^2\mathbf{r}}{ds^2}$$

respectively.

2.5.1. Orthonormal triad of vectors

t, n, b.

We shall now give precise definitions of unit vector **t, n, b** lying along the tangent, principal normal and binormal respectively.

The unit tangent vector **t**. We have seen that the tangent at *P* is parallel to the vector $d\mathbf{r}/ds$. The sense of $d\mathbf{r}/ds$ is the same as that of the curve along which '*s*' increases and, as seen in cor. to § 2.5, its length is unity. We write

$$\mathbf{t} = \frac{d\mathbf{r}}{ds},$$

so that, **t**, is the unit vector along the tangent and in the sense in which the arc, *s*, is measured positively.

The vector, **t**, is called the *unit tangent vector*.

The unit principal **n.** As $\mathbf{t} = d\mathbf{r}/ds$, we have

$$\frac{d\mathbf{t}}{ds} = \frac{d^2\mathbf{r}}{ds^2}.$$

As seen in the cor. to § 2.5, the vector $d^2\mathbf{r}/ds^2$ is parallel to the principal normal. This vector is, however, not necessary of unit length.

The unit vector, **n**, called unit principal normal, is defined as a unit vector having the sense of $d\mathbf{t}/ds$. Clearly '**n**' lies along the principal normal.

The unit binormal **b**. The vector, **b**, is defined by the equation

$$\mathbf{b} = \mathbf{t} \times \mathbf{n},$$

so that, **b**, is a unit vector lying along the binormal such that the triad of vectors

$$\mathbf{t}, \mathbf{n}, \mathbf{b}$$

is right handed.

2.6. CURVATURE AND TORSION

(Meerut 98, Rohilkhand 99, Garhwal 2000)

Def. 1. *Arc-rate of rotation of tangent is defined as curvature.*

Def. 2. *Arc-rate of rotation of binormal is defined is torsion.*

The curvature and torsion are usually denoted by κ and τ respectively. Their reciprocals, called *radius of curvature* and *radius of torsion* are denoted by ρ and σ respectively. Thus

$$\rho = \frac{1}{\kappa}, \quad \sigma = \frac{1}{\tau}.$$

It will be seen that, ρ, is always regarded as positive but σ may be positive or negative so that σ is a signed scalar.

2.6.1. We shall now cstablish the following results:

$$(i) \left|\frac{d\mathbf{t}}{ds}\right| = \kappa, \qquad (ii) \left|\frac{d\mathbf{b}}{ds}\right| = |\tau|.$$

(*i*) Let, $\delta\theta$, denote the angle, lying between 0 and π, between the tangents at P and Q, *i.e.*, between the vectors **t** and $\mathbf{t} + \delta\mathbf{t}$.

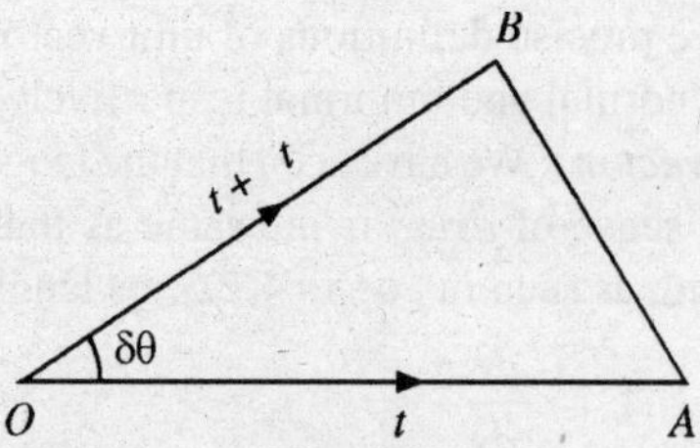

Fig. 2.2

Now $\quad \mathbf{t} \times (\mathbf{t} + \delta\mathbf{t}) = \mathbf{t} \times \delta\mathbf{t},$

represents twice the vector area of the triangle OAB. The magnitude of the area of this triangle being

$$\frac{1}{2} OA \,.\, OB \sin \delta\theta = \frac{1}{2} \sin \delta\theta,$$

we have

$$|\mathbf{t} \times \delta\mathbf{t}| = \sin \delta\theta.$$

$$\therefore \quad \left| \mathbf{t} \times \frac{\delta \mathbf{t}}{\delta s} \right| = \frac{\sin \delta\theta}{\delta s} = \frac{\sin \delta\theta}{\delta\theta} \frac{\delta\theta}{\delta s}.$$

Taking limit, we obtain

$$\left| \mathbf{t} \times \frac{\delta \mathbf{t}}{ds} \right| = 1 \,.\, \kappa = \kappa..$$

As, **t**, is of constant length, it is perpendicular to $d\mathbf{t}/ds$. Hence

$$\left| \mathbf{t} \times \frac{\delta \mathbf{t}}{ds} \right| = \left| \frac{d\mathbf{t}}{ds} \right|.$$

Thus

$$\left| \frac{d\mathbf{t}}{ds} \right| = \kappa..$$

Taking, **b**, instead of, **t**, we may similarly show that

$$\left| \frac{d\mathbf{b}}{ds} \right| = |\tau|.$$

2.6.2. Theorem

A necessary and sufficient condition for the curve to be a straight line is that the curvature $\kappa = 0$ *at all points of the curve.* **(Berhampur 1993)**

The straight line be $\mathbf{r} = s\mathbf{a} + \mathbf{b}$ where **a** and **b** are constant vectors.

$$\therefore \quad \mathbf{t} = \frac{d\mathbf{r}}{ds} = \mathbf{a} \quad \text{and} \quad \frac{d\mathbf{t}}{ds} = \frac{d^2\mathbf{r}}{ds^2} = 0$$

$$\therefore \quad \kappa = \left| \frac{d^2\mathbf{r}}{ds^2} \right| = 0,$$

i.e., if the curve is a straight line, then $\kappa = 0$.

Conversely, if $\kappa = 0$ for all points on the curve, then

$$\frac{d^2\mathbf{r}}{ds^2} = 0 \qquad \text{...(1)}$$

$$\Rightarrow \quad \frac{d\mathbf{r}}{ds} = \mathbf{a} \quad \Rightarrow \quad \mathbf{r} = \mathbf{a}s + \mathbf{b} \qquad \text{...(2)}$$

Equation (2) represents a straight line for all values of **a** and **b**.

2.6.3. Theorem

A necessary and sufficient condition that a given curve is a plane curve is that $\tau = 0$ *at all points.* **(Berhampur 1993)**

If given curve is a plane curve, then the tangent and normal at all points of the curve lie in the plane of the curve. It implies that the plane of the curve is the osculating plane at all points on the curve. Hence

$$\frac{d\mathbf{b}}{ds} = 0 \quad \text{or} \quad \tau = 0.$$

Conversely, if $\tau = 0$ at ali points of the curve, then

$$\frac{d\mathbf{b}}{ds} = 0, \textit{ i.e., } \mathbf{b} \text{ is a constant vector.}$$

Again, $$\frac{d}{ds}(\mathbf{r}\cdot\mathbf{b}) = \frac{d\mathbf{r}}{ds}\cdot\mathbf{b}+\mathbf{r}\cdot\frac{d\mathbf{b}}{ds} = \mathbf{t}\cdot\mathbf{b}+\mathbf{r}\cdot\frac{d\mathbf{b}}{ds}.$$

As **t** and **b** are orthogonal, $\mathbf{t}\cdot\mathbf{b} = 0$.

Also $$\frac{d\mathbf{b}}{ds} = 0$$

$$\therefore \quad \mathbf{r}\cdot\mathbf{b} = \text{constant.}$$

As **b** is constant vector of magnitude unity, **r** . **b** is the projection of the position vector **r** on **b** and is same at all points of the curve by the condition **r** . **b** = constant. This implies that the curve must lie in a plane.

2.6.4. Screw-Curvature

Def. *The arc rate at which principal normal changes direction is called the screw-curvature vector.*

$$\therefore \text{ Screw-curvature vector } = \frac{d\mathbf{n}}{ds} \text{ and}$$

$$\left|\frac{d\mathbf{n}}{ds}\right| = \sqrt{(\kappa^2+\tau^2)}$$

2.7. FRENET'S FORMULAE

We shall now establish the following important results, known as Frenet's formulae. **(Garwal 2002)**

$$\frac{d\mathbf{t}}{ds} = \kappa\mathbf{n}, \quad \text{...(1)}$$

$$\frac{d\mathbf{n}}{ds} = -\kappa\mathbf{t}-\tau\mathbf{b}, \quad \text{...(2)}$$

$$\frac{d\mathbf{b}}{ds} = -\tau\mathbf{n}. \quad \text{...(3)}$$

We have seen that, **n**, is the unit vector having the sense and direction of $d\mathbf{t}/ds$. Also

$$\left|\frac{d\mathbf{t}}{ds}\right| = \kappa,$$

Combining these two facts, we obtain

$$\frac{d\mathbf{t}}{ds} = \kappa\,\mathbf{n};$$

Now,

$$\mathbf{t}\cdot\mathbf{b} = 0.$$

$$\therefore \quad \frac{d\mathbf{t}}{ds} \cdot \mathbf{b} + \mathbf{t} \cdot \frac{d\mathbf{b}}{ds} = 0,$$

or $$\kappa \mathbf{n} \cdot \mathbf{b} + \mathbf{t} \cdot \frac{d\mathbf{b}}{ds} = 0 \quad \text{by (1)}$$

or $$\mathbf{t} \cdot \frac{d\mathbf{b}}{ds} = 0; \text{ for } \mathbf{n} \cdot \mathbf{b} = 0.$$

Thus $d\mathbf{b}/ds$ is perpendicular to $\mathbf{t}$. Also, $\mathbf{b}$, being of constant magnitude, is perpendicular to its derivative $d\mathbf{b}/ds$.

Hence $d\mathbf{b}/ds$ is parallel to the vector $\mathbf{n}$.

Moreover we have seen that

$$\left|\frac{d\mathbf{b}}{ds}\right| = |\tau|$$

Thus, we have

$$\frac{d\mathbf{b}}{ds} = \tau \mathbf{n},$$

which is (3).

Here the torsion τ is positive or negative according as the vectors $\mathbf{n}$ and $d\mathbf{b}/ds$, as defined above, have the same or opposite senses.

Finally, we have

$$\mathbf{n} = \mathbf{b} \times \mathbf{t}.$$

$$\begin{aligned}
\therefore \quad \frac{d\mathbf{n}}{ds} &= \frac{d\mathbf{b}}{ds} \times \mathbf{t} + \mathbf{b} \times \frac{d\mathbf{t}}{ds} \\
&= \tau\mathbf{n} \times \mathbf{t} + \mathbf{b} \times \kappa\mathbf{n} \qquad \text{[by (1) and (3)]} \\
&= -\tau\mathbf{b} - \kappa\mathbf{t} \\
&= -\kappa\mathbf{t} - \tau\mathbf{b},
\end{aligned}$$

which is (2).

2.8. EXPRESSIONS FOR CURVATURE AND TORSION IN TERMS OF THE DERIVATIVES OF, R, WITH RESPECT TO *S*.

We have

$$\frac{d\mathbf{r}}{ds} = \mathbf{t}. \qquad \text{...(1)}$$

$$\therefore \quad \frac{d^2\mathbf{r}}{ds^2} = \frac{d\mathbf{t}}{ds} = \kappa\mathbf{n}. \qquad \text{...(2)}$$

[Frenet's formula (1)]

$$\therefore \quad \kappa = \left|\frac{d^2\mathbf{r}}{ds^2}\right|. \qquad \text{...(3)}$$

Multiplying (1) and (2) vectorially, we obtain

$$\frac{d\mathbf{r}}{ds} \times \frac{d^2\mathbf{r}}{ds^2} = \kappa\mathbf{t} \times \mathbf{n} = \kappa\mathbf{b}.$$

$$\therefore \qquad \kappa = \left| \frac{d\mathbf{r}}{ds} \times \frac{d^2\mathbf{r}}{ds^2} \right|. \qquad ...(4)$$

The equations (3) and (4) give two expressions for κ in the required forms.

Differentiating (2) and making use of Frenet's second formula, we obtain

$$\frac{d^2\mathbf{r}}{ds^2} = \frac{d\kappa}{ds}\mathbf{n} + \kappa\frac{d\mathbf{n}}{ds}$$

$$= \frac{d\kappa}{ds}\mathbf{n} - \kappa\,(\kappa\,\mathbf{t} + \tau\mathbf{b})$$

$$= -\kappa^2\,\mathbf{t} + \frac{d\kappa}{ds}\mathbf{n} - \kappa\tau\,\mathbf{b}. \qquad ...(5)$$

$$\therefore \qquad \frac{d\mathbf{r}}{ds} \times \frac{d^3\mathbf{r}}{ds^3} = \frac{d\kappa}{ds}\mathbf{b} + \kappa\tau\,\mathbf{n}. \qquad ...(6)$$

Multiplying (2) and (6) scalarly, we have

$$\frac{d\mathbf{r}}{ds} \times \frac{d^3\mathbf{r}}{ds^3} \cdot \frac{d^2\mathbf{r}}{ds^2} = \kappa^2\tau.$$

Hence

$$\left[\frac{d\mathbf{r}}{ds}\ \frac{d^2\mathbf{r}}{ds^2}\ \frac{d^3\mathbf{r}}{ds^3}\right] = -\,\kappa^2\tau.$$

2.8.1. Necessary and Sufficient Condition for the curve to be a plane curve

From § 2.8

$$\left[\frac{d\mathbf{r}}{ds}\ \frac{d^2\mathbf{r}}{ds^2}\ \frac{d^3\mathbf{r}}{ds^3}\right] = \kappa^2\tau.$$

In case $\left[\frac{d\mathbf{r}}{ds}\ \frac{d^2\mathbf{r}}{ds^2}\ \frac{d^3\mathbf{r}}{ds^3}\right] = 0$, then either $\kappa = 0$ or $\tau = 0$.

Let $\tau \neq 0$, at some point of the curve, then in the neighbourhood of this point $\tau \neq 0$, therefore $\kappa = 0$ in the neighbourhood of this point. Hence the arc is a straight line (§ 2.6.2) and therefore $\tau = 0$ on this line, which contradicts the hypothesis. Hence $\tau = 0$ at all points and the curve is plane (§ 2.6.3).

Conversely, if $\tau = 0$, *i.e.*, the curve is a plane curve, then

$$\left[\frac{d\mathbf{r}}{ds}\ \frac{d^2\mathbf{r}}{ds^2}\ \frac{d^3\mathbf{r}}{ds^3}\right] = 0$$

Hence the necessary and sufficient condition.

2.8.2. Direction consines of the principal Normal and Binormal

From Fenet's formula.

$$\frac{d\mathbf{t}}{ds} = \kappa\mathbf{n} \qquad \Rightarrow \qquad \mathbf{n} = \frac{d\mathbf{t}/ds}{\kappa}$$

$$\Rightarrow \qquad \mathbf{n} = \frac{d^2\mathbf{r}/ds^2}{\kappa} = \frac{1}{\kappa}\left(\frac{d^2x}{ds^2}\hat{\mathbf{i}} + \frac{d^2y}{ds^2}\hat{\mathbf{j}} + \frac{d^2z}{ds^2}\hat{\mathbf{k}}\right) \qquad \text{...(1)}$$

Since $\mathbf{n}$ is a unit vector, the direction cosines of the principal normal are

$$\frac{d^2x/ds^2}{\kappa}, \quad \frac{d^2y/ds^2}{\kappa}, \quad \frac{d^2z/ds^2}{\kappa}.$$

Squaring (1), we get

$$1 = \frac{1}{\kappa^2}\left[\left(\frac{d^2x}{ds^2}\right)^2 + \left(\frac{d^2y}{ds^2}\right)^2 + \left(\frac{d^2z}{ds^2}\right)^2\right]$$

or
$$\kappa = \left[\left(\frac{d^2x}{ds^2}\right)^2 + \left(\frac{d^2y}{ds^2}\right)^2 + \left(\frac{d^2z}{ds^2}\right)^2\right]^{1/2}$$

We have
$$\mathbf{b} = \mathbf{t}\times\mathbf{n} = \frac{d\mathbf{r}}{ds}\times\frac{d^2\mathbf{r}/ds^2}{\kappa}$$

$$= \frac{1}{\kappa}\left[\left(\frac{dy}{ds}\frac{d^2z}{ds^2} - \frac{dz}{ds}\frac{d^2y}{ds^2}\right)\hat{\mathbf{i}}\right.$$

$$+ \left(\frac{dz}{ds}\frac{d^2x}{ds^2} - \frac{dx}{ds}\frac{d^2z}{ds^2}\right)\hat{\mathbf{j}}$$

$$\left. + \left(\frac{dx}{ds}\frac{d^2y}{ds^2} - \frac{dy}{ds}\frac{d^2x}{ds^2}\right)\hat{\mathbf{k}}\right]$$

Since $\mathbf{b}$ is a unit vector, the direction cosines of the binormal are

$$\frac{1}{\kappa}\left(\frac{dy}{ds}\cdot\frac{d^2z}{ds^2} - \frac{dz}{ds}\cdot\frac{d^2y}{ds^2}\right), \quad \frac{1}{\kappa}\left(\frac{dz}{ds}\frac{d^2x}{ds^2} - \frac{dx}{ds}\frac{d^2z}{ds^2}\right),$$

$$\frac{1}{\kappa}\left(\frac{dx}{ds}\frac{d^2y}{ds^2} - \frac{dy}{ds}\frac{d^2x}{ds^2}\right).$$

2.9. EXPRESSIONS FOR Ρ AND Σ IN TERMS OF THE DERIVATIVES OF R WITH RESPECT TO AN ARBITRAY PARAMETER *T*.

We have

$$\frac{d\mathbf{r}}{ds} = \frac{d\mathbf{r}}{dt}\frac{dt}{ds},$$

$$\frac{d^2\mathbf{r}}{ds^2} = \frac{d^2\mathbf{r}}{dt^2}\left(\frac{dt}{ds}\right)^2 + \frac{d\mathbf{r}}{dt}\frac{d^2t}{ds^2},$$

$$\frac{d^2\mathbf{r}}{ds^2} = \frac{d^2\mathbf{r}}{dt^2}\left(\frac{dt}{ds}\right)^2 + 3\,\frac{d^2\mathbf{r}}{dt^2}\cdot\frac{dt}{ds}\,\frac{d^2t}{ds^2} + \frac{d\mathbf{r}}{dt}\,\frac{d^3t}{ds^3}.$$

$$\therefore \quad \frac{d\mathbf{r}}{ds}\times\frac{d^2\mathbf{r}}{ds^2} = \frac{d\mathbf{r}}{dt}\times\frac{d^2\mathbf{r}}{dt^2}\left(\frac{dt}{ds}\right)^2,$$

or

$$\kappa = \left|\frac{d\mathbf{r}}{ds}\times\frac{d^2\mathbf{r}}{ds}\right| = \frac{\left|\dfrac{d\mathbf{r}}{dt}\times\dfrac{d^2\mathbf{r}}{dt^2}\right|}{\left(\dfrac{ds}{dt}\right)^2}.$$

Also

$$\left|\frac{d\mathbf{r}}{ds}\times\frac{d^2\mathbf{r}}{ds^2}\cdot\frac{d^3\mathbf{r}}{ds^3}\right| = \frac{d\mathbf{r}}{dt}\times\frac{d^2\mathbf{r}}{dt^2}\cdot\frac{d^3\mathbf{r}}{dt^3}\left(\frac{dt}{ds}\right)^2.$$

$$\therefore \quad \kappa^2\tau = -\left[\frac{d\mathbf{r}}{ds}\,\frac{d^2\mathbf{r}}{ds^2}\,\frac{d^3\mathbf{r}}{ds^3}\right] = -\frac{\left[-\dfrac{d\mathbf{r}}{dt}\,\dfrac{d^2\mathbf{r}}{dt^2}\,\dfrac{d^3\mathbf{r}}{dt^3}\right]}{\left(\dfrac{ds}{dt}\right)^2}.$$

Since

$$\left|\frac{d\mathbf{r}}{dt}\right| = \left|\frac{d\mathbf{r}}{ds}\,\frac{ds}{dt}\right| = \left|\frac{d\mathbf{r}}{ds}\right|\left|\frac{ds}{dt}\right| = \frac{ds}{dt}, \text{ for } \left|\frac{d\mathbf{r}}{ds}\right| = 1,$$

we can re-write the results obtained above as follows:

$$\kappa = \left|\frac{d\mathbf{r}}{dt}\times\frac{d^2\mathbf{r}}{dt^2}\right| \Big/ \left|\frac{d\mathbf{r}}{dt}\right|^2,$$

$$\kappa^2\tau = -\left[\frac{d\mathbf{r}}{dt}\,\frac{d^2\mathbf{r}}{dt^2}\,\frac{d^3\mathbf{r}}{dt^3}\right]\left|\frac{d\mathbf{r}}{dt}\right|^6.$$

2.10. EXPRESSIONS FOR P AND Σ OBTAINED INDEPENDENTLY OF FRENET'S FORMULAE

We shall now obtain the expressions for, ρ, and σ, in terms of the derivatives of, $\mathbf{r}$, with respect to directly from the definitions without employing the Frenet's formulae. This derivation will keep the sign of, σ, undetermined.

2.10.1. The tangents at $P(t)$ and $Q(t + \delta t)$ to the curve

$$\mathbf{r} = \mathbf{f}(t)$$

are parallel to the vectors

$$\mathbf{f}'(t),\ \mathbf{f}'(t+\delta t),$$

so that if, $\delta\theta$, lying between 0 and π, be the angle between the same, we have

$$\sin \delta\theta = \frac{|\mathbf{f}'(t) \times \mathbf{f}'(t+\delta t)|}{|\mathbf{f}'(t)||\mathbf{f}'(t+\delta t)|} = \frac{|\mathbf{f}'(t) \times [\mathbf{f}'(t+\delta t) - \mathbf{f}'(t)|}{|\mathbf{f}'(t)||\mathbf{f}'(t+\delta t)|}$$

$$\therefore \quad \left|\frac{\delta\theta}{\delta s}\right| = \left|\frac{\delta\theta}{\sin\delta\theta}\,\frac{\sin\delta\theta}{\delta t}\,\frac{\delta t}{\theta s}\right|$$

$$= \left|\frac{\delta\theta}{\sin\delta\theta}\right|\left|\frac{\sin\delta\theta}{\delta t}\right|\left|\frac{\delta t}{\theta s}\right|$$

$$= \left|\frac{\delta\theta}{\sin\delta\theta}\right|\frac{|\mathbf{f}'(t) \times [\mathbf{f}'(t+\delta t) - \mathbf{f}'(t)]/\delta t|}{|\mathbf{f}'(t)|\ |\mathbf{f}'(t+\delta t)|}\left|\frac{\delta t}{\delta s}\right|.$$

Let $Q \to P$.

$$\therefore \quad \kappa = \frac{|\mathbf{f}'(t) \times \mathbf{f}''(t)|}{|\mathbf{f}'(t)|^3} \times \frac{\left|\frac{d\mathbf{r}}{dt} \times \frac{d^2\mathbf{r}}{dt^2}\right|}{\left|\frac{d\mathbf{r}}{dt}\right|^3}$$

We note that

$$\left|\frac{dt}{ds}\right| = \frac{1}{|ds/dt|} = \frac{1}{|d\mathbf{f}/dt|} = \frac{1}{|\mathbf{f}'(t)|}.$$

2.10.2. The binormals at $P(t)$ and $Q(t + \delta t)$ are parallel to the vectors

$$\mathbf{f}'(t) \times \mathbf{f}''(t),\ \mathbf{f}'(t+\delta t) \times \mathbf{f}''(t+\delta t),$$

respectively so that if $\delta\psi$ be the angle between the same, we have

$$|\sin \delta\phi| = \frac{|[\mathbf{f}'(t) \times \mathbf{f}''(t)] \times [\mathbf{f}'(t+\delta t) \times \mathbf{f}''(t+\delta t)]|}{|\mathbf{f}'(t) \times \mathbf{f}''(t)||[\mathbf{f}'(t+\delta t) \times \mathbf{f}''(t+\delta t)]|}.$$

The numerator of $|\sin \delta\phi|$

$$= |\ [\mathbf{f}'(t) \times \mathbf{f}''(t)\ .\ \mathbf{f}''(t+\delta t)]\,\mathbf{f}'(t+\delta t) - [\mathbf{f}'(t) \times \mathbf{f}''(t)\ .\ \mathbf{f}'(t+\delta t)]\,\mathbf{f}''(t+\delta t)\ |$$

$$= \left[\mathbf{f}'(t) \times \mathbf{f}''(t) \cdot \frac{\mathbf{f}''(t+\delta t) - \mathbf{f}''(t)}{\delta t}\right]\mathbf{f}'(t+\delta t) -$$

$$\left[\mathbf{f}'(t) \times \mathbf{f}''(t) \cdot \frac{\mathbf{f}'(t+\delta t) - \mathbf{f}'(t)}{\delta t}\right](t+\delta t)\ |\delta t|.$$

It is now easy to see that

$$\lim\left|\frac{\sin\delta\psi}{\delta t}\right| = \frac{|[\mathbf{f}' \times \mathbf{f}''\ .\ \mathbf{f}''']\mathbf{f}'' - [\mathbf{f}' \times \mathbf{f}''\ .\ \mathbf{f}''']\mathbf{f}''|}{|\mathbf{f}' \times \mathbf{f}''|^2}$$

$$= \frac{|[\mathbf{f}'\ \mathbf{f}''\ \mathbf{f}''']\mathbf{f}'|}{|\mathbf{f}' \times \mathbf{f}''|^2}.$$

Now
$$\left|\frac{\delta\psi}{\delta s}\right| \doteq \left|\frac{\delta\psi}{\sin\delta\psi}\right|\left|\frac{\sin\delta\psi}{\delta t}\right|\left|\frac{\delta t}{\delta s}\right|.$$

$\therefore$
$$|\tau| = \lim\left|\frac{\delta\psi}{\delta s}\right| = \frac{|[\mathbf{f}'\ \mathbf{f}''\ \mathbf{f}''']\mathbf{f}'|}{|\mathbf{f}'\times\mathbf{f}''|^2}\cdot\frac{1}{|\mathbf{f}'|^2}$$

for $\left|\dfrac{dt}{ds}\right| = \dfrac{1}{|\mathbf{f}'|}$

$$= \frac{|[\mathbf{f}'\ \mathbf{f}''\ \mathbf{f}''']|}{|\mathbf{f}'\times\mathbf{f}''|^2} = \frac{\left|\left[\dfrac{d\mathbf{r}}{dt}\ \dfrac{d^2\mathbf{r}}{dt^2}\ \dfrac{d^3\mathbf{r}}{dt^3}\right]\right|}{\left|\dfrac{d\mathbf{r}}{dt}\times\dfrac{d^2\mathbf{r}}{dt^2}\right|^2}$$

$\therefore$
$$\kappa^2|\tau| = \frac{\left|\left[\dfrac{d\mathbf{r}}{dt}\ \dfrac{d^2\mathbf{r}}{dt^2}\ \dfrac{d^3\mathbf{r}}{dt^3}\right]\right|}{\left|\dfrac{d\mathbf{r}}{dt}\right|^6}.$$

2.11. REPRESENTATION OF AN ARC IN THE NEIGHBOURHOOD OF A POINT

Let O be any given point on a curve and let arc $OP = s$; P being any point on the curve. Let

$$\mathbf{r} = \mathbf{f}(s),$$

be the equation of the curve. We suppose that $\mathbf{f}(s)$ is expressible as a Maclaurin's series in the neighbourhood of $s = 0$. Thus, we have

$$\mathbf{r} = \mathbf{f}(s) = \mathbf{f}(0) + s\mathbf{f}'(0) + \frac{s^2}{2!}\mathbf{f}''(0) + \frac{s^3}{3!}\mathbf{f}'''(0) + \dots \quad \dots(1)$$

This is an easy consequence of the corresponding theorem in Scalar Calculus. We denote the unit vectors **t, n, b** at O by **i, j, k** respectively.

Now

$$\mathbf{f}'(s) = \frac{d\mathbf{r}}{ds} = \mathbf{t},$$

$$\mathbf{f}''(s) = \frac{d^2\mathbf{r}}{ds^2} = \frac{d\mathbf{t}}{ds} = \kappa\mathbf{n} = \frac{1}{\rho}\mathbf{n},$$

$$\mathbf{f}'''(s) = \frac{d^2\mathbf{r}}{ds^3} = \frac{1}{\rho}\frac{d\mathbf{n}}{ds} - \frac{\rho'}{\rho^2}\mathbf{n} = -\frac{1}{\rho}\left(\frac{1}{\rho}\mathbf{t} + \frac{1}{\sigma}\mathbf{b}\right) - \frac{\rho'}{\rho^2}\mathbf{n}.$$

Here

$$\mathbf{f}'(0),\ \mathbf{f}''(0),\ \mathbf{f}'''(0)$$

will be obtained by regarding

$$\mathbf{t}, \mathbf{n}, \mathbf{b}, \rho, \rho', \sigma$$

to have the values as at O. Also we write

$$\mathbf{r} = x\mathbf{i} + y\mathbf{j} + z\mathbf{k},$$

so that x, y, z are the co-ordinates of P referred to the tangent, principal normal and binormal at O as co-ordinate axes.

Substituting in (1), and rejecting powers of, s, higher than three, we obtained

$$x\mathbf{i} + y\mathbf{j} + z\mathbf{k} = \mathbf{0} + s\mathbf{i} + \frac{s^2}{2!}\frac{1}{\rho}\mathbf{j} + \frac{s^3}{3!}\left(-\frac{1}{\rho^2}\mathbf{i} - \frac{1}{\rho\sigma}\mathbf{k} - \frac{\rho'}{\rho^2}\mathbf{j}\right)$$

$$\therefore \quad x = s - \frac{s^3}{6\rho^2}, \quad y = \frac{s^2}{2\rho} - \frac{s^3\rho'}{6\rho^2}, \quad z = -\frac{s^2}{2\rho\sigma}.$$

These give the co-ordinates x, y, z of a point P on the curve in a neighbourhood of O.

2.12. SPHERICAL INDICATRICES

The locus of the point whose position vector is equal to the unit tangent vector at any point of a given curve is called the spherical indicatrix of the tangent. **(Garhwal 2000, 2002)**

We may similarly have the spherical indicatrices of the principal normal and binormal.

Clearly the three spherical indicatrices lie on the sphere whose centre is the origin and radius unity.

We shall now proceed to obtain the radii of curvature and torsion of the spherical indicatrices in terms of those of the given curve.

The position vector, $\mathbf{R}$, of any point on the spherical indicatrix of the tangent for the curve $\mathbf{r} = \mathbf{f}(s)$ is given by

$$\mathbf{R} = \mathbf{t}, \text{ where } \mathbf{t} = \frac{d\mathbf{r}}{ds}$$

$$\therefore \quad \frac{d\mathbf{R}}{ds} = \frac{d\mathbf{t}}{ds} = \kappa\mathbf{n},$$

and

$$\frac{d^2\mathbf{R}}{ds^2} = \kappa'\mathbf{n} - \kappa(\kappa\mathbf{t} + \tau\mathbf{b}) = -\kappa^2\mathbf{t} - \kappa\tau\mathbf{b} + \kappa'\mathbf{n}.$$

The curvature of the spherical indicatrix

$$= \frac{\left|\dfrac{d\mathbf{R}}{ds} \times \dfrac{d^2\mathbf{R}}{ds^2}\right|}{\left|\dfrac{d\mathbf{R}}{ds}\right|^3} = \frac{\left|\kappa^3\mathbf{b} - \kappa^2\tau\mathbf{t}\right|}{|\kappa\mathbf{n}|^3}$$

$$= \frac{\sqrt{(\kappa^6 + \kappa^4 + \tau^2)}}{3} = \frac{\sqrt{(\kappa^2 + \tau^2)}}{\kappa}.$$

Also, we have

$$\frac{d^3\mathbf{R}}{ds^3} = -3\kappa\kappa'\mathbf{t} - (2\kappa'\tau + \kappa\tau')\mathbf{b} + (\kappa t^2).$$

$\therefore$ torsion of the spherical indicatrix

$$= -\frac{\left[\frac{d\mathbf{R}}{ds}\ \frac{d^2\mathbf{R}}{ds^2}\ \frac{d^3\mathbf{R}}{ds^3}\right]}{\left|\frac{d\mathbf{R}}{ds}\right|^6} \cdot \frac{\kappa^2}{(\kappa^2+\tau^2)} = \frac{\kappa'\tau - \kappa\tau'}{\kappa(\kappa^2+\tau^2)}.$$

We may similarly treat the case of the spherical indicatrices of the principal normal and binormal. It can be shown that the curvature and torsion of the spherical indicatrix of binormal are

$$\frac{\sqrt{(\kappa^2+\tau^2)}}{\tau},\ \frac{\kappa'\tau - \kappa\tau'}{\tau(\kappa^2+\tau^2)}.$$

2.13. SCALAR FORMULATION IN TERMS OF RECTANGULAR CARTESIAN CO-ORDINATES

This section is devoted to giving the scalar equivalents, in terms of Rectangular Co-ordinate axes, of the results in the Differential Geometry of curves which we have obtained.

As usual, **i, j, k** denote unit vectors along the axes. Let the curve be

$$r = \mathbf{f}(t) = \mathbf{i}x + \mathbf{j}y + \mathbf{k}z,$$

so that x, y, z are functions of t. Also we shall denote the derivatives by dashes. Thus we write

$$\frac{dx}{dt} = x',\ \frac{d^2x}{dt} = x'',\ \frac{d^3x}{dt^3} = x''' \text{ etc.}$$

Also for the current position vector **R** of any point on the locus associated with the given curve, we write

$$\mathbf{R} = \mathbf{i}X + \mathbf{j}Y + \mathbf{k}Z.$$

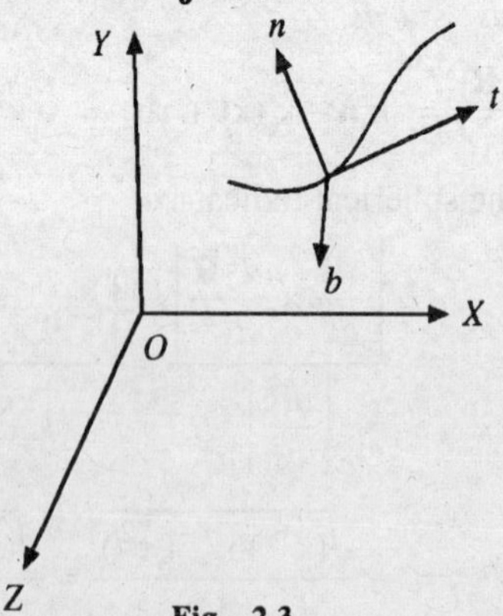

Fig. 2.3

The *tangent* at (x, y, z) is

$$\mathbf{R} = \mathbf{r} + u\frac{d\mathbf{r}}{dt},$$

i.e.,

$$\mathbf{i}X + \mathbf{j}Y + \mathbf{k}Z = (\mathbf{i}x + \mathbf{j}y + \mathbf{k}z) + u(\mathbf{i}x' + \mathbf{j}y' + \mathbf{k}z') \qquad \text{§ 2.1}$$

or

$$\frac{X-x}{x'} = \frac{Y-y}{y'} = \frac{Z-z}{z'}.$$

The *osculating piece* at (x, y, z) is

$$(\mathbf{R}-2)\ .\ \frac{d\mathbf{r}}{dt} \times \frac{d^2\mathbf{r}}{dt^2} = 0. \qquad \text{§ 2.2}$$

i.e.,

$$\begin{vmatrix} X-x & Y-y & Z-z \\ x' & y' & z' \\ x'' & y'' & z'' \end{vmatrix} = 0.$$

The *normal plane at* (x, y, z) is

$$(\mathbf{R}-\mathbf{r})\ .\ \frac{d\mathbf{r}}{dt} = 0, \qquad \text{§ 2.3}$$

i.e., $\Sigma(X-x)x' = 0.$

The *binormal* is parallel to the vector

$$\frac{d\mathbf{r}}{dt} \times \frac{d^2\mathbf{r}}{dt^2}, \qquad \text{§ 2.4}$$

i.e., $(\mathbf{i}x' + \mathbf{j}y' + \mathbf{K}z') \times (\mathbf{i}x'' + \mathbf{j}y'' + \mathbf{z}z''),$

so that the *direction ratios of the binormal are*

$$y'z'' - y''z',\ z'x'' - z''x',\ x'y'' - x''y'.$$

The *principal normal* is parallel to the vector

$$\left(\frac{d\mathbf{r}}{dt}\right)^2 \frac{d^2\mathbf{r}}{dt^2} - \left(\frac{d\mathbf{r}}{dt} \cdot \frac{d^2\mathbf{r}}{dt^2}\right)\frac{d\mathbf{r}}{dt}, \qquad \text{§ 2.5}$$

i.e., $(\Sigma x'^2)\ \Sigma\mathbf{i}x'' - (\Sigma x'x'')\ (\Sigma\mathbf{i}x'),$

so that the *direction ratios of the principal normal are*

$$y'(x''y' - x'y'') + z'(x''z' - x'z''), \text{ etc.}$$

If, s, be taken as the parameter so that dashes denote derivatives with respect to s, then the *direction ratios of the tangent, principal normal and binormal take the form*

$$x', y', z';\ x'', y'', z'';\ y'z'' - y''z',\ z'x'' - z''x',\ x'y'' - x''y'. \qquad (\text{§ } 2.7)$$

Frenet's formulae. If

$$l_1, m_1, n_1;\ l_2, m_2, n_2;\ l_3, m_3, n_3$$

denote the direction cosines of the tangent, principal normal and binormal so that

$$\mathbf{t} = \mathbf{i}l_1 + \mathbf{j}m_1 + \mathbf{k}n_1; \quad \mathbf{n} = \mathbf{i}l_2 + \mathbf{j}m_2 + \mathbf{k}n_2; \quad \mathbf{b} = \mathbf{i}l_3 + \mathbf{j}m_3 + \mathbf{k}n_3.$$

then, making substitutions in the results of § 15, we obtain

$$\frac{dl_1}{ds} = \frac{l_2}{\rho}, \quad \frac{dm_1}{ds} = \frac{m_2}{\rho}, \quad \frac{dn_1}{ds} = \frac{n_2}{\rho},$$

$$\frac{dl_2}{ds} = -\frac{l_1}{\rho} - \frac{l_3}{\sigma}, \quad \frac{dm_2}{ds} = -\frac{m_1}{\rho} - \frac{m_3}{\sigma}, \quad \frac{dn_2}{ds} = -\frac{n_1}{\rho} - \frac{n_3}{\rho}.$$

$$\frac{dl_3}{ds} = \frac{l_2}{\sigma}, \quad \frac{dm_3}{ds} = \frac{m_2}{\sigma}, \quad \frac{dn_3}{ds} = \frac{n_2}{\sigma},$$

as Frenet's Formulae formulated scalarly.

Curvature. We have § 2.9

$$\kappa = \left|\frac{d\mathbf{r}}{dt} \times \frac{d^2\mathbf{r}}{dt^2}\right| \Big/ \left|\frac{d\mathbf{r}}{dt}\right|^3$$

$$= \frac{|\Sigma \mathbf{i}x' \times \Sigma \mathbf{i}x''|}{|\Sigma \mathbf{i}x'|^3}$$

$$= \frac{\sqrt{\Sigma (y'z'' - y''z')^2}}{\Sigma (x'^2)^{3/2}}.$$

Torsion. We have § 2.9

$$\tau = -\frac{1}{\kappa^2} \frac{\left[\dfrac{d\mathbf{r}}{dt} \; \dfrac{d^2\mathbf{r}}{dt^2} \; \dfrac{d^3\mathbf{r}}{dt^3}\right]}{\left|\dfrac{d\mathbf{r}}{dt}\right|^6}$$

$$= -\frac{1}{\kappa^2} \frac{\begin{vmatrix} x' & y' & z' \\ x'' & y'' & z'' \\ x''' & y''' & z''' \end{vmatrix}}{(x'^2 + y'^2 + z'^2)^3}.$$

EXAMPLES

1. *Find the osculating plane, curvature and torsion at any point of the curve*

$$x = a \cos 2t, \quad y = a \sin 2t, \quad z = 2a \sin t.$$

For the position vector $\mathbf{r}$ of any point of the curve,

$$\mathbf{r} = \mathbf{i}a \cos 2t + \mathbf{j}a \sin 2t + \mathbf{k}^2 a \sin t.$$

We write

$$\mathbf{r} = (a \cos 2t, \; a \sin 2t, \; 2a \sin t)$$

where the meaning of the notation is self evident.

$$\therefore \quad \frac{d\mathbf{r}}{dt} = (-2a \sin 2t, \; 2a \cos 2t, \; 2a \cos t),$$

$$\frac{d^2\mathbf{r}}{dt^2} = (-4a\cos 2t, -4a\sin 2t, -2a\sin t),$$

$$\frac{d^3\mathbf{r}}{dt^3} = (-8a\sin 2t, -8a\cos 2t, -2a\cos t).$$

$$\therefore \quad \frac{d\mathbf{r}}{dt}\times\frac{d^2\mathbf{r}}{dt^2} = [4a^2(\sin t + \sin 2t\cos t),$$
$$-4a^2(\cos t + \cos t\cos 2t), 8a^2]$$

Thus the osculating plane is

$$(\mathbf{R}-\mathbf{r}) \cdot \frac{d\mathbf{r}}{dt}\times\frac{d^2\mathbf{r}}{dt^2} = 0.$$

i.e., $(x - a\cos 2t)(\sin t + \sin 2t\cos t) -$
$$(y - a\sin 2t)(\cos t + \cos t\cos 2t) + (z - 2a\sin t)2 = 0,$$

or $(\sin t + \sin 2t\cos t)x - (\cos t + \cos t + \cos t\cos 2t)y + 2z = 3a\sin t.$

Also curvature

$$= \left|\frac{d\mathbf{r}}{dt}\times\frac{d^2\mathbf{r}}{dt^2}\right| \Big/ \left|\frac{d\mathbf{r}}{dt}\right|^3$$

$$= \frac{4a^2[(\sin t + \sin 2t\cos t)^2 + (\cos t + \cos 2t\cos t)^2 + 4]^{\frac{1}{2}}}{8a^3[\sin^2 2t + \cos^2 2t + \cos^2 t]^{\frac{3}{2}}}$$

$$= \frac{1}{2a}\sqrt{\left[\frac{5+3\cos^2 t}{(1+\cos^2 t)^3}\right]}.$$

$$\text{Torsion} = \frac{\dfrac{d\mathbf{r}}{dt}\times\dfrac{d^2\mathbf{r}}{dt^2}\cdot\dfrac{d^3\mathbf{r}}{dt^3}}{\left|\dfrac{d\mathbf{r}}{dt}\times\dfrac{d^2\mathbf{r}}{dt^2}\right|^2} = \frac{3}{a(5\sec t + 3\cos t)}.$$

2. *For the curve* $x = a(3t - t^3),\ y = 3at^2,\ z = a(3t + t^3),$

prove that $$\kappa = \tau = \frac{1}{3a(1+t^2)^2}.$$

In this case $\mathbf{r} = (3at - at^3,\ 3at^2,\ 3at + at^3)$

$$\therefore \quad \frac{d\mathbf{r}}{dt} = (3a - 3at^2,\ 6at,\ 3a + 3at^2)$$

$$\frac{d^2\mathbf{r}}{dt^2} = (-6at,\ 6a,\ 6at)$$

and $$\frac{d^3\mathbf{r}}{dt^3} = (-6a, 0, 6a)$$

$$\therefore \quad \frac{d\mathbf{r}}{dt} = 3a\sqrt{\left[(1-t^2)^2 + 4t^2 + (1+t^2)^2\right]}$$

$$= 3\sqrt{2}\, a\,(1+t^2)$$

$$\frac{d\mathbf{r}}{dt} \times \frac{d^2\mathbf{r}}{dt^2} = (18\, a^2t^2 - 18a^2,\ -36a^2t,\ 18a^2 + 18a^2t^2)$$

$$\therefore \quad \left|\frac{d\mathbf{r}}{dt} \times \frac{d^2\mathbf{r}}{dt^2}\right| = 18\sqrt{2}\; a^2\,(1+t^2)$$

$$\therefore \quad \kappa = \frac{\left|\frac{d\mathbf{r}}{dt} \times \frac{d^2\mathbf{r}}{dt^2}\right|}{\left|\frac{d\mathbf{r}}{dt}\right|^3} = \frac{18\sqrt{2}\, a^2\,(1+t^2)}{54\sqrt{2}\, a^3\,(1+t^2)^3}$$

$$= \frac{1}{3a(1+t^2)^2}$$

Again,

$$\left[\frac{d\mathbf{r}}{dt}\ \frac{d^2\mathbf{r}}{dt^2}\ \frac{d^3\mathbf{r}}{dt^3}\right] = \begin{vmatrix} 3a-3at^2 & 6at & 3a+3at^2 \\ -6at & 6a & 6at \\ -6a & 0 & 6a \end{vmatrix} = 216\, a^3$$

$$\therefore \quad \tau = \frac{\left[\frac{d\mathbf{r}}{dt}\ \frac{d^2\mathbf{r}}{dt^2}\ \frac{d^3\mathbf{r}}{dt^3}\right]}{\left|\frac{d\mathbf{r}}{dt} \times \frac{d^2\mathbf{r}}{dt^2}\right|^2} = \frac{216\, a^3}{\left\{18\sqrt{2}\, a^2\,(1+t^2)\right\}^2}$$

$$= \frac{1}{3a(1+t^2)^2}.$$

3. *Prove that at the point of intersection of the surfaces*

$$x^2 + y^2 = z^2, \quad z = a \tan^{-1}(y/x),$$

where $$y = x\tan\theta,$$

$$\rho = \frac{a(2+\theta^2)^{3/2}}{(8+5\theta^2+\theta^4)} \quad \text{and} \quad \sigma = \frac{a(8+5\theta^2+\theta^4)}{6+\theta^2}.$$

For the point of intersection

$$y = x\tan\theta \Rightarrow z = a\tan^{-1}(y/x) = a\theta$$

and $x^2 + y^2 = z^2 \Rightarrow x = a\theta \cos\theta,\ y = a\,\theta \sin\theta$

$$\therefore \quad \mathbf{r} = (a\,\theta \cos\theta,\ a\,\theta \sin\theta,\ a\,\theta)$$

Then
$$\frac{d\mathbf{r}}{d\theta} \times \frac{d^2\mathbf{r}}{d\theta^2} = a^2\,[\theta \sin\theta - 2\cos\theta,\ -\theta\cos\theta - 2\sin\theta,$$
$$(\theta\sin\theta - \cos\theta)\ (\theta\sin\theta - 2\cos\theta)$$
$$+ (\theta\cos\theta + \sin\theta)\,(\theta\cos\theta + 2\sin)]$$
$$= a^2\,[\theta\sin\theta - 2\cos\theta, -\theta\cos\theta - 2\sin\theta, \theta^2 + 2]$$

$$\Rightarrow \quad \left|\frac{d\mathbf{r}}{d\theta} \times \frac{d^2\mathbf{r}}{d\theta^2}\right|^2 = a^4\,(8 + 5\,\theta^2 + \theta^4)$$

$$\left|\frac{d\mathbf{r}}{d\theta}\right|^2 = a^2\,(2 + \theta^2)$$

$$\therefore \quad \rho = \frac{1}{\kappa} = \frac{\left|\dfrac{d\mathbf{r}}{d\theta}\right|^3}{\left|\dfrac{d\mathbf{r}}{d\theta} \times \dfrac{d^2\mathbf{r}}{d\theta^2}\right|} = \frac{a^3\,(2+\theta^2)^{3/2}}{a^2\,(8+5\theta^2+\theta^4)}$$

$$= \frac{a(2+\theta^2)^{3/2}}{(8+5\theta^2+\theta^4)}$$

Also
$$\left|\frac{d\mathbf{r}}{d\theta}, \frac{d^2\mathbf{r}}{d\theta^2}, \frac{d^3\mathbf{r}}{d\theta^3}\right| = a^3\,(6 + \theta^2)$$

Then
$$\sigma = \frac{1}{\tau} = \frac{\left|\dfrac{d\mathbf{r}}{d\theta} \times \dfrac{d^2\mathbf{r}}{d\theta^2}\right|^2}{\left[\dfrac{d\mathbf{r}}{d\theta}\ \dfrac{d^2\mathbf{r}}{d\theta^2}\ \dfrac{d^3\mathbf{r}}{d\theta^3}\right]}$$

$$= \frac{a^4\,(8+5\theta^2+\theta^4)}{a^3\,(6+\theta^2)} = \frac{a(8+5\theta^2+\theta^4)}{6+\theta^2}.$$

4. *If the tangent to a curve makes a constant angle, α, with a fixed line, then*

$$\sigma = \pm\,\rho \tan\alpha.$$

Conversely, show that if σ/ρ is constant, the tangent makes a constant angle with a fixed direction.

Let, **e**, be the unit vector parallel to the given fixed line so that, as given,

$$\mathbf{t}.\,\mathbf{e} = \cos\alpha. \qquad ...(1)$$

Differentiating, we get

$$\frac{d\mathbf{t}}{ds} \cdot \mathbf{e} = 0,$$

or $$\kappa\, \mathbf{n} \cdot \mathbf{e} = 0. \qquad \text{(Frenet's first)}$$

Thus $$\mathbf{n} \cdot \mathbf{e} = 0 \qquad ...(2)$$

Hence $\mathbf{n}$ is perpendicular to $\mathbf{e}$. Thus the vectors $\mathbf{b}, \mathbf{t}, \mathbf{e}$ are coplanar.

$$\therefore \quad \mathbf{b} \cdot \mathbf{e} = \pm \sin \alpha. \qquad ...(3)$$

Differentiating (2) and employing Frenet's second, we get

$$\frac{d\mathbf{n}}{ds} \cdot \mathbf{e} = 0$$

i.e., $$-(\kappa \mathbf{t} + \tau \mathbf{b}) \cdot \mathbf{e} = 0$$

$$\therefore \quad \kappa \cos \alpha \pm \tau \sin \alpha = 0 \qquad \text{[From (1) and (3)]}$$

or $$\sigma = \pm \rho \tan \alpha.$$

Converse. Let $\sigma = a\rho$ where a, is some constant scalar.

Now

$$\frac{d\mathbf{t}}{ds} = \frac{1}{\rho}\mathbf{n} \text{ and } \frac{d\mathbf{b}}{ds} = \frac{1}{\sigma}\mathbf{n}.$$

$$\therefore \quad \rho \frac{d\mathbf{t}}{ds} = \mathbf{n} = \sigma \frac{d\mathbf{b}}{ds},$$

or $$\frac{d\mathbf{t}}{ds} = \frac{\sigma}{\rho} \frac{d\mathbf{b}}{ds} = a \frac{d\mathbf{b}}{ds}.$$

Integrating, we get

$$\mathbf{t} = a\mathbf{b} + \mathbf{c},$$

where $\mathbf{c}$ is a constant vector.

Multiplying scalarly with $\mathbf{t}$, we get

$$\mathbf{t} \cdot \mathbf{c} = 1.$$

Hence the tangent makes a constant angle with the direction of the fixed vector $\mathbf{c}$.

5. *Prove that the Serret-Frenet formulae can be written in the form*

$$\frac{d\mathbf{t}}{ds} = \mathbf{w} \times \mathbf{t}, \quad \frac{d\mathbf{n}}{ds} = \mathbf{w} \times \mathbf{n}, \quad \frac{d\mathbf{b}}{ds} = \mathbf{w} \times \mathbf{b}$$

and determine $\mathbf{w}$. (*Here* $\mathbf{w}$ *is called Darboux vector of the curve.*)

We have

$$\frac{d\mathbf{t}}{ds} = \kappa \mathbf{n} = \tau\, \mathbf{t} \times \mathbf{t} + \kappa\, \mathbf{b} \times \mathbf{t}$$

$$(\because \ \mathbf{t} \times \mathbf{t} = \mathbf{0} \text{ and } \mathbf{b} \times \mathbf{t} = \mathbf{n})$$

$$= (\tau \mathbf{t} + \kappa \mathbf{b}) \times \mathbf{t} \qquad ...(1)$$

$$\frac{d\mathbf{n}}{ds} = \tau \mathbf{b} - \kappa \mathbf{t} = \tau(\mathbf{t} \times \mathbf{n}) + \kappa\,(\mathbf{b} \times \mathbf{n})$$

$$= (\tau \mathbf{t} + \kappa \mathbf{b}) \times \mathbf{n} \qquad ...(2)$$

and $$\frac{d\mathbf{b}}{ds} = -\tau \mathbf{n} = \tau(\mathbf{t} \times \mathbf{b}) + \kappa\,(\mathbf{b} \times \mathbf{b})$$

$$= (\tau \mathbf{t} + \kappa \mathbf{b}) \times \mathbf{f} \qquad ...(3)$$

From eqns. (1), (2) and (3), we get

$$\frac{d\mathbf{t}}{ds} = \mathbf{w}\times\mathbf{t},\quad \frac{d\mathbf{n}}{ds} = \mathbf{w}\times\mathbf{n},\quad \frac{d\mathbf{b}}{ds} = \mathbf{w}\times\mathbf{b}$$

where $\mathbf{w} = \tau\mathbf{t} + \kappa\mathbf{b}$.

6. *Prove that*

$$\frac{d^3\mathbf{r}}{ds^3} = \frac{d\mathbf{k}}{ds}\,\mathbf{n} - \kappa^2\, t + \kappa\,\tau\,\mathbf{b} \text{ and hence show that}$$

$$\frac{d^4\mathbf{r}}{ds^4} = \left(\frac{d^2\kappa}{ds^2} - \kappa^3 - \kappa\tau^2\right)\mathbf{n} - 3\frac{d\mathbf{k}}{ds}\kappa\mathbf{t} + \left(2\frac{d\kappa}{ds}\tau + \frac{d\tau}{ds}\kappa\right)\mathbf{b}$$

we have

$$\frac{d^2\mathbf{r}}{ds^2} = \kappa\,\mathbf{n}$$

$$\therefore \quad \frac{d^3\mathbf{r}}{ds^3} = \kappa\frac{d\mathbf{n}}{ds} + \frac{d\kappa}{ds}\mathbf{n} = \kappa(\tau\mathbf{b} - \kappa\mathbf{t}) + \frac{d\kappa}{ds}\mathbf{n}$$

$$= \frac{d\kappa}{ds}\mathbf{n} - \kappa^2\,\mathbf{t} + \kappa\tau\,\mathbf{b}.$$

Again,

$$\frac{d^4\mathbf{r}}{ds^4} = \frac{d^2\kappa}{ds^2}\mathbf{n} + \frac{d\kappa}{ds}\frac{d\mathbf{n}}{ds} - 2k\frac{d\kappa}{ds}\mathbf{t} - \kappa^2\frac{d\mathbf{t}}{ds} + \left(k\frac{d\tau}{ds} + \frac{d\kappa}{ds}\tau\right)\mathbf{b} + \kappa\tau\frac{d\mathbf{b}}{ds}$$

$$= \frac{d^2k}{ds^2}\mathbf{n} + \frac{d\kappa}{ds}(\tau\mathbf{b} - \kappa\mathbf{t}) - 2\kappa\frac{d\kappa}{ds}\mathbf{t} - \kappa^3\,\mathbf{n} + \left(k\frac{d\tau}{ds} + \frac{d\kappa}{ds}\tau\right)\mathbf{b} - \kappa\tau^2\,\mathbf{n}$$

$$\therefore \quad = \left(\frac{d^2\kappa}{ds^2} - \kappa^3 - \kappa\tau^2\right)\mathbf{n} - 3\kappa\frac{d\kappa}{ds}\mathbf{t} + \left(2\frac{d\kappa}{ds}\tau + \kappa\frac{d\tau}{ds}\right)\mathbf{b}.$$

7. ***The osculating plane at every point of a curve touches a fixed sphere; show that the plane through the tangent perpendicular to the principal normal passes through the centre of the sphere.***

Let, **a**, be the centre and, p, the radius of the given sphere.

The osculating plane

$$(\mathbf{R} - \mathbf{r})\,.\,\mathbf{b} = 0,$$

will touch the sphere if

$$(\mathbf{a}-\mathbf{r}) \cdot \mathbf{b} = p.$$

Differentiating *w.r.* to *s*, we obtain

$$(\mathbf{a}-\mathbf{r}) \cdot \frac{d\mathbf{b}}{ds} - \frac{d\mathbf{r}}{ds} \cdot \mathbf{b} = 0; \ \mathbf{a}, p \text{ being constants.}$$

$$\therefore \qquad (\mathbf{a}-\mathbf{r}) \cdot \tau\mathbf{n} - \mathbf{t} \cdot \mathbf{b} = 0$$

i.e.,

$$(\mathbf{a}-\mathbf{r}) \cdot \mathbf{n} = 0,$$

which shows that the plane $(\mathbf{R}-\mathbf{r}) \cdot \mathbf{n} = 0$ through the tangent and perpendicular to the principal normal passes through the centre, **a**, of the given sphere.

8. *If x, y, z are the rectangular cartesian co-ordinates of any point on a curve, show that*

$$x'''^2 + y'''^2 + z'''^2 = \frac{1}{\rho^2\sigma^2} + \frac{1+\rho'^2}{\rho^4},$$

where dashes denote differentiation with respect to arc lengths.

(Garhwal 2002)

Clearly $\qquad x'''^2 + y'''^2 + z'''^2 = \left|\frac{d^3\mathbf{r}}{ds^3}\right|^2.$

Now

$$\frac{d\mathbf{r}}{ds} = \mathbf{t}.$$

$$\frac{d^2\mathbf{r}}{ds^2} = \frac{dt}{ds} = \frac{1}{\rho}\,\mathbf{n}.$$

$$\frac{d^3\mathbf{r}}{ds^3} = -\frac{\rho'}{\rho^2}\mathbf{n} + \frac{1}{\rho}\frac{d\mathbf{n}}{ds}$$

$$= -\frac{\rho'}{\rho^2}\mathbf{n} - \frac{1}{\rho^2}\,\mathbf{t} - \frac{1}{\rho\sigma}\,\mathbf{b}.$$

$$\therefore \qquad \left|\frac{d^3\mathbf{r}}{ds^3}\right|^2 = \frac{\rho'^2}{\rho^4} + \frac{1}{\rho^4} + \frac{1}{\rho^2\sigma^2} = \frac{1}{\rho^2\sigma^2} + \frac{1+\rho'^2}{\rho^4}.$$

9. *If the tangent and binormal at a point of the curve make angles* θ *and* ϕ *respectively with a fixed direction, prove that*

$$\frac{\sin\theta}{\sin\phi} \cdot \frac{d\theta}{d\phi} = -\frac{\kappa}{\tau}.$$

Let the tangent **t** and binormal **b** at a point of the curve make angles θ and ϕ respectively with the fixed direction, say **a** in space, then

then $\qquad \mathbf{t} \cdot \mathbf{a} = a\cos\theta \qquad (|\mathbf{a}| = a)$

and $\qquad \mathbf{b} \cdot \mathbf{a} = a\cos\phi$

Then

$$\frac{d\mathbf{t}'}{ds} \cdot \mathbf{a} = -a\sin\theta\,\frac{d\theta}{ds}$$

$$\Rightarrow \qquad \kappa\,\mathbf{n}\,.\,\mathbf{a} = -\,a\sin\theta\,\frac{d\theta}{ds} \qquad ...(1)$$

Also, $$\frac{d\mathbf{b}}{ds}\,.\,\mathbf{a} = -\,a\sin\phi\,\frac{d\phi}{ds},$$

$$\Rightarrow \qquad \square - \kappa\,\mathbf{n}\,.\,\mathbf{a} = -\,a\sin\phi\,\frac{d\phi}{ds} \qquad ...(2)$$

Dividing (1) by (2), we get

$$\frac{\sin\theta}{\sin\phi}\cdot\frac{d\theta}{d\phi} = -\frac{\kappa}{\tau}.$$

10. *If* $\mathbf{m}_1$, $\mathbf{m}_2$, $\mathbf{m}_3$, *are the moments about the origin of unit vectors* **t**, **n**, **b** *localised in the tangent, principal normal and binormal, prove that*

$$\mathbf{m}_1' = \kappa\,\mathbf{m}_2,\ \mathbf{m}_2' = \mathbf{b} - \kappa\,\mathbf{m}_1 + \tau\,\mathbf{m}_3,$$
$$\mathbf{m}_3' = -\,\mathbf{n} - \tau\,\mathbf{m}_2.$$

Here dash denotes $\dfrac{d}{ds}$.

Let **r** be the position vector of any current point, then

$$\mathbf{m}_1 = \mathbf{r}\times\mathbf{t},\ \mathbf{m}_2 = \mathbf{r}\times\mathbf{n},\ \mathbf{m}_3 = \mathbf{r}\times\mathbf{b}$$

$$\therefore \qquad \mathbf{m}_1' = \mathbf{r}'\times\mathbf{t} + \mathbf{r}\times\kappa\mathbf{n} = \kappa\mathbf{r}\times\mathbf{n} \qquad (\because\ \mathbf{r}' = \mathbf{t})$$
$$= \kappa\,\mathbf{m}_2.$$

Also $$\mathbf{m}' = \mathbf{r}'\times\mathbf{n} + \mathbf{r}\times\mathbf{n}'$$
$$= \mathbf{t}\times\mathbf{n} + \mathbf{r}\times(\tau\mathbf{b} - \kappa\mathbf{t})$$
$$= \mathbf{b} + \tau\,\mathbf{m}_3 - \kappa\,\mathbf{m}_1.$$

Similarly the remaining result.

11. *Prove that the length of the common perpendicular (shortest distance) d of the tangents at two nearer points distant s apart is approximately given by*

$$d = \frac{1}{12}\,\kappa\,\tau\,s^3.$$

Let P and Q be two neighbouring points on the curve. Take P as origin and Q at a distance s from P. Thus the position vectors of P and Q may be taken as $\mathbf{r}(0)$ and $\mathbf{r}(s)$ respectively. The unit tangent vectors at P and Q are $\mathbf{r}'(0)$ and $\mathbf{r}'(s)$, so the unit vector of the common perpendicular is along the vector $\mathbf{r}'(s)\times\mathbf{r}'(0)$. $\left(\textbf{Note. } \mathbf{r}' \equiv \dfrac{d\mathbf{r}}{ds}\right)$

$$\therefore \qquad d = \text{projection of } \overrightarrow{PQ} \text{ in the direction of } \mathbf{r}'(s)\times\mathbf{r}'(0)$$

$$= [\mathbf{r}(s) - \mathbf{r}(0)]\cdot\frac{r'(s)\times r'(0)}{|r'(s)\times\mathbf{r}(0)|}$$

$$\mathbf{r}(s) = \mathbf{r}(0) + s\,\mathbf{r}'(0) + \frac{s^2}{2!}\,\mathbf{r}''(0) + ...$$

[By Taylor's Theorem]

Also, $\mathbf{r}'(0) = t,\ \mathbf{r}''(0) = \kappa\,\mathbf{n}$

and $\mathbf{r}'''(0) = -\kappa^2\mathbf{t} + \kappa'\mathbf{n} + \kappa\,\tau\,\mathbf{b}$

On substitution, we get

$$\mathbf{r}(s) = \mathbf{r}(0) + s\mathbf{t} + \frac{1}{2}s^2\kappa\,\mathbf{n} + \frac{1}{6}s^3(-\kappa^2\mathbf{t} + \kappa'\,\mathbf{n} + \tau\,\kappa\,\mathbf{b}) + ...$$

$$= \mathbf{t} + s\,\kappa\,\mathbf{n} + \frac{1}{2}s^2(-\kappa^2\,\mathbf{t} + \kappa'\,\mathbf{n} + \kappa\,\tau\,\mathbf{b}) + ...$$

$$\therefore\quad \mathbf{r}'(s)\times\mathbf{r}'(0) = -s\,\kappa\,\mathbf{b} + \frac{1}{2}s^2(-\kappa'\,\mathbf{b} + \kappa\,\tau\,\mathbf{n}) + ...$$

$$\therefore\quad [\mathbf{r}(s) - \mathbf{r}(0)]\ .\ \mathbf{r}'(s)\times\mathbf{r}'(0)$$

$$= \left[s\mathbf{t} + \frac{1}{2}s^2\kappa\,\mathbf{n} + \frac{1}{6}s^2\left(-\kappa^2\,\mathbf{t} + \kappa'\mathbf{n} + \kappa\tau\mathbf{b}\right)\right]$$

$$\cdot\left[\frac{1}{2}s^2\kappa\tau\mathbf{n} - \left(s\kappa + \frac{1}{2}s^2\kappa'\right)\mathbf{b}\right]$$

$$= \left(\frac{1}{2}s^2\kappa + \frac{1}{6}s^3\kappa'\right).\ \frac{1}{2}\ s^2\kappa\,\tau - \frac{1}{9}s^3\kappa\,\tau$$

$$\cdot\left(s\kappa + \frac{1}{2}s^2\kappa'\right)$$

$$= \frac{1}{4}\kappa\,\tau\,s^4 - \frac{1}{6}\ \kappa^2\,\tau\,s^4 = \frac{1}{12}\ \kappa^2\,\tau\,s^4 \qquad \text{(approx.)}$$

Also,

$$\left|\mathbf{r}'(s)\times\mathbf{r}'(0)\right|^2 = \frac{1}{4}\kappa^2\,\tau^2\,s^4 + \left(s\kappa' + \frac{1}{2}s^2\kappa'\right)^2$$

$$= s^2\,\kappa^2 + s^3\,\kappa\kappa'$$

$$\therefore\quad \left|\mathbf{r}'(s) - \mathbf{r}'(0)\right|^2 = s\,\kappa \qquad \text{(approx.)}$$

$$\therefore\quad d = \frac{1}{12}\frac{\kappa^2\tau s^4}{s\kappa} = \frac{1}{12}\ \kappa\,\tau\,s^3 \qquad \text{(approx.)}$$

12. *A curve is drawn on a cylinder of radius 'a' and the cylinder is developed into a plane. If* ρ *be the radius of curvature of the curve and* ρ_1 *the radius of curvature of the developed curve of the corresponding points,*

$$\frac{1}{\rho^2} = \frac{1}{\rho_1^2} = \frac{\sin^4\phi}{a^2},$$

where ϕ *is the angle that the tangent to the curve makes with the generator of the cylinder through the point.*

The point on the given curve is

$$\mathbf{r} = (a\cos t, -a\sin t, z)$$

Differentiating w.r.t. 't', we get

$$\dot{\mathbf{r}} = (-a\sin t, -a\cos t, \dot{z}), \quad \ddot{\mathbf{r}} = (-a\cos t, a\sin t, \ddot{z})$$

$$\therefore \quad \dot{\mathbf{r}}\times\ddot{\mathbf{r}} = -[(a\sin t\,\dot{z} + a\cos t\,\ddot{z}), (a\cos t\,\dot{z} - a\sin t\,\ddot{z}), a^2]$$

$$\therefore \quad |\dot{\mathbf{r}}\times\ddot{\mathbf{r}}| = [(a\sin t\,\dot{z} + a\cos t\,\ddot{z})^2 + (a\cos t\,\dot{z} - a\sin t\,\ddot{z})^2 + a^2]^{1/2}$$

$$= (a^2\dot{z}^2 + a^2\ddot{z}^2 + a^4)^{1/2}$$

and $\quad |\dot{\mathbf{r}}| = (a^2 + \dot{z}^2)^{1/2}$

$$\therefore \quad \kappa = \frac{|\dot{\mathbf{r}}\times\ddot{\mathbf{r}}|}{|\dot{\mathbf{r}}|^3} = \frac{(a^2\dot{z}^2 + a^2\ddot{z}^2 + a^4)^{1/2}}{(a^2+\dot{z}^2)^{3/2}}$$

$$\Rightarrow \quad \frac{1}{\rho^2} = \frac{(a^2\dot{z}^2 + a^2\ddot{z}^2 + a^4)}{(a^2+\dot{z}^2)^3} \qquad \ldots(1)$$

Now, when the cylinder is developed into the plane, the point is

$$\mathbf{r}_1 = [a(\pi + t), -a, z]$$

$$\therefore \quad \dot{\mathbf{r}}_1 = (a, 0, \dot{z}), \quad \ddot{\mathbf{r}}_1 = (0, 0, \ddot{z})$$

$$\therefore \quad \dot{\mathbf{r}}\times\ddot{\mathbf{r}} = (0, -a\,\ddot{z}, 0)$$

$$\therefore \quad |\dot{\mathbf{r}}\times\ddot{\mathbf{r}}| = (a^2\ddot{z}^2)^{1/2} \text{ and } |\dot{\mathbf{r}}_1| = (a^2 + \dot{z}^2)^{1/2}$$

$$\therefore \quad \kappa_1^2 = \frac{1}{\rho_1^2} = \frac{|\dot{\mathbf{r}}_1\times\ddot{\mathbf{r}}_1|}{|\dot{\mathbf{r}}_1|^6} = \frac{a^2\ddot{z}^2}{(a^2+\dot{z}^2)^3} \qquad \ldots(2)$$

From (1) and (2), we have

$$\frac{1}{\rho^2} - \frac{1}{\rho_1^2} = \frac{a^2\dot{z}^2 + a^2\ddot{z}^2 + a^4 - a^2\ddot{z}^2}{(a^2+\dot{z}^2)^3}$$

$$= \frac{a^2}{(a^2+\dot{z}^2)^3} \qquad \ldots(3)$$

Also, ϕ is the angle between generator, *i.e.*, z-axis which is parallel to unit vector $\mathbf{k}$ and tangent to the curve.

$$\therefore \quad \cos\phi = \frac{\mathbf{k}.\dot{\mathbf{r}}}{|\dot{\mathbf{r}}|} = \frac{\dot{z}}{(a^2+\dot{z}^2)^{1/2}}$$

$$\Rightarrow \quad (a^2 + \dot{z}^2)\cos^2\phi = \dot{z}^2$$

$$\Rightarrow \quad \dot{z}^2 = a^2\cot^2\phi$$

Substituting in (3)

$$\frac{1}{\rho^2} - \frac{1}{\rho_1^2} = \frac{a^2}{(a^2 + a^2\cot^2\phi)^2} = \frac{\sin^4\phi}{a^2}.$$

EXERCISES

1. Find the equation of the osculating plane and the direction cosines of the tangent, principal normal and binormal at any point of the curve

$$x = 2 \log t, \quad y = 4t, \quad z = 2t^2 + 1.$$

Find also the curvature and torsion.

Show also that the arc, s, is given by

$$t \frac{ds}{dt} = 2x.$$

2. Find the curvature at the point, q, of the curve given by

$$x = \sqrt{3} \sec \theta, \quad y = 2 \cos \theta - \sec \theta, \quad z = 2 \sin \theta,$$

and prove that the principal normal cuts the x-axis at a constant angle.

3. Find the curvature of the curve

$$x = a(u - \sin u), \quad y = a(1 - \cos u), \quad z = tu.$$

Also show that the torsion is given by

$$-b[b^2 + a^2(1 - \cos u)^2]^{-1}.$$

4. Calculate the curvature and torsion for the curve

$$x = ae^u, \quad y = ae^{-u}, \quad z = \sqrt{2}\, au,$$

and show that the tangents to the curve make a constant angle with a certain fixed direction.

5. Show that the curve

$$x = a \cos 2\theta, \quad y = a \sin 2\theta, \quad z = 2a \sin \theta$$

passes twice in perpendicular directions through the point $(a, 0, 0)$.

6. Show that the locus of the feet of the perpendiculars from the origin to the tangents to the curve

$$\mathbf{r} = (a \cos \theta,\ a \sin \theta,\ a\,\theta)$$

is a curve which lies completely on the hyperboloid $x^2 + y^2 - z^2 = a^2$.

Further prove that for this curve

$$\rho = \frac{a(2+\theta^2)^{\frac{3}{2}}}{2\theta(1+\theta^2)^{\frac{1}{2}}} \text{ and } \sigma = \frac{a(1+\theta^2)}{2}.$$

7. Prove that a variable tangent to the curve

$$x = 2t^2, \quad y = 3t^2, \quad z = 6t$$

makes a fixed angle with a certain fixed direction and that the locus of the point where the tangent cuts the plane $x = 0$ is a conic.

8. Prove the relations

$$\frac{d\mathbf{r}}{ds} \cdot \frac{d^2\mathbf{r}}{ds^2} = 0, \quad \mathbf{r} \cdot \frac{d^3\mathbf{r}}{ds^3} = -\kappa^2, \quad \mathbf{r} \cdot \frac{d^4\mathbf{r}}{ds^4}$$

$$= -3\kappa \frac{d\kappa}{ds}, \quad \frac{d^2\mathbf{r}}{ds^2} \cdot \frac{d^3\mathbf{r}}{ds^3} = \kappa \frac{d\kappa}{ds}, \quad \frac{d^3\mathbf{r}}{ds^3} \cdot \frac{d^4\mathbf{r}}{ds^4}$$

$$= \kappa\left(\frac{d^2\kappa}{ds^2} - \kappa^3 - \kappa\tau^2\right),$$

$$\frac{d^3\mathbf{r}}{ds^3}\cdot\frac{d^4\mathbf{r}}{ds^4} = \frac{d\kappa}{ds}\frac{d^2\mathbf{r}}{ds^2} + 2\kappa^3\frac{d\kappa}{ds} + \kappa^3\tau\frac{d\tau}{ds} + \kappa\frac{d\kappa}{ds}\tau^2.$$

9. If there is a one-one correspondence between the points of two curves and the tangents at corresponding points are parallel, prove that the principal normals are parallel and therefore also the binormals.

Also, prove that

$$\frac{\tau_1}{\tau} = \frac{ds}{ds_1} = \frac{\kappa_1}{\kappa}.$$

10. Determine the function $f(u)$, so that the curve given by $\mathbf{r} = (a\cos u, a\sin u, f(u))$ shall be plane.

11. Prove that the shortest distance between the principal normals at consecutive points, distance s apart is $\dfrac{s\rho}{\sqrt{(\rho^2+\sigma^2)}}$ and that it divides the radius of curvature in the ratio $\rho^2 : \sigma^2$.

12. Prove that if P and Q are adjacent points on a curve and the arc $PQ = s$, the arc length, then angle that the shortest distance between the tangents at P and Q makes with the binormal at P is $s/2\sigma$.

13. A curve cuts the generators of the cylinder

$$y = c\cosh\frac{x}{c}$$

at a constant angle α. Find expressions for its curvature and torsion in terms of arc.

14. Prove that the position vector of the current point on a curve satisfies the differential equation

$$\frac{d}{ds}\left[\sigma\frac{d}{ds}\left(\rho\frac{d^2\mathbf{r}}{ds^2}\right)\right] + \frac{d}{ds}\left(\frac{\sigma}{\rho}\frac{d\mathbf{r}}{ds}\right) + \frac{\rho}{\sigma}\frac{d^2\mathbf{r}}{ds^2} = \mathbf{0}.$$

15. Prove that there exists a vector **d** such that

$$\frac{d\mathbf{t}}{ds} = \mathbf{d}\times\mathbf{t},\quad \frac{d\mathbf{n}}{ds} = \mathbf{d}\times\mathbf{n},\quad \frac{d\mathbf{b}}{ds} = \mathbf{d}\times\mathbf{b},$$

where **t, n, b,** are unit vectors along the tangent, principal normal and binormal to a curve.

16. Prove that if the principal normal remains perpendicular to a fixed direction, the tangent makes a constant angle with that direction.

17. Prove that for two points P, Q on a curve

$$\overrightarrow{PQ} = \mathbf{t}\,\delta s + \frac{1}{2}\kappa\,\mathbf{n}\,(\delta s)^2 + \frac{1}{6}\left[-\kappa\,\mathbf{t} + \frac{d\kappa}{ds}\mathbf{n} + \kappa\tau\,\mathbf{b}\right](\delta s)^3 +$$

$$\left[-3\kappa\frac{d\kappa}{ds}\mathbf{t}+\left(\frac{d^2\kappa}{ds^2}-\kappa^2-\kappa\tau^2\right)\mathbf{n}+\left(2\frac{d\kappa}{ds}\tau+\kappa\frac{d\tau}{ds}\right)\mathbf{b}\right](\delta s)^3$$

where $\mathbf{t}, \mathbf{n}, \mathbf{b}, \kappa, \tau, d\kappa/ds, d^2\kappa/ds^2, d\tau/ds$ refer to P and arc $PQ = \delta s$.

Deduce that the shortest distance between

(*i*) principal normals at P, Q is $\tau\delta s\,(\tau^2 + \kappa^2)$,

(*ii*) tangents at P, Q is $\frac{1}{12}\kappa\tau\,(\delta s)^3$.

2.14. INVOLUTES AND EVOLUTES

Def. *If there is a one-one correspondence between the points of two curves C, C_1 such that the tangent at any point of C is a normal at the corresponding point of C_1 then C is called an evolute of C_1 and C_1, an involute of C.* In this case the line joining the corresponding points of C and C_1 is the tangent to C and a normal to C_1 at the corresponding points.

2.14.1. Determination of the Involutes of a given curve. (*Kanpur 2002*)

Let

$$\mathbf{r} = \mathbf{f}(s), \qquad ...(1)$$

be a given curve C and let C_1 be an involute of C.

The point P_1 of C_1 corresponding to any point P of C is a suitably chosen point on the tangent at P and dependent on P. Thus the position vector, $\mathbf{R}$, of P_1, is given by

$$\mathbf{R} = \mathbf{r} + u\mathbf{t},$$

where $$\mathbf{r} = \mathbf{f}(s), \quad \mathbf{t} = d\mathbf{r}/ds, \qquad ...(2)$$

and $u = PP_1$ is a suitably chosen function of s.

Then the problem of the determination of C_1 is equivalent to that of the determination of, u, as a suitable function of s.

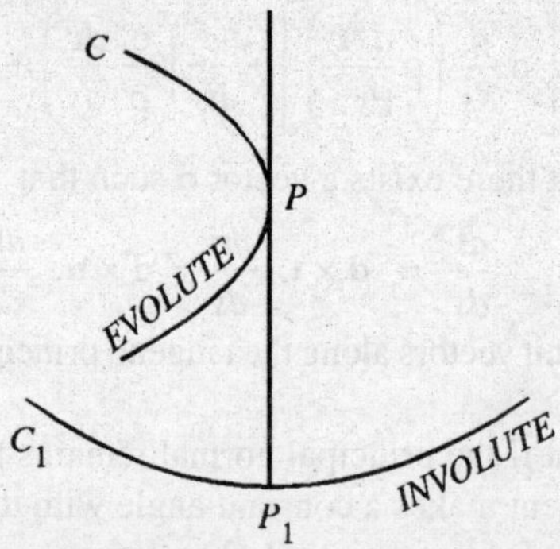

Fig. 2.4

Now we have
$$\frac{d\mathbf{R}}{ds} = \frac{d\mathbf{r}}{ds} + u\frac{d\mathbf{t}}{ds} + \mathbf{t}\frac{du}{ds}$$
$$= \mathbf{t} + u\kappa\,\mathbf{n} + \mathbf{t}\frac{du}{ds}$$
$$= \left(1 + \frac{du}{ds}\right)\mathbf{t} + u\,\kappa\mathbf{n}. \qquad ...(3)$$

The vector, $d\mathbf{R}/ds$, is parallel to the tangent to the curve C_1 at P_1 and accordingly it must be perpendicular to PP_1 which is a normal at P_1. Also the line PP_1 is tangent to C at P and as such it is parallel to the vector $\mathbf{t}$.

$$\therefore \qquad \mathbf{t} \cdot \frac{d\mathbf{R}}{ds} = 0. \qquad ...(4)$$

From (3) and (4), we obtain
$$1 + \frac{du}{ds} = 0,$$
which, on integration, gives
$$u = c - s;$$
c being any constant. Thus the curve, C_1, is given by
$$\mathbf{R} = \mathbf{r} + (c - s)\mathbf{t}. \qquad ...(5)$$

It can also be easily shown that for each value of the constant c, the curve (5) is an involute of the curve (1). In fact, from (5), we obtain
$$\frac{d\mathbf{R}}{ds} = \frac{d\mathbf{r}}{ds} - \mathbf{t} + (c - s)\frac{d\mathbf{t}}{ds} = (c - s)\,\kappa\mathbf{n},$$
so that
$$\mathbf{t} \cdot \frac{d\mathbf{R}}{ds} = 0,$$
i.e., the tangent to C is a normal to C_1.

Thus, we state that
$$\mathbf{r} = \mathbf{f}(s) + (c - s)\,\mathbf{f}'(s)$$
is an involute of $\quad \mathbf{r} = \mathbf{f}(s),$
for every value of the constant c.

2.14.2. Determination of the evolutes of a given curve

Let $\quad \mathbf{r} = \mathbf{f}(s),$
be a given curve C_1 and let C be an evolute of C_1.

A normal P_1P to C_1 at P_1 is the tangent to C at P.

The line P_1P, being a normal to C_1, is coplanar with the unit principal normal vector $\mathbf{n}$ and the unit binormal vector $\mathbf{b}$.

Thus the position vector $\mathbf{R}$ of P is given by an expression of the form
$$\overrightarrow{P_1P} = \mathbf{R} - \mathbf{r} = u\mathbf{b} + v\mathbf{n}, \qquad ...(1)$$

where u, v are suitable functions of s. We have, from (1),

$$\frac{d\mathbf{R}}{ds} = \frac{d\mathbf{r}}{ds} + u\frac{d\mathbf{r}}{ds} + v\frac{d\mathbf{n}}{ds} + \mathbf{b}\frac{du}{ds} + \mathbf{n}\frac{dv}{ds}$$

$$= \mathbf{t} + u\,\tau\,\mathbf{n} + v\,(-\,\kappa\mathbf{t} - \tau\,\mathbf{b}) + \mathbf{b}\,\frac{du}{ds} + \mathbf{n}\frac{dv}{ds}$$

$$= (1 - v\kappa)\,\mathbf{t} + \left(-\,v\tau + \frac{du}{ds}\right)\mathbf{b} + \left(u\tau + \frac{dv}{ds}\right)\mathbf{n}. \qquad ...(2)$$

The vector, $d\mathbf{R}/ds$, is parallel to the tangent at P and as such it is parallel to the line P_1P which is known to be parallel to the vector

$$u\mathbf{b} + v\mathbf{n}.$$

$$\therefore \qquad (u\mathbf{b} + v\mathbf{n}) \times \frac{d\mathbf{R}}{ds} = 0. \qquad ...(3)$$

From (2) and (3), we obtain

$$\left[v\left(-\,v\tau + \frac{du}{ds}\right) - u\left(u\tau + \frac{dv}{ds}\right)\right]\mathbf{t} + u\,(1 - v\kappa)\,\mathbf{n} - v\,(1 - v\kappa)\,\mathbf{b} = 0, \quad ...(4)$$

making use of the relations

$$\mathbf{t} \times \mathbf{n} = \mathbf{b}, \;\; \mathbf{n} \times \mathbf{b} = \mathbf{t}, \;\; \mathbf{b} \times \mathbf{t} = \mathbf{n}.$$

The vectors, **t, n, b** being non-coplanar, we obtain from (4)

$$v\left(-\,v\tau + \frac{du}{ds}\right) - u\left(u\tau + \frac{dv}{ds}\right) = 0$$

i.e.,
$$(u^2 + v^2)\,\tau = v\,\frac{du}{ds} - u\frac{dv}{ds},$$

and
$$u\,(1 - v\kappa) = 0,$$

$$v\,(1 - v\kappa) = 0.$$

Suppose $u^2 + v^2 \neq 0$. We then obtain

$$1 - v\,\kappa = 0, \;\; i.e., \;\; v = 1/\kappa = \rho, \qquad ...(5)$$

and
$$\tau = \frac{v\dfrac{du}{ds} - u\dfrac{dv}{ds}}{u^2 + v^2} = \frac{d}{ds}\left(\tan^{-1}\frac{u}{v}\right)$$

$$\therefore \qquad \tan^{-1}\frac{u}{v} = \int \tau ds + c = \phi(s) + c, \text{ say;}$$

c, being any arbitrary constant.

$$\therefore \qquad u = v\tan\,[\phi(s) + c] = \rho\tan\,[\phi\,(s) + c]. \qquad ...(6)$$

Thus the locus of P is

$$\mathbf{R} = \mathbf{r} + \rho\tan\,[\phi\,(s) + c]\;\mathbf{b} + \rho\mathbf{n}. \qquad ...(7)$$

Also we may show that the curve (7) is an evolute of the given curve for every value of the constant c.

Thus every curve possesses a one parameter family of evolutes.

2.15. BERTRAND CURVES

In the preceding section we considered pairs of curves such that there exists a one-one correspondence between points on any pair and the line joining corresponding points is a normal to one and tangent to the other. We shall now consider the case of pairs of curves such that there exists a one-one correspondence between the points on any pair and the line joining any pair of corresponding points is the principal normal to both. In this connection, we shall prove the following theorem.

Theorem. *If there exists a one-one correspondence between the points of two curves such that the line joining any pair of corresponding points is the principal normal to both, then*

(*i*) *the distance between corresponding points is constant,*

(*ii*) *the tangents at the corresponding points are inclined at a constant angle,*

(*iii*) *there exists a linear relation between the curvature and torsion of either curve.*

Let
$$\mathbf{r} = \mathbf{f}(s) \qquad ...(1)$$

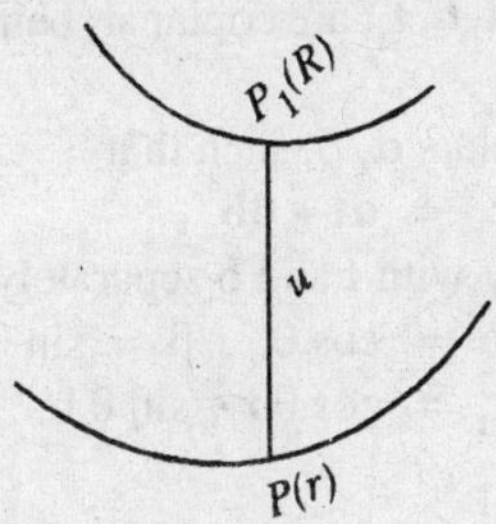

Fig. 2.5

be the equation of one curve. The equation of the second curve must then be of the form
$$\mathbf{R} = \mathbf{r} + u\mathbf{n}, \qquad ...(2)$$
where u is a suitable function of s and denotes the distance between corresponding points of the two curves.

Let s_1 denote arc length and $\mathbf{t}_1$, $\mathbf{n}_1$, $\mathbf{b}_1$, denote unit tangent, principal normal and binormal for the second curve.

We are given
$$\mathbf{n} = \mathbf{n}_1. \qquad ...(3)$$
Differentiating (2) with respect to s, we get
$$\frac{d\mathbf{R}}{ds} = \frac{d\mathbf{r}}{ds} + du\,\mathbf{n} + u\frac{d\mathbf{n}}{ds}$$
$$= \mathbf{t} + \frac{du}{ds}\mathbf{n} + (-\kappa\mathbf{t} - \tau\mathbf{b})u. \qquad ...(4)$$
The vector $d\mathbf{R}/ds$ being parallel to the tangent to the second curve, we have

$$0 = \frac{d\mathbf{R}}{ds} \cdot \mathbf{n}_1 = \frac{d\mathbf{R}}{ds} \cdot \mathbf{n}. \qquad ...(5)$$

From (4) and (5), we obtain

$$\frac{du}{ds} = 0.$$

Thus, u, is constant. Hence (*i*).

If, θ, be the angle between the tangents at corresponding points, we have

$$\cos\theta = \mathbf{t} \cdot \mathbf{t}_1.$$

$$\therefore \quad -\sin\theta \frac{d\theta}{ds} = \frac{d\mathbf{t}}{ds} \cdot \mathbf{t}_1 + \mathbf{t} \cdot \frac{d\mathbf{t}_1}{ds_1}\frac{ds_1}{ds}$$

$$= \kappa\,\mathbf{n} \cdot \mathbf{t}_1 + \mathbf{t} \cdot \kappa_1\,\mathbf{n}_1 \frac{ds_1}{ds}$$

$$= \kappa\,\mathbf{n}_1 \cdot \mathbf{t}_1 + \mathbf{t} \cdot n\kappa_1 \frac{ds_1}{ds} = 0$$

$$\therefore \quad \frac{d\theta}{ds} = 0.$$

Thus θ is constant. Hence we have (*ii*).

Finally, the vectors $\mathbf{t}, \mathbf{b}, \mathbf{t}_1$, are coplanar; being perpendicular to the same vector $\mathbf{n}$.

Thus there exist scalars α, β, such that

$$\mathbf{t}_1 = \alpha\mathbf{t} + \beta\mathbf{b}.$$

Multiplying scalarly with $\mathbf{t}$ and $\mathbf{b}$ separately, we obtain

$$\alpha = \cos\theta, \quad \beta = \sin\theta.$$

$$\therefore \quad \mathbf{t}_1 = \cos\theta\,\mathbf{t} + \sin\theta\,\mathbf{b}. \qquad ...(6)$$

Also we have

$$\mathbf{t}_1 = \frac{d\mathbf{R}}{ds_1}$$

$$= \frac{d\mathbf{R}}{ds}\frac{ds}{ds_1}$$

$$= \left[(1-u\kappa)\mathbf{t} - u\tau\mathbf{b} + \frac{du}{ds}\mathbf{n}\right]\frac{ds}{ds_1} \qquad \text{[From (4)]}$$

$$= [(1-u\kappa)\mathbf{t} - u\tau\mathbf{b}]\frac{ds}{ds_1}, \text{ for } \frac{du}{ds} = 0. \qquad ...(7)$$

From (6) and (7), we obtain

$$\cos\theta\mathbf{t} + \sin\theta\mathbf{b} = [(1-u\kappa)\,\mathbf{t} - u\tau\mathbf{b}]\,\frac{ds}{ds_1}.$$

$$\therefore \quad \cos\theta = (1-u\kappa)\frac{ds}{ds_1}, \; \sin\theta = -u\tau\,\frac{ds}{ds_1}.$$

These give $\dfrac{\cos\theta}{1-u\kappa} = \dfrac{\cos\theta}{-u\tau}$

or $u\kappa - u\tau\cot\theta = 1,$

which is the required linear relation between κ and τ. Hence (*iii*).

2.16. OSCULATING SPHERE

Def. *The osculating sphere at any point P of a curve is the limiting position of the sphere through the four points P, Q, R, S of the curve when the points Q, R, S tend to coincide with P.*

We shall now obtain the radius and the position vector of the centre of the osculating sphere of any point P of a given curve,

$$\mathbf{r} = \mathbf{f}(s).$$

Let the points P, Q, R, S correspond to the value s_1, s_2, s_3, s_4 of s. Let, p, be the radius and, $\mathbf{a}$, the position vector of the sphere through P, Q, R, S.

Consider, now, the scalar function

$$\phi(s) = (\mathbf{r}-\mathbf{a})^2 - p^2, \text{ where } \mathbf{r} = \mathbf{f}(s).$$

Here $\phi(s)$ vanishes for $s = s_1, s_2, s_3, s_4$. Thus, by Rolle's Theorem in the Scalar Calculus, $\phi'(s)$ vanishes for three values $\alpha_1, \alpha_2, \alpha_3$ of s such that

$$s_1 < \alpha_1 < s_2 < \alpha_2 < s_3 < \alpha_3 < s_4.$$

Again, by Rolle's Theorem, $\phi''(s)$, vanishes for two values β_1, β_2 of s such that

$$\alpha_1 < \beta_1 < \alpha_2 < \beta_2 < \alpha_3.$$

Finally, $\phi'''(s)$, vanishes for a value γ such that

$$\beta_1 < \gamma < \beta_2.$$

Let, now, Q, R, S tend to P. Then

$$s_2, s_3, s_4, \alpha_1, \alpha_2, \alpha_3, \beta_1, \beta_2, \gamma.$$

all tend to s_1. Thus if p_1 denote the radius and $\mathbf{a}_1$, the position vector of the centre of the osculating sphere at $P(s_1)$, and

$$\psi(s) = (\mathbf{r}-\mathbf{a}_1)^2 - p_1^2. \qquad ...(1)$$

then $\psi(s), \psi'(s), \psi''(s), \psi'''(s)$ must all be zero for $s = s_1$.

Equating to zero the first three derivatives of (1) and employing Frenet's formulae, we get

$$(\mathbf{r}-\mathbf{a}_1)\cdot\mathbf{t} = 0 \qquad ...(2)$$

$$(\mathbf{r}-\mathbf{a}_1)\cdot\kappa\mathbf{n} + 1 = 0 \qquad ...(3)$$

$$-\kappa^2(\mathbf{r}-\mathbf{a}_1)\cdot\mathbf{t} - \kappa\tau(\mathbf{r}-\mathbf{a}_1)\cdot\mathbf{b} + \frac{d\kappa}{ds}(\mathbf{r}-\mathbf{a}_1)\cdot\mathbf{n} = 0. \qquad ...(4)$$

The equations (2), (3) and (4) are equivalent to

$$(\mathbf{r}-\mathbf{a}_1)\cdot\mathbf{t} = 0 \qquad ...(2')$$

$$(\mathbf{r}-\mathbf{a}_1)\cdot\mathbf{n} = -\frac{1}{\kappa} = -\rho. \qquad ...(3')$$

$$(\mathbf{r}-\mathbf{a}_1)\cdot\mathbf{b} = -\frac{1}{\kappa^2\tau}\frac{d\kappa}{ds} = \sigma\frac{d\rho}{ds} \qquad ...(4')$$

To solve these for $\mathbf{a}_1$, we write

$$\mathbf{r} - \mathbf{a}_1 = \alpha\,\mathbf{t} + \beta\mathbf{n} + \gamma\mathbf{b},$$

and multiply scalarly with **t, n, b** respectively. Thus we obtain

$$\alpha = 0, \;\; \beta = -\rho, \;\; \gamma = \sigma\frac{d\rho}{ds},$$

$$\therefore \quad \mathbf{a}_1 = \mathbf{r} + \rho\,\mathbf{n} - \sigma\frac{d\rho}{ds}\,\mathbf{b}.$$

Substituting in (1), we obtain

$$p_1^{\,2} = \rho^2 + \sigma^2\left(\frac{d\rho}{ds}\right)^2.$$

Thus

$$\sqrt{\left[\rho^2 + \sigma^2\left(\frac{d\rho}{ds}\right)^2\right]}$$

is the radius, and

$$\mathbf{r} + \rho\,\mathbf{n} - \sigma\,\frac{d\rho}{ds}\,\mathbf{b},$$

the position vector of the centre of the osculating sphere at $P(s_1)$; **r, n, t,** $\rho, \sigma,$ $d\rho/ds$ *referring to the point* $P(s_1)$.

2.17. OSCULATING CIRCLE

Def. *The osculating circle at any point P of a curve is the limiting position of the circle through the three points P, Q, R when Q and R tend to P.*

It is clear that the osculating circle at any point of a curve lies in the osculating plane at the point.

We shall now determine the radius and the centre of the osculating circle at any point $P(s_1)$ on a curve $\mathbf{r} = \mathbf{f}(s)$.

Proceeding as in the case of the osculating sphere, starting with three values s_1, s_2, s_3 instead of four, we may show that the radius p_1 and the centre $\mathbf{a}_1$ of the osculating circle satisfy the equations

$$(\mathbf{r} - \mathbf{a}_1)^2 - p_1^{\,2} = 0 \qquad \text{...(1)}$$

$$(\mathbf{r} - \mathbf{a}_1) \cdot \mathbf{t} = 0, \qquad \text{...(2)}$$

$$(\mathbf{r} - \mathbf{a}_1) \cdot \kappa\,\mathbf{n} + 1 = 0. \qquad \text{...(3)}$$

In addition to these, we have

$$(\mathbf{r} - \mathbf{a}_1) \cdot \mathbf{b} = 0, \qquad \text{...(4)}$$

for the line joining the point P to the centre, being in the osculating plane, is perpendicular to the vector **b**.

As in § 24, we may now show that

$$\mathbf{a}_1 = \mathbf{r} + \frac{1}{\kappa}\,\mathbf{n} = \mathbf{r} + \rho\mathbf{n}.$$

$$p_1 = \rho$$

Thus $\mathbf{r} + \rho\mathbf{n}$ *is the centre and* ρ *the radius of the osculating circle at any point;* $\mathbf{r}, \rho, \mathbf{n}$ referring to the point.

Clearly the centre of the osculating circle lies on the principal normal at P and is at a distance ρ from P.

It will also be seen that the centre, $\mathbf{r} + \rho\mathbf{n} - \dfrac{d\rho}{ds}\,\mathbf{b}$, of the osculating sphere lies on the line through the centre $\mathbf{r} + \rho\mathbf{n}$, of the osculating circle drawn parallel and in opposite direction to the binormal vector $\mathbf{b}$ and at distance, $\sigma\, d\rho/ds$, from the same.

EXAMPLES

1. *Prove that the radius R of the sphere of curvature is given by*

$$R^2 = \rho^4 \sigma^2 \left(\frac{d^3\mathbf{r}}{ds^3}\right)^2 - \sigma^2.$$

We have

$$\frac{d^2\mathbf{r}}{ds^2} = \kappa\,\mathbf{n} = \frac{1}{\rho}\,\mathbf{n}$$

$$\therefore \quad \frac{d^3\mathbf{r}}{ds^3} = \frac{1}{\rho}(\tau\,\mathbf{b} - \kappa\,\mathbf{t}) - \frac{1}{\rho^2}\left(\frac{d\rho}{ds}\right)\mathbf{n}$$

$$= -\frac{1}{\rho^4}\,\mathbf{t} - \frac{1}{\rho^4}\left(\frac{d\rho}{ds}\right)\mathbf{n} - \frac{1}{\sigma\rho}\,\mathbf{b}$$

Squaring, we have

$$\left(\frac{d^3\mathbf{r}}{ds^3}\right)^2 = \frac{1}{\rho^4} + \frac{1}{\rho^4}\left(\frac{d\rho}{ds}\right)^2 + \frac{1}{\sigma^2\rho^2}$$

$$\Rightarrow \quad \rho^4\sigma^2\left(\frac{d^3\mathbf{r}}{ds^3}\right)^2 = \sigma^2 + \left(\frac{d\rho}{ds}\right)^2 \sigma^3 + \rho^2$$

$$= \sigma^2 + R^2$$

$$\Rightarrow \quad R^2 = \rho^4\sigma^2\left(\frac{d^3\mathbf{r}}{ds^3}\right)^2 - \sigma^2.$$

2. *If R is the radius of the spherical curvature, prove that*

$$R = \left|\left(\mathbf{t} \times \frac{d^2\mathbf{t}}{ds^2}\right)\Big/ \kappa^2\,\tau\right| \qquad \textit{(Berhampur 1993)}$$

We know that

$$\frac{d\mathbf{t}}{ds} = \kappa\,\mathbf{n} = \frac{1}{\rho}\,\mathbf{n}$$

$$\therefore \quad \frac{d^2\mathbf{t}}{ds^2} = \frac{1}{\rho}(\tau\,\mathbf{b} - \kappa\,\mathbf{t}) - \frac{1}{\rho^2}\left(\frac{d\rho}{ds}\right)\mathbf{n}$$

$$= -\frac{1}{\rho^2}\,\mathbf{t} - \frac{1}{\rho^2}\left(\frac{d\rho}{ds}\right)\mathbf{n} + \frac{1}{\rho\sigma}\,\mathbf{b}$$

$$\therefore \quad \mathbf{t}\times\frac{d^2\mathbf{t}}{ds^2} = -\frac{1}{\rho^2}\left(\frac{d\rho}{ds}\right)\mathbf{b} - \frac{1}{\rho\sigma}\,\mathbf{n}$$

$$[\because \quad \mathbf{t}\times\mathbf{t} = 0,\ \mathbf{t}\times\mathbf{n} = \mathbf{b},\ \mathbf{t}\times\mathbf{b} = -\mathbf{n}]$$

Therefore,

$$\left|\mathbf{t}\times\frac{d^2\mathbf{t}}{ds^2}\right| = \sqrt{\frac{1}{\rho^4}\left(\frac{d\rho}{ds}\right)^2 + \frac{1}{\rho^2\sigma^2}}$$

$$= \sqrt{\left\{\frac{\sigma^2\left(\frac{d\rho}{ds}\right)^2 + \rho^2}{\rho^4\sigma^2}\right\}}$$

$$= \frac{R}{\rho^2\sigma}$$

$$\therefore \quad R = \left|\rho^2\,\sigma\mathbf{t}\times\frac{d^2\mathbf{t}}{ds^2}\right| = \left|\left(\mathbf{t}\times\frac{d^2\mathbf{t}}{ds^2}\right)\Big/\kappa^2\,\tau\right|.$$

3. *For spherical curve, prove that*

$$\rho + \frac{d^2\rho}{d\psi^2} = 0,$$

where ψ *is such that* $d\psi = \tau\, ds$.

It can easily be shown for a spherical curve,

$$\frac{d}{ds}\left(\sigma\,\frac{d\rho}{ds}\right) + \frac{\rho}{\sigma} = 0$$

$$\Rightarrow \quad \frac{d}{ds}\left(\frac{ds}{d\psi}\cdot\frac{d\psi}{ds}\right) + \frac{\rho}{\sigma} = 0 \qquad \left[\because \quad \sigma = \frac{1}{\tau} = \frac{ds}{d\psi}\right]$$

$$\Rightarrow \quad \frac{d}{d\psi}\left(\frac{d\rho}{d\psi}\right)\cdot\frac{d\psi}{ds} + \rho\,\frac{d\psi}{ds} = 0$$

$$\Rightarrow \quad \frac{d^2\rho}{d\psi^2} + \rho = 0.$$

4. *Show that the locus of the centre of curvature is an evolute only when the curve is plane.* *(Berhampur 93)*

The position vector of a current point on the evalute is

$$\mathbf{r}_1 = \mathbf{r} + \rho\mathbf{n} - \rho \tan(\phi + c)\,\mathbf{b} \qquad \text{...(1)}$$

[Equation (7) of § 2.14.2]

where c is an arbitrary constant and for its various values, we get an infinite system of evolutes.

Also, the locus of the centre of curvature is given by

$$\mathbf{c} = \mathbf{r} + \rho\mathbf{n} \qquad \text{...(2)}$$

Equations (1) and (2) will coincide, *i.e.*, the locus of the centre of curvature c is an evolute $\mathbf{r}_1$, then we must have

$$\tan(\phi + c) = 0,$$

$$\Rightarrow \qquad \phi + c = m\pi \quad \Rightarrow \quad \frac{d\phi}{ds} = 0, \text{ i.e., } \tau = 0.$$

Hence the curve must be plane.

5. *Show that the tangent to the locus of the centre of the osculating sphere passes through the centre of the osculating circle.*

The position vector, $\mathbf{R}$, of any point on the locus of the centre of the osculating sphere of the curve $\mathbf{r} = \mathbf{f}(s)$ is given by

$$\mathbf{R} = \mathbf{r} + \rho\mathbf{n} - \sigma \frac{d\rho}{ds}\mathbf{b}.$$

$$\therefore \quad \frac{d\mathbf{R}}{ds} = \frac{d\mathbf{r}}{ds} + \rho\frac{d\mathbf{n}}{ds} - \sigma\frac{d\rho}{ds}\frac{d\mathbf{b}}{ds} + \mathbf{n}\frac{d\rho}{ds} - \frac{d}{ds}\left(\sigma\frac{d\rho}{ds}\right)\mathbf{b}$$

$$= \mathbf{t} + \rho(-\kappa\,\mathbf{t} - \tau\,\mathbf{b}) - \sigma\frac{d\rho}{ds}\tau\,\mathbf{n}$$

$$+ \mathbf{n}\frac{d\rho}{ds} - \frac{d}{ds}\left(\sigma\frac{d\rho}{ds}\right)\mathbf{b}$$

$$= -\left[\rho\tau + \frac{d}{ds}\left(\sigma\frac{d\rho}{ds}\right)\right]\mathbf{b}.$$

Any point on the tangent at the point, $\mathbf{R}$, to the locus of the centre of the osculating sphere is given by

$$r + \rho\mathbf{n} - \sigma\frac{d\rho}{ds}\mathbf{b} + p\left[\rho\tau + \frac{d}{ds}\left(\sigma\frac{d\rho}{ds}\right)\right]\mathbf{b};$$

p, being different for different points on the tangent.

Taking $$p = \sigma\frac{d\rho}{ds}\Big/\left[\rho\tau + \frac{d}{ds}\left(\sigma\frac{d\rho}{ds}\right)\right],$$

we see that the centre of osculating circle $\mathbf{r} + \rho\mathbf{n}$ lies on the tangent in question.

EXERCISES

1. The binormals of a curve C are normals to another curve C_1. If the binormal at the point P on C meets C_1 at P_1, prove that PP_1 is constant.

2. The principal normals of a curve are the binormals of another. Prove that the radii of curvature and torsion for the first curve satisfy the relation

$$a\left(\frac{1}{\rho^2}+\frac{1}{\sigma^2}\right) = \frac{1}{\rho},$$

where, a, is constant.

3. The radii of curvature and torsion ρ, σ at any point of a curve C are connected by the relation

$$\alpha\rho^{-1} + \beta\sigma^{-1} = 1,$$

where α, β are constants. Show that the locus of points on the principal normals of C, distant α from C, is a curve with the same principal normals as C.

4. On the tangent at any point P of a given curve a point Q is taken so that PQ is in the positive direction of the tangent and is of constant length c. Prove that the normal at Q to the locus of Q passes through the centre of curvature of the given curve at P.

Show also that, if ρ is the radius of curvature of the given curve at P and if 's' is the arc of this curve measured from a fixed point of it to P, then the radius of curvature at Q of the locus of Q is

$$(\rho^2+c^2)^{\frac{3}{2}} \Big/ \left(c^2+\rho^2-c\rho\frac{d\rho}{ds}\right).$$

5. Prove that a necessary and sufficient condition for a curve to be on a sphere is that

$$\rho\tau + \frac{d}{ds}\left(\frac{1}{\tau}\frac{d\rho}{ds}\right) = 0,$$

where ρ^{-1} and τ are curvature and torsion respectively of the curve.

[Equate to zero the derivative *w.r.* to 's' of the radius of the osculating sphere.]

6. Prove that if C is the circular helix

$$x = a\cos\theta,\ y = a\sin\theta,\ z = a\theta\tan\alpha,$$

then the locus C_1 of the centres of spherical curvature of C is another circular helix and C is the locus of the centres of spherical curvature of C_1.

7. In any curve, if R be the radius of spherical curvature, prove that at any point (x, y, z)

$$\rho^4\left\{\left(\frac{d^3x}{ds^3}\right)^1+\left(\frac{d^3y}{ds^3}\right)^2+\left(\frac{d^3z}{ds^3}\right)^3\right\} = 1+\frac{R^2}{\sigma^2}.$$

8. Prove that the tangent, principal normal and binormal to the locus C_1 of the centre of spherical curvature of a curve C are paralleled respectively to the binormal, principal normal and tangent to C at the corresponding point.

Deduce that $\rho\rho_1 = \sigma\sigma_1$, where ρ, σ are the radii of curvature and torsion at a point of C and ρ_1, σ_1 those at the corresponding point of C_1.

9. If C is a curve of constant curvature κ, prove that the locus C' of its centre of curvature is also a curve of constant curvature.

10. If ψ is such that $d\psi = \tau\, ds$, show that

(*i*) $\rho_1 = \dfrac{1}{\kappa_1} = \rho + \dfrac{d^2\rho}{d\psi^2}$

(*ii*) $\sigma_1 = \dfrac{1}{\tau_1} + \dfrac{\rho}{\sigma}\left(\rho + \dfrac{d^2\rho}{d\psi^2}\right) = \dfrac{R\rho}{\sigma} \cdot \dfrac{dR}{d\rho}$

(*iii*) $R^2 = \rho^2 + \left(\dfrac{d\rho}{d\psi}\right)^2$

(*iv*) $R\dfrac{dR}{d\rho} = \rho + \dfrac{d^2\rho}{d\psi^2} = \dfrac{1}{\kappa_1} = \rho + \sigma\dfrac{d}{ds}\left(\sigma\dfrac{d\rho}{ds}\right)$

(*v*) $R = \dfrac{dR}{ds} = \sqrt{(R^2 - \rho^2)} \cdot \dfrac{ds_1}{ds}$,

where ρ_1 and σ_1 are the radii of curvature and torsion of the locus of centre of spherical curvature and R is the radius of osculating sphere.

11. A curve is drawn on a sphere, prove that the centre of curvature at any point is the foot of the perpendicular from the centre of the sphere upon the osculating plane at that point.

12. Prove that the torsion of an involute of a curve is equal to

$$\frac{\rho\left(\sigma\dfrac{d\rho}{ds} - \rho\dfrac{d\sigma}{ds}\right)}{(\rho^2 + \sigma^2)(c - s)}.$$

OBJECTIVE QUESTIONS

For each of the following questions, four alternatives are given for the answer. Only one of them is correct. Choose the correct alternative.

1. Vector $\dfrac{d\mathbf{r}}{ds} \times \dfrac{d^2\mathbf{r}}{ds^2}$

(*a*) lies in osculating plane (*b*) is normal to the osculating plane
(*c*) is inclined at a fixed angle with osculating plane
(*d*) does not exist.

2. The normal lying in the osculating plane at a point P on the space curve is called

(*a*) Principal normal (*b*) Binormal
(*c*) Secondary normal (*d*) Osculating normal

3. A normal perpendicular to the principal normal is

(*a*) Binormal (*b*) Osculating normal
(*c*) Evolute (*d*) Involute

4. The plane containing the tangent and binormal is called
(*a*) Normal plane (*b*) Osculating plane
(*c*) Rectifying plane (*d*) Tangent plane

5. Osculating plane at a point P of a space curve contains
(*a*) $\hat{\mathbf{b}}, \hat{\mathbf{n}}$ (*b*) $\hat{\mathbf{t}}, \hat{\mathbf{n}}$
(*c*) $\hat{\mathbf{b}}, \hat{\mathbf{t}}$ (*d*) $\hat{\mathbf{b}}, \hat{\mathbf{n}}$ and $\hat{\mathbf{t}}$

6. Arc rate of rotation of tangent is
(*a*) Curvature (*b*) Radius of Curvature
(*c*) Torsion (*d*) Radius of Torsion

7. For a curve to be a straight line is that at all points of the curve
(*a*) $\tau = 0$ (*b*) $\dfrac{1}{\tau} = 0$
(*c*) $\kappa = 0$ (*d*) $\dfrac{1}{\kappa} = 0$

8. For a curve to be a plane curve is that
(*a*) $\sigma = 0$ (*b*) $\rho = 0$
(*c*) $\tau = 0$ (*d*) $\kappa = 0$

9. Value of $\dfrac{d\mathbf{b}}{ds}$ is
(*a*) $\kappa\,\mathbf{n}$ (*b*) $\tau\,\mathbf{n}$
(*c*) $-\tau\,\mathbf{n}$ (*d*) $-\kappa\,\mathbf{n}$

10. Direction cosines of the principal normal are
(*a*) $\dfrac{x''}{\kappa}, \dfrac{y''}{\kappa}, \dfrac{z''}{\kappa}$ (*b*) $\dfrac{x'}{\kappa}, \dfrac{y'}{\kappa}, \dfrac{z'}{\kappa}$
(*c*) $\dfrac{x''}{\tau}, \dfrac{y''}{\tau}, \dfrac{z''}{\tau}$ (*d*) $\dfrac{x'}{\tau}, \dfrac{y'}{\tau}, \dfrac{z'}{\tau}$

11. Principal normals at consecutive points do not intersect unless
(*a*) $\kappa = 0$ (*b*) $\rho = 0$
(*c*) $\sigma = 0$ (*d*) $\tau = 0$

12. Centre of osculating sphere coincides with the centre of osculating circle if
(*a*) $\rho = 0$ (*b*) $\rho =$ constant
(*c*) $\sigma = 0$ (*d*) $\sigma =$ constant

13. Principal normal to the evolute is
(*a*) perpendicular to **t** (*b*) parallel to **t**
(*c*) perpendicular to **b** (*d*) parallel to **b**

14. For a plane curve the locus of the centre of curvature is
(*a*) evolute (*b*) involute
(*c*) circle of curvature (*d*) Radius of curvature

15. The locus of a point whose position vector is equal to the unit principal normal **n** at any point of a given curve is called

(*a*) Evolute (*b*) Involute

(*c*) Spherical indicatrix of the principal normal

(*d*) Spherical indicatrix of the binormal

ANSWERS

1. (*b*)	**2.** (*a*)	**3.** (*a*)	**4.** (*c*)	**5.** (*b*)
6. (*a*)	**7.** (*c*)	**8.** (*c*)	**9.** (*c*)	**10.** (*a*)
11. (*d*)	**12.** (*b*)	**13.** (*b*)	**14.** (*a*)	**15.** (*c*)

3

Differential Geometry II

SURFACES

Introduction. This chapter will be concerned with some elementary aspects of the Differential Geometry of Surfaces, whereas the previous chapter was concerned with those of curves. It may be noted that the basic distinction between a curve and surface is that a curve is a locus with **one** *degree of freedom,* whereas a surface is a locus with **two** *degrees of freedom.* Thus a curve is given scalarly by equations of the form

$$x = f(t), \; y = g(t), \; z = h(t)$$

or vectorially by

$$\mathbf{r} = \mathbf{F}(t)$$

and a surface scalarly by equations of the form

$$x = f(u, v), \; y = g(u, v), \; z = h(u, v)$$

or vectorially by

$$\mathbf{r} = \mathbf{F}(u, v),$$

there being *one parameter t,* for the curve and *two parameters, u, v* for the surface.

3.1. SURFACE, TANGENT PLANE, NORMAL

Consider the surface

$$\mathbf{r} = \mathbf{f}(u, v). \qquad ...(1)$$

Here u, v are two parameters and, $\mathbf{r}$, is the position vector of any point on the surface corresponding to the values u, v of the parameters.

Any curve on the surface is given by equations of the form

$$u = \phi(t), \; v = \psi(t) \qquad ...(2)$$

where, in particular, we may have arc length, s, in place of t.

In terms of the parameter, t, the same curve is also given by

$$\mathbf{r} = \mathbf{f}[\phi(t), \; \psi(t)]. \qquad ...(3)$$

We shall suppose that $\mathbf{f}(u, v)$ possesses continuous partial derivatives of every order which appear in any investigation. In general, we shall not need derivatives of order higher than the second.

3.1.1. Tangent line to any curve on the surface, Tangent plane.

Differentiating (3), with respect to t, we obtain

$$\frac{d\mathbf{r}}{dt} = \frac{\partial \mathbf{r}}{\partial u}\frac{du}{dt} + \frac{\partial \mathbf{r}}{\partial v}\frac{dv}{dt}. \qquad \text{(Refer § 1.13.3} \qquad ...(4)$$

From (4), we see that the vector $d\mathbf{r}/dt$ which is parallel to the tangent to the curve at the point $P(u, v)$ is coplanar with the two vectors

$$\frac{\partial \mathbf{r}}{\partial u}, \frac{\partial \mathbf{r}}{\partial v},$$

which depend only upon the point P and are independent of any particular curve on the surface through P.

Thus the tangent at the point $P(u, v)$ to every curve on the surface through P lies on the plane through P normal to the vector

$$\frac{\partial \mathbf{r}}{\partial u} \times \frac{\partial \mathbf{r}}{\partial v}. \qquad \text{[Refer note 1, below]}$$

Hence the locus of the tangents at $P(u, v)$ to different curves on the surface through P is the plane

$$(\mathbf{R} - \mathbf{r}) \cdot \frac{\partial \mathbf{r}}{\partial u} \times \frac{\partial \mathbf{r}}{\partial v} = 0,$$

called the *tangent plane* at P.

Normal. *The line through any point perpendicular to the tangent plane at the point is called the normal at the point.*

The equation of the normal at $P(u, v)$ is

$$\mathbf{R} = \mathbf{r} + p\frac{\partial \mathbf{r}}{\partial u} \times \frac{\partial \mathbf{r}}{\partial v}.$$

3.1.2. UNIT NORMAL VECTOR

The vector

$$\frac{\dfrac{\partial \mathbf{r}}{\partial u} \times \dfrac{\partial \mathbf{r}}{\partial v}}{\left|\dfrac{\partial \mathbf{r}}{\partial u} \times \dfrac{\partial \mathbf{r}}{\partial v}\right|}$$

whose length is unity and which is parallel to the normal at (u, v) is called *unit normal vector*.

Notes. 1. The above discussion will not be applicable to points for which

$$\frac{\partial \mathbf{r}}{\partial u} \times \frac{\partial \mathbf{r}}{\partial v} = 0.$$

Such a point is called a **Singular point.** The consideration of surfaces with singular points is not within the scope of this book.

2. The curves u = constant, v = constant are called *parametric* curves. There passes one member of either family through each point of the surface.

Thus the parametric curves through a point (u_0, v_0) are

$$\mathbf{r} = \mathbf{f}(u, v_0); \quad u \text{ is a parameter;}$$

and

$$\mathbf{r}_3 = \mathbf{f}(u_0, v); \quad v \text{ is a parameter.}$$

The vectors $\partial\mathbf{r}/\partial u$, $\partial\mathbf{r}/\partial v$ are parallel to the tangents to these two curves. *The two parametric curves through every point will be orthogonal, if*

$$\frac{\partial\mathbf{r}}{du} \cdot \frac{\partial\mathbf{r}}{dv} = 0.$$

for every pair of values u, v.

3. A curve on a surface can also be given by a relation of the form $v = \phi(u)$.

This curve is the same as

$$\mathbf{r} = \mathbf{f}[u, \phi(u)]; \qquad u \text{ is a parameter.}$$

The vector

$$\frac{d\mathbf{f}}{du} = \frac{\partial\mathbf{r}}{du} + \frac{\partial\mathbf{r}}{dv}\frac{dv}{du}, \qquad [\S\ 1.13]$$

is parallel to the tangent to this curve.

4. Taking, s, as the parameter, instead of t, we have

$$\frac{\partial\mathbf{r}}{ds} = \frac{\partial\mathbf{r}}{ds}\frac{du}{ds} + \frac{\partial\mathbf{r}}{dv}\frac{dv}{ds}.$$

Here $d\mathbf{r}/ds$ is a unit tangent vector coplanar with the vectors $\partial\mathbf{r}/\partial u$ and $\partial\mathbf{r}/\partial v$. As a vector in a plane is *uniquely* expressible in terms of two given vectors in the same plane, we see that for two curves through a point with the same tangent,

$$\frac{du}{ds}, \frac{dv}{ds}$$

have the same values.

5. The numbers u, v which determine uniquely a point of the surface are called *curvilinear* co-ordinates of the point.

3.2. THE TWO FUNDAMENTAL FORMS

We shall now define two *differential quadratic forms* involving six functions

$$E, F, G; \; L, M, N$$

of u, v which play a fundamental role in the study of surfaces.

3.2.1. The First form

Take any point and consider any curve on the surface through the point. Let, s, denote the arc length along the curve so that the co-ordinates u, v of any point on the curve are functions of s.

Now, we have

$$\frac{d\mathbf{r}}{ds} = \frac{\partial\mathbf{r}}{\partial s}\frac{du}{ds} + \frac{\partial\mathbf{r}}{\partial v}\frac{dv}{ds},$$

where $d\mathbf{r}/ds$ is the unit tangent vector to the curve. Thus

$$I = \left(\frac{d\mathbf{r}}{ds}\right)^2$$

$$= \left(\frac{\partial \mathbf{r}}{\partial u}\frac{du}{ds} + \frac{\partial \mathbf{r}}{\partial v}\frac{dv}{ds}\right)^2$$

$$= \left(\frac{\partial \mathbf{r}}{\partial u}\right)^2 \left(\frac{du}{ds}\right)^2 + 2\frac{\partial \mathbf{r}}{\partial u}\cdot\frac{\partial \mathbf{r}}{\partial v}\frac{du}{ds}\frac{dv}{ds} + \left(\frac{\partial \mathbf{r}}{\partial v}\right)^2 \left(\frac{dv}{ds}\right)^2.$$

As is customary, we now write

$$E = \left(\frac{\partial \mathbf{r}}{\partial u}\right)^2, \; F = \frac{\partial \mathbf{r}}{\partial u}\cdot\frac{\partial \mathbf{r}}{\partial v}, \; G = \left(\frac{\partial \mathbf{r}}{\partial v}\right)^2,$$

so that E, F, G are scalar functions of u, v.

Thus
$$I = E\left(\frac{du}{ds}\right)^2 + 2F\frac{du}{ds}\frac{dv}{ds} + G\left(\frac{dv}{ds}\right)^2 \qquad ...(1)$$

or, in terms of differentials

$$(ds)^2 = E\,(du)^2 + 2F\,du\,dv + G\,(dv)^2. \qquad ...(2)$$

The expression $E\,(du)^2 + 2F\,du\,dv + G\,(dv)^2$,
on the right of (2), is known as the *First fundamental form* and the functions E, F, G are known as *Magnitudes of first order for* the surface.

Cor. For the parametric curve, v = constant, we have

$$dv/ds = 0 \text{ and, from (1), } du/ds = 1/\sqrt{E}.$$

Similarly for u = constant, we have

$$du/ds = 0 \text{ and } dv/ds = 1\sqrt{G}.$$

3.2.2. The Second form

We have

$$\frac{d\mathbf{r}}{ds} = \frac{\partial \mathbf{r}}{du}\frac{du}{ds} + \frac{\partial \mathbf{r}}{\partial v}\frac{du}{ds}.$$

$$\frac{d^2\mathbf{r}}{ds^2} = \left(\frac{\partial^2 \mathbf{r}}{du^2}\frac{du}{ds} + \frac{\partial^2 \mathbf{r}}{\partial v\,\partial u}\frac{dv}{ds}\right)\frac{du}{ds} + \left(\frac{\partial^2 \mathbf{r}}{\partial u}\frac{du}{ds} + \frac{\partial^2 \mathbf{r}}{\partial v^2}\frac{dv}{ds}\right)\frac{dv}{ds} + \frac{\partial \mathbf{r}}{\partial u}\frac{d^2\mathbf{r}}{ds^2} + \frac{\partial \mathbf{r}}{\partial v}\frac{d^2 v}{ds^2}$$

$$= \frac{\partial^2 \mathbf{r}}{\partial u^2}\left(\frac{du}{ds}\right)^2 + 2\frac{\partial^2 \mathbf{r}}{\partial u\,\partial v}\frac{du}{ds}\frac{dv}{ds} +$$

$$\frac{\partial^2 \mathbf{r}}{\partial v^2}\left(\frac{dv}{ds}\right)^2 + \frac{\partial \mathbf{r}}{\partial u}\frac{d^2 u}{ds^2} + \frac{\partial \mathbf{r}}{\partial v}\frac{d^2 v}{ds^2}.$$

Multiplying scalarly with the unit normal vector **n**, we obtain

$$\mathbf{n} \,.\, \frac{d^2\mathbf{r}}{ds^2} = \mathbf{n} \,.\, \frac{\partial^2 \mathbf{r}}{\partial u^2}\left(\frac{du}{ds}\right)^2 + 2\mathbf{n} \,.\, \frac{\partial^2 \mathbf{r}}{\partial u\, \partial v}\frac{du}{ds}\frac{dv}{ds} +$$

$$\mathbf{n} \,.\, \frac{\partial^2 \mathbf{r}}{\partial v^2}\left(\frac{dv}{ds}\right)^2, \qquad \text{...(1)}$$

$$\mathbf{n} \,.\, \frac{\partial \mathbf{r}}{\partial u} = 0 \text{ and } \mathbf{n} \,.\, \frac{\partial \mathbf{r}}{\partial v} = 0.$$

As is customary, we write

$$L = \mathbf{n} \,.\, \frac{\partial^2 \mathbf{r}}{\partial u^2},\ M = \mathbf{n} \,.\, \frac{\partial^2 \mathbf{r}}{\partial u\, \partial v},\ N = \mathbf{n} \,.\, \frac{\partial^2 \mathbf{r}}{\partial v^2},$$

so that, we may re-write (1) as

$$\mathbf{n} \,.\, \frac{d^2\mathbf{r}}{ds^2} = L\left(\frac{du}{ds}\right)^2 + 2M\frac{du}{ds}\frac{dv}{ds} + N\left(\frac{dv}{ds}\right)^2.$$

The quadratic differential form

$$L\,(du)^2 + 2M\,du\,dv + N\,(dv)^2,$$

is known as the **Second fundamental form** and L, M, N are known as *magnitudes of the second order.*

3.3. DISTRIBUTION OF CURVATURES OF CURVES THROUGH A POINT ON A SURFACE

3.3.1. Curvature at any point of a curve on a surface

Let κ denote the curvature and, $\mathbf{n}_c$, the unit principal normal vector at any point of the curve so that, by Frenet's first formula, we have

$$\frac{d^2\mathbf{r}}{ds^2} = \kappa\,\mathbf{n}_c. \qquad \text{...(1)}$$

Also, as seen in the preceeding section,

$$\mathbf{n} \,.\, \frac{d^2\mathbf{r}}{ds^2} = L\left(\frac{du}{ds}\right)^2 + 2M\frac{du}{ds}\frac{dv}{ds} + N\left(\frac{dv}{ds}\right)^2. \qquad \text{...(2)}$$

From (1) and (2), we obtain the relation

$$\kappa\,\mathbf{n} \,.\, \mathbf{n}_c = L\left(\frac{du}{ds}\right)^2 + 2M\frac{du}{ds}\frac{dv}{ds} + N\left(\frac{dv}{ds}\right)^2. \qquad \text{...(3)}$$

which gives the curvature κ. This relation also shows that the *curvature at any point of a curve drawn on a surface depends upon the tangent line* [*refer note* (*4*) *after* § *3.1*] *and the principal normal to the curve.* Thus two curves on the surface through a point have the same curvature at the point, if the two urves have the same tangent line and the same principal normal at the point, *e.,* if they have the same osculating plane at the point. In other words, we see hat *the curvature of any curve on the surface at a point is the same as that f the section of the surface by the osculating plane of the curve at the point.* hus it is enough to investigate the curvatures of plane sections of a surface nly.

.3.2. Relation between the curvatures of normal and oblique plane sections through the same tangent line. Meunier's Theorem.

If ρ, ρ_0 *denote the radii of curvatures of oblique and normal sections hrough the same tangent line, and* θ *be the angle between the sections, then*

$$\rho = \rho_0 \cos\theta.$$

onsider the section of the surface by any plane π through a point P of the urface. The tangent to this section is the line of intersection, l, of the plane π vith the tangent plane at P.

Consider also the section of the surface by the plane π_1 through the line and the normal, **n**, to the surface. The unit vector, **n**, is the unit principal ormal vector to this section.

These two sections are respectively oblique and normal sections through he same tangent line. Let κ, κ_0 be the curvatures of these sections respecively and let θ be the angle between the planes of the two sections. The angle between the planes is the same as the angle between the vectors $\mathbf{n}$, $\mathbf{n}_c$. We ave

$$\mathbf{n} \cdot \mathbf{n}_c = \cos\theta.$$

Applying (3, § 3.3.1) to the normal section, we get

$$\kappa_0 = L\left(\frac{du}{ds}\right)^2 + 2M\frac{du}{ds}\frac{dv}{ds} + N\left(\frac{dv}{ds}\right)^2. \quad ...(4)$$

We may note that *du*/*ds* and *dv*/*ds* have the same values for the two secons. From (3) and (4), we have

$$\kappa \cos\theta = \kappa_0$$

r

$$\rho = \rho_0 \cos\theta.$$

Hence the result which is known as *Meunier's Theorem.*

Note. As a result of the Meunier's Theorem, curvature of any plane ection of a surface can be expressed in terms of the normal section having he same tangent line. The next section will deal with the totality of all noral sections through a point.

.3.3. Principal sections. Principal directions. Principal radii of curvature

Consider different normal sections through a point P of the surface. The lane of each of these sections passes through the normal to the surface at P nd the tangent line to any section is the line of intersection of its plane with he tangent plane at P.

If κ, denote the curvature of any normal section through P, we have

$$\kappa = L\left(\frac{du}{ds}\right)^2 + 2M\frac{du}{ds}\frac{dv}{ds} + N\left(\frac{dv}{ds}\right)^2,$$

where du/ds, dv/ds are different for different normal sections.

$$\therefore \quad L(du)^2 + 2M\,du\,dv + N(dv)^2 = \kappa(ds)^2$$

$$= \kappa\,[E\,(du)^2 + 2F\,du\,dv + G\,(dv)^2]$$

or $$(\kappa E - L)(du)^2 + 2(\kappa F - M)\,du\,dv + (\kappa G - N)(dv)^2 = 0,$$

or $$(\kappa E - L)\left(\frac{du}{dv}\right)^2 + 2(\kappa F - M)\frac{du}{dv} + (\kappa G - N) = 0. \quad ...(5)$$

This shows that to each value of du/dv, *i.e.*, to each normal section corresponds a value of, κ, but to each value of, κ, there correspond two values of du/dv. Thus *there are two normal sections having a given curvature* κ. We have

$$\kappa = \frac{Lp^2 + 2Mp + N}{Ep^2 + 2Fp + G}, \text{ where } p = \frac{du}{dv}.$$

Equating to zero the derivative of, κ, with respect to p, we see that, κ, is stationary for the values of p, given by

$$(LF - ME)p^2 + (LG - NE)p + (MG - NF) = 0.$$

Replacing, p, by du/dv, we see that the directions of the normal sections corresponding to the stationary values of κ, are given by

$$(LF - ME)\left(\frac{du}{dv}\right)^2 + (LG - NE)\frac{du}{dv} + (MG - NF) = 0. \quad ...(6)$$

The stationary values of κ will be obtained by substituting in (5) the values of du/dv as given by (6), *i.e.*, by eliminating du/dv from (5) and (6). Thus the stationary values of, κ, are the roots of the equation.

$$(F^2 - EG)\kappa^2 + (LG + NE - 2MF)\kappa + (M^2 - LN) = 0. \quad ...(7)$$

The values of, κ, obtained as the roots of (7), are called the *Principal curvatures* and the corresponding normal sections are called *Principal sections.* The directions of the tangents to the principal sections which are given by (6) are called *Principal Directions.*

The product

$$\frac{M^2 - LN}{F^2 - EG}$$

of the two principal curvatures is called the *Gaussian Curvature.*

Lines of curvature. Def. *A curve drawn on a surface is called a line of curvature if the tangent line at any point thereof lies along one of the two principal directions of the surface at the point.*

It follows from the equation (6) which gives the differential equation, in the curvilinear co-ordinates, of the lines of curvature, that there exist two families of such lines and that through each point there passes one member of each family.

It also follows (6) that if $F = 0$ and $M = 0$, then the differential equation of the lines of curvature is

$$du\,dv = 0,$$

so that the parametric curves are themselves the lines of curvature.

Equations of line of curvature

Since the direction of line of curvature at any point is along the principal direction at that point, hence the differential equations of the two systems of line of curvature is the same as the equation of the principal direction, so their equation is

$$(EM - FL)\,du^2 + (EN - GL)\,du\,dv + (FN - GM)\,dv^2 = 0 \quad \text{...(1)}$$

Since $E(FN - GM) - F(EN - GL) + G(EM - FL) = 0$, therefore, the two systems of line of curvature are orthogonal.

Equation (1) can be expressed in the following from:

$$\begin{vmatrix} dv^2 & -du\,dv & du^2 \\ E & F & G \\ L & M & N \end{vmatrix} = 0.$$

3.3.4. Rodrigue's Formula

A necessary and sufficient condition that a curve on a surface be line of curvature is that

$$\frac{d\mathbf{N}}{ds} \propto \frac{d\mathbf{r}}{ds} \quad \text{or } d\mathbf{N} + \kappa\, d\mathbf{r} = 0.$$

at each of its points, where κ *denotes the normal curvature.*

Let (du, dv) be a line of curvature on the surface, then it is a principal direction at point (u, v) to the surface, so we have from equations

$$(L - \kappa E)\,du + (M - \kappa F)\,dv = 0 \quad \text{...(1)}$$

and

$$(M - \kappa F)\,du + (N - \kappa G)\,dv = 0 \quad \text{...(2)}$$

where κ is one of the principal curvatures.

Substituting the values of L, M, N, E, F, G in (1) and (2), we have

$$(-\mathbf{N}_1,\ \mathbf{r}_1 - \kappa\,\mathbf{r}_1{}^2)\,du + (-\mathbf{N}_2 \,.\, \mathbf{r}_1 - \kappa\,\mathbf{r}_1 \,.\, \mathbf{r}_2)\,dv = 0$$

$$\Rightarrow \quad (\mathbf{N}_1\,du + \mathbf{N}_2\,dv) \,.\, \mathbf{r}_1 + \kappa\,(\mathbf{r}_1\,du + \mathbf{r}_2\,dv) \,.\, \mathbf{r}_1 = 0$$

$$\Rightarrow \quad (d\mathbf{N}) \,.\, \mathbf{r}_1 + (\kappa\, d\mathbf{r}) \,.\, \mathbf{r}_1 = 0 \quad \text{...(3)}$$

Similarly, from (2)

$$(d\mathbf{N} + \kappa\, d\mathbf{r}) \,.\, \mathbf{r}_2 = 0 \quad \text{...(4)}$$

we have, $\quad \mathbf{N} \,.\, \mathbf{N} = 1 \Rightarrow 2\mathbf{N} \,.\, d\mathbf{N} = 0,$

i.e., $d\mathbf{N}$ is normal to $\mathbf{N}$ or $d\mathbf{N}$ is a tangential vector to the surface. Also, $\mathbf{r}_1$ and $\mathbf{r}_2$ are tangential vectors, therefore, in order that equations (3) and (4) are satisfied, we must have

$$d\mathbf{N} + \kappa\, d\mathbf{r} = 0$$

$$\Rightarrow \quad \frac{d\mathbf{N}}{ds} \propto \frac{d\mathbf{r}}{ds},$$

i.e., the condition is necessary.

Again, let there be a curve on the surface at each point of which

$$\frac{d\mathbf{N}}{ds} \propto \frac{d\mathbf{r}}{ds} \quad \Rightarrow \quad \kappa\, d\mathbf{r} + d\mathbf{N} = 0.$$

Now, reversing the order of steps we get the equations (3) and (4).

Thus, the curve is line of curvature in case κ is normal curvature to the surface.

$$\text{Given} \quad \kappa d\mathbf{r} + d\mathbf{N} = 0 \Rightarrow \kappa\, d\mathbf{r} = -d\mathbf{N}$$

$$\Rightarrow \kappa(\mathbf{r}_1\, du + \mathbf{r}_2\, dv) = -(\mathbf{N}_1\, du + \mathbf{N}_2\, dv)$$

$$\Rightarrow \kappa(\mathbf{r}_1\, du + \mathbf{r}_2\, dv) \cdot (\mathbf{r}_1\, du + \mathbf{r}_2\, dv) = -(\mathbf{N}_1\, du + \mathbf{N}_2\, dv) \cdot (\mathbf{r}_1\, du + \mathbf{r}_2\, dv)$$

$$\Rightarrow \quad \kappa = \frac{L\, du^2 + 2M\, du\, dv + N\, dv^2}{E\, du^2 + 2F\, du\, dv + G\, dv^2}$$

$\Rightarrow$ κ is a normal curvature at the point (u, v) in the direction (du, dv).

Hence, the direction at each point of the curve is a principal direction and thus the curve is a line of curvature on the surface.

3.3.5. Orthogonality of Principal Directions

Theorem. *The two principal directions at any point are orthogonal.*

Let $\qquad u = \phi_1(v), \quad v = \phi_2(v)$

be the two lines of curvature through any point P. The $d\phi_1/dv$, $d\phi_2/dv$ are the two roots of the quadratic (6) in du/dv.

$$\therefore \quad \frac{d\phi_1}{dv} + \frac{d\phi_2}{dv} = -\frac{LG - NE}{LF - ME}, \quad \frac{d\phi_1}{dv}\frac{d\phi_2}{dv} = \frac{MG - NF}{LF - ME}.$$

The vectors

$$\frac{\partial \mathbf{r}}{\partial u}\frac{d\phi_1}{dv} + \frac{\partial \mathbf{r}}{\partial v}, \quad \frac{\partial \mathbf{r}}{\partial u}\frac{d\phi_2}{dv} + \frac{\partial \mathbf{r}}{\partial v}$$

are parallel to the tangents to these curves. We have

$$\left(\frac{\partial \mathbf{r}}{\partial u}\frac{d\phi_1}{dv} + \frac{\partial \mathbf{r}}{\partial v}\right) \cdot \left(\frac{\partial \mathbf{r}}{\partial u}\frac{d\phi_2}{dv} + \frac{\partial \mathbf{r}}{\partial v}\right)$$

$$= E\frac{d\phi_1}{dv}\frac{d\phi_2}{dv} + F\left(\frac{d\phi_1}{dv} + \frac{d\phi_2}{dv}\right) + G$$

$$= E\frac{MG - NF}{LF - ME} - F\frac{LG - NE}{LF - ME} + G$$

$$= 0.$$

Hence the theorem.

3.4. ASYMPTOTIC LINES

The system of curves on the surface $\mathbf{r} = \mathbf{f}(u, v)$ whose differential equation is

$$L(du)^2 + 2M\,du\,dv + N\,(dv)^2 = 0$$

is called the system of *Asymptotic lines.* Clearly there are two families of asymptotic lines such that though each point on the surface there passes one member of each family.

From the relation

$$\mathbf{n}\,.\,\frac{d^2\mathbf{r}}{ds^2} = L\left(\frac{du}{ds}\right)^2 + 2M\,\frac{du}{ds}\,\frac{dv}{ds} + N\left(\frac{dv}{ds}\right)^2,$$

and the fact that the vector $d^2\mathbf{r}/ds^2$ is parallel to the principal normal of the curve, we see that *for an asymptotic line, principal normal at any point is perpendicular to the normal to the surface at the point.* The tangent is also necessarily perpendicular to the normal to the surface. Thus we deduce that the *osculating plane to an asymptotic line at any point coincides with the tangent plane to the surface at the point.* This may be thought of as the characteristic property of any asymptotic line.

3.4.1. Condition for Asymptotic lines to be orthogonal

We know that the two directions given by

$$P\,du^2 + 2Q\,du\,dv + R\,dv^2 = 0$$

are orthogonal, iff

$$ER - 2\,FQ + GP = 0 \qquad ...(1)$$

The differential equation giving asymptotic direction is

$$L\,du^2 + 2M\,du\,dv + N\,dv^2 = 0$$

Thus, the asymptotic directions are orthogonal, iff

$$EN - 2\,FM + GL = 0$$

which gives first curvature = 0.

3.4.2. The necessary and sufficient condition that the parametric curves be asymptotic lines are

$$L = 0, \quad N = 0, \quad M \neq = 0$$

The differential equation of the parametric curve is

$$du\,dv = 0 \qquad ...(1)$$

and the differential equation of the asymptotic line is

$$L\,du^2 + 2M\,du\,dv + N\,dv^2 = 0 \qquad ...(2)$$

In case, the parametric curves are asymptotic lines, equations (1) and (2) must be identical. On comparing the coefficients of du^2, $du\,dv$ and dv^2, we get

$$L = 0, \quad N = 0, \quad M \neq 0.$$

Conversely, if $L = 0$, $N = 0$, $M \neq 0$, the equation (2) reduces to equation (1), *i.e.,* the asymptotic lines are the parametric curves. Hence the proposition.

3.5. CARTESIAN FORMULATION OF FUNDAMENTAL MAGNITUDES

Let $\qquad \mathbf{r} = \mathbf{f}(u, v) = \mathbf{i}x + \mathbf{j}y + \mathbf{k}z,$

so that x, y, z are functions of u, v.

$$\therefore \quad \frac{\partial \mathbf{r}}{\partial u} = \Sigma\, \mathbf{i} x_u, \quad \frac{\partial \mathbf{r}}{\partial v} = \Sigma\, \mathbf{i} x_v,$$

$$\frac{\partial^2 \mathbf{r}}{\partial s^2} = \Sigma\, \mathbf{i} x_u^{\ 2}, \quad \frac{\partial^2 \mathbf{r}}{\partial u\, \partial v} = \Sigma\, \mathbf{i} x_{uv}, \quad \frac{\partial^2 \mathbf{r}}{\partial v^2} = \Sigma\, \mathbf{i} x_v^{\ 2}.$$

$$\therefore \quad E = \left(\frac{\partial \mathbf{r}}{\partial u}\right)^2 = \Sigma\, (x_u)^2, \quad F = \frac{\partial \mathbf{r}}{\partial u}, \frac{\partial \mathbf{r}}{\partial v} = \Sigma\, x_u x_v.$$

$$G = \left(\frac{\partial \mathbf{r}}{\partial v}\right)^2 = \Sigma\, (x_v)^2.$$

Again

$$\mathbf{n} = \frac{\dfrac{\partial \mathbf{r}}{\partial u} \times \dfrac{\partial \mathbf{r}}{\partial v}}{\left| \dfrac{\partial \mathbf{r}}{\partial u} \times \dfrac{\partial \mathbf{r}}{\partial v} \right|} = \frac{\mathbf{i}(y_u z_v - y_v z_u) + \mathbf{j}(z_u x_v - z_v x_u) + \mathbf{k}(x_u y_v - x_v y_u) +}{\sqrt{\Sigma (y_u z_v - y_v z_v)^2}},$$

so that

$$y_u z_v - y_v z_u,\ z_u x_v - z_v x_u,\ x_u y_v - x_v z_v$$

are the direction ratios of the normal. Also we have

$$L = \mathbf{n} \,.\, \frac{\partial^2 \mathbf{r}}{\partial u^2} = \frac{\Sigma\, x_u^2\ (y_u z_v - y_v z_u)}{\sqrt{\Sigma (y_u z_v - y_v z_u)^2}},$$

$$M = \mathbf{n} \,.\, \frac{\partial^2 \mathbf{r}}{\partial u \partial v} = \frac{\Sigma\, x_{uv}\ (y_u z_v - y_v z_u)}{\sqrt{\Sigma (y_u z_v - y_v z_u)^2}},$$

$$N = \mathbf{n} \,.\, \frac{\partial^2 \mathbf{r}}{\partial v^2} = \frac{\Sigma\, x_v^2\ (y_u z_v - y_v z_u)}{\sqrt{\Sigma (y_u z_v - y_v z_u)^2}}.$$

Also it may be seen that

$$\Sigma\, (y_u z_v - y_v z_u)^2 = \Sigma\, (x_u)^2\, \Sigma\, (x_v)^2 - (\Sigma x_u x_v)^2$$
$$= EG - F^2.$$

EXAMPLES

1. *Show that the parametric curves on the sphere*

$$x = a \sin u \cos v,\ y = a \sin u \sin v,\ z = a \cos u$$

form an orthogonal system.

Eliminating u, v, we obtain

$$x^2 + y^2 + z^2 = a^2,$$

so that the surface is a sphere.

We have

$$\mathbf{r} = (a \sin u \cos v,\ a \sin u \sin v,\ a \cos u).$$

$$\therefore \quad \frac{\partial \mathbf{r}}{\partial u} = (a \cos u \cos v,\ a \cos u \sin v,\ - a \sin u),$$

and $$\frac{\partial \mathbf{r}}{\partial v} = (-a \sin u \sin v, a \sin u \cos v, 0).$$

$$\therefore \quad \frac{\partial \mathbf{r}}{\partial u}, \frac{\partial \mathbf{r}}{\partial v} = 0$$

Hence the result. [Refer Note 2 after § 3.1.2]

2. *Show that the normal to the surface*

$$x = (u+v)/\sqrt{2}, \quad y = (u-v)/\sqrt{2}, \quad z = uv$$

at any point (u, v) is described by any unit sector.

$$\mathbf{n} = \frac{x, -y, -1}{\sqrt{(1+x^2+y^2)}} = \frac{(u+v), (v-u), -\sqrt{2}}{\sqrt{\{2(1+u^2+v^2)\}}}.$$

Also evaluate curvature at the origin for the normal section in any direction du, dv and show that curvature is zero for the normal sections which have the same tangents as the parametric curves through the origin.

We have

$$\mathbf{r} = \left(\frac{u+v}{\sqrt{2}}, \frac{u-v}{\sqrt{2}}, uv\right).$$

$$\therefore \quad \frac{\partial \mathbf{r}}{\partial u} = \left(\frac{1}{\sqrt{2}}, \frac{1}{\sqrt{2}}, v\right), \quad \frac{\partial \mathbf{r}}{\partial v}\left(\frac{1}{\sqrt{2}}, -\frac{1}{\sqrt{2}}, u\right).$$

$$\therefore \quad \frac{\partial \mathbf{r}}{\partial u} \times \frac{\partial \mathbf{r}}{\partial v} = \left(\frac{u+v}{\sqrt{2}}, \frac{v-u}{\sqrt{2}}, -1\right).$$

$$\therefore \quad \mathbf{n} = \frac{\dfrac{\partial \mathbf{r}}{\partial u} \times \dfrac{\partial \mathbf{r}}{\partial v}}{\left|\dfrac{\partial \mathbf{r}}{\partial u} \times \dfrac{\partial \mathbf{r}}{\partial v}\right|} = \frac{(u+v), (v-u), -\sqrt{2}}{\sqrt{\{2(1+u^2+v^2)\}}}.$$

The curvature, κ, of the normal section in any direction *du, dv* is given by

$$\kappa = \frac{L(du)^2 + 2M\,du\,dv + N(dv)^2}{E(du)^2 + 2F\,du\,dv + G(dv)^2}.$$

We shall, therefore, now proceed to determine

E, F, G; L, M, N.

We have

$$\frac{\partial^2 \mathbf{r}}{\partial u^2} = (0, 0, 0), \quad \frac{\partial^2 \mathbf{r}}{\partial u\,\partial v} = (0, 0, 1), \quad \frac{\partial^2 \mathbf{r}}{\partial v^2} = (0, 0, 0)$$

$$E = \left(\frac{\partial \mathbf{r}}{\partial u}\right)^2 = \frac{1}{2} + \frac{1}{2} + v^2 = 1 + v^2,$$

$$F = \frac{\partial \mathbf{r}}{\partial u} \cdot \frac{\partial \mathbf{r}}{\partial v} = \frac{1}{2} - \frac{1}{2} + uv = uv,$$

$$G = \left(\frac{\partial \mathbf{r}}{\partial v}\right)^2 = \frac{1}{2} + \frac{1}{2} + u^2 = 1 + u^2.$$

$$L = \mathbf{n} \,.\, \frac{\partial^2 \mathbf{r}}{\partial u^2} = 0,$$

$$N = \mathbf{n} \,.\, \frac{\partial^2 \mathbf{r}}{\partial v^2} = 0,$$

$$M = \mathbf{n} \,.\, \frac{\partial^2 \mathbf{r}}{\partial u\, \partial v} = \frac{\sqrt{2}}{\sqrt{\{2(1+u^2+v^2)\}}} = -\frac{\mathbf{1}}{\sqrt{(\mathbf{1}+u^2+v^2)}}$$

$$\therefore \quad \kappa = \frac{-2du\,dv}{\sqrt{(1+u^2+v^2)}\;[(1+v^2)(du^2)+2uv\,du\,dv+(1+u^2)(dv)^2]}.$$

Clearly $\kappa = 0$ when $du = 0$ or $dv = 0$ so that the curvature is zero for normal sections touching the parametric lines.

$$u = \text{constant}, \quad v = \text{constant}.$$

3. *Find the lines of curvature and principal curvatures for the surface generated by the tangents to a skew curve.*

Let $\mathbf{r} = \mathbf{f}(s)$ be the given curve. Then the equation of the surface generated by the tangents to this curve is

$$\mathbf{R} = \mathbf{r} + u\,\frac{d\mathbf{r}}{ds},$$

where the position vector $\mathbf{R}$ of any point on the surface is a function of s, u and $\mathbf{r} = \mathbf{f}(s)$. We have

$$\frac{\partial \mathbf{R}}{\partial u} = \frac{\partial \mathbf{r}}{\partial s} = \mathbf{t}, \quad \frac{\partial \mathbf{R}}{\partial s} = \frac{d\mathbf{r}}{ds} + u\,\frac{d^2\mathbf{r}}{ds^2} = \mathbf{t} + u\kappa\,\mathbf{n}.$$

$$\therefore \quad E = \left(\frac{\partial \mathbf{R}}{\partial u}\right)^2 = 1, \quad F = \frac{\partial \mathbf{R}}{\partial u} \cdot \frac{\partial \mathbf{R}}{\partial s} = 1.$$

$$G = \left(\frac{\partial \mathbf{R}}{\partial s}\right)^2 = 1 + u^2\kappa^2 = 1 + \frac{u^2}{\rho^2}.$$

Also
$$\mathbf{n}_s = \frac{\dfrac{\partial \mathbf{R}}{\partial u} \times \dfrac{\partial \mathbf{R}}{\partial s}}{\left|\dfrac{\partial \mathbf{R}}{\partial u} \times \dfrac{\partial \mathbf{R}}{\partial s}\right|} = \frac{u\kappa\mathbf{b}}{u\kappa} = \mathbf{b}, \qquad \text{...(1)}$$

where $\mathbf{n}_s$ denotes the unit vector normal to the surface. From (1) it also follows that the normal at any point to the surface generated by tangents is parallel to the binormal to the curve at the corresponding point.

Further we have

$$\frac{\partial^2 \mathbf{R}}{\partial u^2} = 0, \quad \frac{\partial^2 \mathbf{R}}{\partial u\, \partial s} = \frac{d\mathbf{t}}{ds} = \kappa \mathbf{n}$$

$$\frac{\partial^2 \mathbf{R}}{\partial s^2} = \frac{d\mathbf{t}}{ds} + u\frac{d\kappa}{ds}\mathbf{n} + u\kappa \frac{d\mathbf{n}}{ds}$$

$$= \left(\kappa + u\frac{d\kappa}{ds}\right)\mathbf{n} - u\kappa^2 \mathbf{t} - u\kappa\tau \mathbf{b}.$$

$$L = \mathbf{n}_s \cdot \frac{\partial^2 \mathbf{R}}{\partial u^2} = 0, \quad M = \mathbf{n}_s \cdot \frac{\partial^2 \mathbf{R}}{\partial u\, \partial s} = \mathbf{b} \cdot \kappa \mathbf{n} = 0$$

$$N = \mathbf{n}_s \cdot \frac{\partial^2 \mathbf{R}}{\partial s^2} = -u\kappa\tau = -\frac{u}{\rho\sigma}.$$

$\therefore$ the principal curvatures are given by

$$-\frac{u^2}{\rho^2}\kappa^2 + \left(-\frac{u}{\rho\sigma}\right)\kappa = 0,$$

so that the two principal curvatures are

$$0, \quad -\frac{\rho}{\sigma u}.$$

The differential equation of the lines of curvature is

$$0\,(du)^2 + \frac{u}{\rho\sigma}\,du\,ds + \frac{u}{\rho\sigma}(ds)^2 = 0,$$

$$ds = 0, \quad du + ds = 0.$$

or

$$s = \text{constant}, \quad u + s = \text{constant}.$$

The eqaution $s =$ constant shows that the family of tangents to the curve is itself one of the systems of the lines of curvature.

4. *If a plane cuts a surface everywhere at a constant angle, then show that the curve of intersection is a line of curvature on the surface.*

Let **N** be the unit normal to the cutting plane, then $\frac{d\mathbf{N}}{ds} = 0$, thus the equation

$$\frac{d}{ds}\left(\mathbf{N} \cdot \frac{d\mathbf{N}}{ds}\right) = 0, \quad i.e., \quad \frac{d\mathbf{N}}{ds} \cdot \mathbf{N} + \mathbf{N} \cdot \frac{d\mathbf{N}}{ds} = 0$$

reduces to $\mathbf{N} \cdot \frac{d\mathbf{N}}{ds} = 0$. There $\frac{d\mathbf{N}}{ds}$ is the unit normal to the given surface.

Therefore, the following two equations

$$\mathbf{N} \cdot \frac{d\mathbf{N}}{ds} = 0 \quad \text{and} \quad \mathbf{N} \cdot \frac{d\mathbf{N}}{ds} = 0,$$

together gives that $\frac{d\mathbf{N}}{ds}$ is parallel to $N \times \frac{d\mathbf{N}}{ds}$ and hence parallel to **t**. Thus, the curve of intersection is a line of curvature on the surface.

5. *If l_1, m_1, n_1 be the direction cosines of the tangent to a line of curvature at l, m, n are the direction cosines of the normal to the surface at the point, then*

$$\frac{dl}{l_1} = \frac{dm}{m_1} = \frac{dn}{n_1}$$

Let **t** be unit tangent vector at a point P of the line of curvature, then $\mathbf{t} = (l_1, m_1, n_1)$.

If **N** be a unit vector along the normal to the surface at P, then $\mathbf{N} = (l, m, n)$.

Since the curve is a line of curvature, then by Rodrigue's formula, we get

$$\kappa \mathbf{t} + \frac{d\mathbf{N}}{ds} = 0 \Rightarrow \frac{d\mathbf{N}}{ds} \text{ is parallel to } \mathbf{t}.,$$

$$\Rightarrow \left(\frac{dl}{ds}, \frac{dm}{ds}, \frac{dn}{ds}\right) \text{ is parallel to } (l_1, m_1, n_1)$$

$$\Rightarrow \frac{(dl/ds)}{l_1} = \frac{(dm/ds)}{m_1} = \frac{(dn/ds)}{n_1}$$

$$\Rightarrow \frac{dl}{l_1} = \frac{dm}{m_1} = \frac{dn}{n_1}.$$

6. *Show that the osculating plane at any point of an asymptotic line is the tangent plane to the surface.* (*Berhampur 1993*)

Let $\mathbf{r} = \mathbf{r}(s)$ be a curved asymptotic line lying on the surface $\mathbf{r} = \mathbf{r}(u, v)$. The tangent **t** to the asymptotic line is perpendicular to the surface normal **N**, therefore

$$\mathbf{N} \cdot \mathbf{t} = 0$$

$$\Rightarrow \frac{d\mathbf{N}}{ds} \cdot \mathbf{t} + \mathbf{N} \cdot \frac{d\mathbf{t}}{ds} = 0$$

$$\Rightarrow \frac{d\mathbf{N}}{ds} \cdot \frac{d\mathbf{r}}{ds} + N \cdot (\kappa \mathbf{n}) = 0$$

$$\Rightarrow \mathbf{N} \cdot (\kappa \mathbf{n}) = 0 \quad \left[\because \mathbf{r} = \mathbf{r}(s) \text{ is an asymptotic line and for asymptotic line } \frac{d\mathbf{N}}{ds} \cdot \frac{d\mathbf{r}}{ds} = 0\right]$$

$$\Rightarrow \mathbf{N} \cdot \mathbf{n} = 0 \quad (\because \kappa \neq 0)$$

Also, we have $\mathbf{N} \cdot \mathbf{t} = 0$. Therefore, **N** is perpendicular to both **t** and **n** and is parallel to $\mathbf{t} \times \mathbf{n} = \mathbf{b}$. Hence $\mathbf{N} = \pm \mathbf{b}$. Thus, at any point of a curved asymptotic line, the binormal is also normal to the surface. We choose the direction of **b** so that $\mathbf{N} = \mathbf{b}$.

Now, equation of the osculating plane to $\mathbf{r} = \mathbf{r}(s)$ at any point is

$$(\mathbf{R} - \mathbf{r}) \cdot \mathbf{b} = 0$$

and the equation of the tangent plane to the surface at this point is

$$(\mathbf{R}-\mathbf{r}).\mathbf{N} = 0$$

The relation $\mathbf{N}=\mathbf{b}$ makes these two planes coincident at the point on the curve on the surface and hence the proposition.

7. *Find the asymptotic lines and their torsions on the surface generated by the binormals to a twisted curve.*

The position vector of a current point on the surface is

$$\mathbf{R} = \mathbf{R} + v\,\mathbf{b}$$

we have $$L=0,\ M = 0-\frac{\tau}{H},\ N = \frac{\left(\kappa+\kappa\tau^2 v^2 - \frac{d\tau}{ds}v\right)}{H},$$

$$H^2 = 1+\tau^2 v^2.$$

Therefore, asymptotic lines given by

$$L\,du^2 + 2M\,du\,dv + N dv^2 = 0$$

reduces to

$$-2\tau\,dv\,ds + \left(\kappa+\kappa\tau^2 v^2 - \frac{d\tau}{ds}v\right)ds^2 = 0$$

which is the differential equation of the asymptotic line.

Now, $$\tau = \pm\sqrt{(-\kappa)}$$

$$= \pm\sqrt{\left(\frac{M^2-LN}{H^2}\right)} = \pm\frac{\tau}{H^2}$$

$$= \pm\frac{\tau}{(1+\tau^2 v^2)}.$$

8. *Find the asymptotic lines of the surface*

$$x = 3u(1+v^2)-u^3,\ y = 3v(1+u^2)-v^3,$$
$$z = 3(u^2-v^2).$$

We have

$$\mathbf{r} = [3u(1+v^2)-u^3,\ 3v(1+u^2)-v^3,\ 3(u^2-v^2)].$$

$$\therefore\quad \frac{\partial \mathbf{r}}{\partial u} = [3(1+v^2-u^2),\ 6uv,\ 6u],$$

$$\frac{\partial \mathbf{r}}{\partial v} = [6uv,\ 3(1+u^2-v^2),\ -6v],$$

$$\frac{\partial^2 \mathbf{r}}{\partial u^2} = (-6u,\ 6v,\ 6),$$

$$\frac{\partial^2 \mathbf{r}}{\partial u\,\partial v} = (6v,\ 6u,\ 0)$$

$$\frac{\partial^2 \mathbf{r}}{\partial v^2} = (6u,\ -6v,\ -6).$$

$$\frac{\partial \mathbf{r}}{\partial u} \times \frac{\partial \mathbf{r}}{\partial v} = [-18u\,(1+u^2+v^2),\ 18\,v\,(\mathbf{1}+u^2+v^2),$$
$$9\,(1-u^4-v^4-2u^2v^2)].$$

$$\therefore \quad \frac{\partial \mathbf{r}}{\partial u} \times \frac{\partial \mathbf{r}}{\partial v} \cdot \frac{\partial^2 \mathbf{r}}{\partial u^2} = 54\,(u^2+v^2+1)^2,$$

$$\frac{\partial \mathbf{r}}{\partial u} \times \frac{\partial \mathbf{r}}{\partial v} \cdot \frac{\partial^2 \mathbf{r}}{\partial u\,\partial v} = 0.$$

$$\frac{\partial \mathbf{r}}{\partial u} \times \frac{\partial \mathbf{r}}{\partial v} \cdot \frac{\partial^2 \mathbf{r}}{\partial v^2} = -54\,(u^2+v^2+1)^2.$$

$$\therefore \quad L = \frac{54(u^2+v^2+1)^2}{\left|\frac{\partial \mathbf{r}}{\partial u} \times \frac{\partial \mathbf{r}}{\partial v}\right|},\ M = 0,$$

$$N = \frac{-54(u^2+v^2+1)^2}{\left|\frac{\partial \mathbf{r}}{\partial u} \times \frac{\partial \mathbf{r}}{\partial v}\right|}.$$

Thus the differential equation of the asymptotic lines is

$$(du)^2 - (dv)^2 = 0, \text{ i.e., } du \pm dv = 0.$$

$$\therefore \quad u \pm v = \text{constant},$$

are the two systems of asymptotic lines.

EXERCISES

1. Find the fundamental magnitudes of the first and second order for the following surfaces:

(*i*) $x = u,\ y = av,\ z = auv.$

(*ii*) $x = u\cos\theta,\ y = u\sin\theta,\ z = c\,\theta.$

(*iii*) $x = u\cos\theta,\ y = u\sin\theta,\ z = c\log\left(u+\sqrt{u^2-c^2}\right).$

2. Find the lines of curvature and the asymptotic lines for the surface

$$x = a\,(u+v),\ y = b\,(u-v),\ z = uv.$$

3. Find the Gaussian curvature at any point of the surface generated by the binormal to a curve $\mathbf{r} = \mathbf{f}\,(s)$.

4. Find the principal radii at any point of the surface

$$x = 3u\,(1+v^2) - u^3,\ y = 3v\,(1+u^2) - v^2,\ z = 3\,(u^2-v^2)$$

and show that the parametric curves are the lines of curvature of the surface.

5. Prove that the differential equation of the lines of curvature on the surface $\mathbf{r} = \mathbf{r}\,(u, v)$ is

$$[\mathbf{n}\,.\,\mathbf{n}_1, \mathbf{r}_1]\,(du)^2 + \{[\mathbf{n}, \mathbf{n}_1, \mathbf{r}_2] + [\mathbf{n}, \mathbf{n}_2, \mathbf{r}_1]\}\,du\,dv + [\mathbf{n}, \mathbf{n}_1, \mathbf{r}_2]\,(dv)^2 = 0,$$

where, $\mathbf{n}$, is the unit normal to the surface and the suffixes 1, 2 denote partial derivatives with respect to the parameters u and v respectively.

6. Show that the asymptotic lines of the hyperboloid

$$x = a \cos u \sec v,\ y = b \sin u \sec v,\ z = c \tan v$$

are given by

$$u \pm v = \text{constant}.$$

7. Show that the product of the principal radii of curvature at any

$$x = uv,\ \ y = v\,(u-1),\ \ z = u\,(v-1)$$

is

$$-\{2\,(u^2 + uv + v^2) - 2\,(u+v) + 1\}^2.$$

8. Prove that the curvature of an asymptotic line may be expressed as

$$\frac{\left(\mathbf{r}_1 \cdot \dfrac{d\mathbf{r}}{ds}\right)\left(\mathbf{r}_2 \cdot \dfrac{d^2\mathbf{r}}{ds^2}\right) - \left(\mathbf{r}_2 \cdot \dfrac{d\mathbf{r}}{ds}\right)\left(\mathbf{r}_1 \cdot \dfrac{d^2\mathbf{r}}{ds^2}\right)}{H}.$$

9. Prove that all straight lines on surface are asymptotic lines.

OBJECTIVE QUESTIONS

For each of the following questions, four alternatives are given for the answer. Only one of them is correct. Choose the correct alternative.

1. For a surface $\mathbf{r} = \mathbf{F}\,(u, v)$, singular point is the point at which

(*a*) $\dfrac{\partial \mathbf{r}}{\partial u} \cdot \dfrac{\partial \mathbf{r}}{\partial v} = 0$ (*b*) $\dfrac{\partial \mathbf{r}}{\partial u} \times \dfrac{\partial \mathbf{r}}{\partial v} = 0$

(*c*) $\dfrac{\partial \mathbf{r}}{\partial u} + \dfrac{\partial \mathbf{r}}{\partial v} = 0$ (*d*) $\dfrac{\partial^2 \mathbf{r}}{\partial u\, \partial v} = 0$

2. Two parametric curves through every point will be orthogonal, if for every pair of values u, v,

(*a*) $\dfrac{\partial \mathbf{r}}{\partial u} \cdot \dfrac{\partial \mathbf{r}}{\partial v} = 0$ (*b*) $\dfrac{\partial \mathbf{r}}{\partial u} \times \dfrac{\partial \mathbf{r}}{\partial v} = 0$

(*c*) $\dfrac{\partial \mathbf{r}}{\partial u} + \dfrac{\partial \mathbf{r}}{\partial v} = 0$ (*d*) $\dfrac{\partial^2 \mathbf{r}}{\partial u\, \partial v} = 0$

3. For the parametric curve u = constant

(*a*) $\dfrac{dv}{ds} = 0$ (*b*) $\dfrac{du}{ds} = 1/\sqrt{G}$

(*c*) $\dfrac{dv}{ds} = 1/\sqrt{G}$ (*d*) None of these

4. The two principal directions at any point are

(*a*) parallel (*b*) orthogonal

(*c*) collinear (*d*) inclined at certain angle

5. Osculating plane to an asymptotic line at any point
   (*a*) coincides with the tangent plane at the point
   (*b*) coincides with the normal plane at the point
   (*c*) is normal to the tangent plane at the point
   (*d*) None of these
6. The necessary and sufficient condition that the parametric curves be asymptotic lines are
   (*a*) $L \neq 0,\ M = 0,\ N = 0$ (*b*) $L = 0,\ M \neq 0,\ N = 0$
   (*c*) $L = 0,\ M = 0,\ N \neq 0$ (*d*) $L = 0,\ M = 0,\ N = 0$

ANSWERS

1. (*b*) **2.** (*a*) **3.** (*c*) **4.** (*b*) **5.** (*a*)
6. (*b*)

4

Mechanics I

PARTICLES

Introduction. All motion is relative and any given particle can be thought of as fixed or moving in any given manner relative to a suitably chosen body of reference so that the concepts of a fixed body or of a body in motion are, as such, meaningless and acquire a meaning only when a reference body is assigned. Thus in the following study of motion the importance of an underlying reference body should be constantly borne in mind.

If is convenient to split up the study of Mechanics into two parts known as

Kinematics and *Kinetics.*

Kinematics is the study of motion and of the time rates of change in the same without reference to the causes. The notion of *Force* which is the known cause of change in motion does not appear in the study of kinematics. The study of Kinematics is a preliminary to that of *Kinetics* which deals with the action of forces and the resulting change in motion as based on the well-known *Newton's Laws of Motion.*

We shall take up the study of Mechanics of a single particle and of a system of particles in the present chapter, of a rigid body in the following chapter and the study of Mechanics of Fluids will be made in Chapter VIII.

By a *particle* will be meant a material system which, at any given time, can be located at a point so that the position of a particle is known by that of the point occupied by it. Often we refer to a point in motion and mean thereby the motion of a particle located at the point.

A *Rigid body* is defined as a system of particles such that the distance between any pair of particles of the system remains unchanged irrespective of the nature of the forces acting on the system. From this point of view, fluids do not constitute a rigid system and form, what may be called, *Deformable Systems.* The concepts of a particle and of a rigid body are only mathematical abstractions and, in reailty, no such thing as a particle or a rigid body is known to exist. But the notions are of importance, for the motions of various important types of material systems can be approximated to by those of a particle or a rigid body.

The reference body must be thought of as a rigid body which may, however, be replaced by a system of rectangular axes rigidly connected with the body, for the position of the body can be determined by that of the system of axes.

4.1. KINEMATICS OF A PARTICLE. DISPLACEMENT. VELOCITY. ACCELERATION.

Consider a moving point P and a system of rectangular cartesian co-ordinate axes OX, OY, OZ regarded as a Reference System. The reference system may also be called Reference Frame.

Let **j, j, k** denote unit vectors along the axes. We write

$$\overrightarrow{OP} = \mathbf{r},$$

so that if x, y, z, are the projections of OP on the co-ordinate axes, *i.e.*, if x, y, z, are the co-ordinates of P, we have

$$\overrightarrow{OP} = \mathbf{r} = \mathbf{i}x + \mathbf{j}y + \mathbf{k}z.$$

The motion of the point P is given by **r**, as a vector function of the time t, say

$$\mathbf{r} = \mathbf{f}(t) = \mathbf{i}f_1(t) + \mathbf{j}f_2(t) + \mathbf{k}f_3(t).$$

The locus of P, known as its *Trajectory,* is the curve with vector equation $\mathbf{r} = \mathbf{f}(t)$; t being now thought of as a parameter.

Let P, Q be the positions of the moving particle at times t, $t + \delta t$ so that we may write

$$\overrightarrow{OP} = \mathbf{r} = \mathbf{f}(t), \quad \overrightarrow{OQ} = \mathbf{r} + \delta\mathbf{r} = \mathbf{f}(t + \delta t).$$

The displacement of the particle in time δt is the vector $\overrightarrow{PQ}$ where

$$\begin{aligned}\overrightarrow{PQ} &= \overrightarrow{OQ} - \overrightarrow{OP}\\ &= (\mathbf{r} + \delta\mathbf{r}) - \mathbf{r}\\ &= \delta\mathbf{r}\\ &= \mathbf{f}(t + \delta t) - \mathbf{f}(t)\\ &= \mathbf{i}\,[f_1(t+\delta t) - f_1(t)] + \mathbf{j}\,[f_2(t+\delta t) - f_2(t)] + \mathbf{k}\,[f_3(t+\delta t) - f_3(t)].\end{aligned}$$

The *velocity of a particle relative to any given frame of reference is, by definition, the time rate of change of the position vector,* **r**, *of the particle relative to the frame.*

Hence the velocity, **v**, of the particle P, at time t, is given by

$$\begin{aligned}\mathbf{v} &= \frac{d\mathbf{r}}{dt} = \frac{d\mathbf{f}}{dt} = \mathbf{i}\frac{df_1}{dt} + \mathbf{j}\frac{df_2}{dt} + \mathbf{k}\frac{df_3}{dt}\\ &= \mathbf{i}\frac{dx}{dt} + \mathbf{j}\frac{dy}{dt} + \mathbf{k}\frac{dz}{dt}.\end{aligned}$$

Again, *the acceleration being, by definition, the time rate of change of velocity*, we see that the acceleration, **a**, of the particle P is given by

$$\mathbf{a} = \frac{d^2\mathbf{r}}{dt^2} = \frac{d^2\mathbf{f}}{dt^2} = \mathbf{i}\frac{d^2 f_1}{dt^2} + \mathbf{j}\frac{d^2 f_2}{dt^2} + \mathbf{k}\frac{d^2 f_3}{dt^2}.$$

$$= \mathbf{i}\frac{d^2 x}{dt^2} + \mathbf{j}\frac{d^2 y}{dt^2} + \mathbf{k}\frac{d^2 z}{dt^2}.$$

4.1.1. Speed of a particle is the magnitude |v| of its velocity v

If, s, denote the length of an are of the trajectory $\mathbf{r} = \mathbf{f}(t)$ of the particle measured from any point A thereof upto P,

$$\mathbf{v} = \frac{d\mathbf{r}}{dt} = \frac{d\mathbf{f}}{ds}\frac{ds}{dt} = \frac{ds}{dt}\mathbf{t}, \qquad ...(1)$$

where, **t**, denotes unit tangent vector to the curve.

The relation (1) shows that the speed is equal to the absolute value of the scalar ds/dt.

Also we see that the direction of the velocity of P is the same as that of the tangent to the trajectory at P and the sense is the same or opposite to that the unit tangent vector, **t**, according as ds/dt is positive or negative, *i.e.*, according as, s, increases or decreases with the time t.

Again since

$$\mathbf{v} = \mathbf{i}\frac{dx}{dt} + \mathbf{j}\frac{dy}{dt} + \mathbf{k}\frac{dz}{dt},$$

we see that the speed |**v**| is also given by

$$|\mathbf{v}| = \sqrt{\left[\left(\frac{dx}{dt}\right)^2 + \left(\frac{dy}{dt}\right)^2 + \left(\frac{dz}{dt}\right)\right]}.$$

4.1.2. Components of velocity and acceleration

Let **p, q, r** be any three unit non-coplanar vectors so that **v** and **a** can be expressed as linear combinations of the same. Let

$$\mathbf{v} = v_1\mathbf{p} + v_2\mathbf{q} + v_2\mathbf{r};$$

$$\mathbf{a} = a_1\mathbf{p} + a_2\mathbf{q} + a_2\mathbf{r}.$$

Then v_1, v_2, v_3 and a_1, a_2, a_3 are said to be the components of the velocity **v** and the acceleration **a** along the vectors **p, q, r**.

The case of mutually orthogonal vectors **p, q, r** will play a comparatively more important part.

It follows from above that

$$\frac{dx}{dt}, \frac{dy}{dt}, \frac{dz}{dt};$$

$$\frac{d^2x}{dt^2}, \frac{d^2y}{dt^2}, \frac{d^2z}{dt^2},$$

are the components of the velocity and the acceleration of a moving particle along the axes of the frame of reference.

Ex. v_1, v_2, v_3 are the components of velocity along the axes of a cartesian rectangular system and $l_1, m_1, n_1; l_2, m_2, n_2; l_3, m_3, n_3$, are the direction cosines of the axes of another rectangular system with respect to the first. Show that the components of velocity along the axes of the second system are

$$l_1v_1 + m_1v_2 + n_1v_3,\ \ l_2v_1 + m_2v_2 + n_2v_3,\ \ l_3v_1 + m_3v_2 + n_3v_3.$$

State and prove a similar result regarding acceleration.

Notes 1. In the preceding § 4.1, nothing has been said about the nature of the motion of the Reference system of axes OX, OY, OZ itself and we should remember that the statements made are all independent of this motion provided only that it is clearly understood that all remarks about velocity and acceleration pertain to the motion *relative to this reference system.*

2. In case we have two systems of rectangular axes

$$\text{OX, OY, OZ; OX}', \text{OY}', \text{OZ}'$$

each moving relatively to the other, then the velocities and accelerations of a moving point relative to the two systems will, of course, be different and there arises the question of the determination of the relations between the same. A very elegant form of these relations which must obviously depend upon the nature of motion of either system relative to the other will, however, only be obtained later on after we have considered the Kinematics of a Rigid body.

The following remarks may prove useful in this connection.

Let **i, j, k; i′, j′, k′**s denote unit vectors along the axes of the two systems and let $x, y, z;\ x', y', z'$ be the projections of $\overrightarrow{OP}$ on the axes. We have

$$\mathbf{r} = \overrightarrow{OP} = \mathbf{i}x + \mathbf{j}y + \mathbf{k}z = \mathbf{i}'x' + \mathbf{j}'y' + \mathbf{k}'z'. \quad ...(1)$$

The velocities of P relative to the two systems of axes separately are

$$\mathbf{i}\frac{dx}{dt} + \mathbf{j}\frac{dy}{dt} + \mathbf{k}\frac{dz}{dt},\ \ \mathbf{i}'\frac{dx'}{dt} + \mathbf{j}'\frac{dy'}{dt} + \mathbf{k}'\frac{dz'}{dt}$$

Also taking into consideration the motion of the system OX′, OY′, OZ′ relative to OX, OY, OZ, we have from (1)

$$\mathbf{i}\frac{dx}{dt} + \mathbf{j}\frac{dy}{dt} + \mathbf{k}\frac{dz}{dt} = \mathbf{i}'\frac{dx'}{dt} + \mathbf{j}'\frac{dy'}{dt} + \mathbf{k}'\frac{dz'}{dt} + x'\frac{d\mathbf{i}'}{dt} + y'\frac{d\mathbf{j}'}{dt} + z'\frac{d\mathbf{k}'}{dt} \quad ...(2)$$

so that either side gives the velocity relative to the system OX, OY, OZ; the time derivatives $d\mathbf{i}'/dt,\ d\mathbf{j}'/dt,\ d\mathbf{k}'/dt$ denoting the velocities relative to OX, OY, OZ. If we could express these time derivatives in terms of **i′**, **j′**, **k′**, then the coefficients of **i′**, **j′**, **k′** on the R.H.S. of (2) will be the components along OX′, OY′, OZ′ of the velocity relative to OX, OY, OZ.

4.2. CIRCULAR MOTION

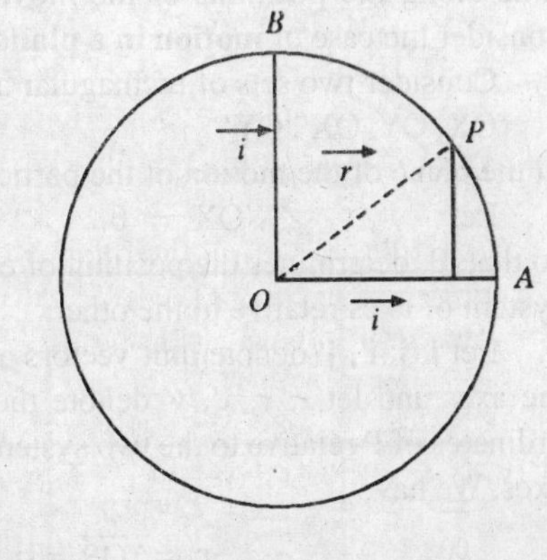

Fig. 4.1

Let ω be the angular velocity of a moving particle P at time t, in a circle of radius r and centre O and A be the starting point. Take **i**, **j** unit vectors along OA and OB taken as axes of x and y respectively.

Let $\overrightarrow{OP} = \mathbf{r}$, $\angle AOP = \theta$.

Then $\theta = \omega t$ ...(1)

Equation of the circular path is

$$\mathbf{r}(t) = r\,\hat{\mathbf{r}}(t), \ |\mathbf{r}| = r \quad ...(2)$$

If PM be perpendicular from P to OA, then

$$\overrightarrow{OP} = \overrightarrow{OM} + \overrightarrow{MP}$$

$$= r\hat{\mathbf{r}}(t) = r\cos\theta\,\mathbf{i} + r\sin\theta\,\mathbf{j}$$

$$\Rightarrow \quad r(t) = r\hat{\mathbf{r}}(t) = r\cos\omega t\,\mathbf{i} + r\sin\omega t\,\mathbf{j} \quad ...(3)$$

$$\Rightarrow \quad \hat{\mathbf{r}}(t) = \cos\omega t\,\mathbf{i} + \sin\omega t\,\mathbf{J} \quad ...(4)$$

Now, velocity $\mathbf{v} = \dfrac{d\mathbf{r}}{dt} = r\dfrac{d\hat{\mathbf{r}}}{ds} = r\left(\mathbf{i}\dfrac{d}{dt}\cos\omega t + \mathbf{j}\dfrac{d}{dt}\sin\omega t\right)$

$$= \omega t\,(-\sin\omega t\,\mathbf{i} + \cos\omega t\,\mathbf{j})$$

∴ Speed $|\mathbf{v}| = \omega r = v$ (say) ...(5)

Acceleration of the particle

$$\mathbf{a} = \frac{d\mathbf{v}}{dt} = \frac{d^2 r}{dt^2}$$

$$= \omega r\,\frac{d}{dt}(-\sin\omega t\,\mathbf{i} + \cos\omega t\,\mathbf{j})$$

$$= -\omega^2\,\mathbf{r} \quad ...(6)$$

Its magnitude $|\mathbf{a}| = |-\omega^2\,\mathbf{r}| = \omega^2 t = a$ (say).

Eliminating ω from (5) and (6), we get

$$a = \frac{v^2}{r}.$$

This acceleration is known as the *Centripetal* (*i.e.*, centre seeping) *acceleration.*

If f be the frequency of the particle, $2\pi f = \omega$.

So, the time period T of the motion is given by

$$T = \frac{2\pi}{\omega} = \frac{1}{f}.$$

4.3. COMPONENTS ALONG MOVING AXES

It is often found useful to determine components along axes which are moving relative to the Reference frame so that we find components at any time along the positions of the moving axes at the time. Here below we consider the case of **motion in a plane.**

Consider two sets of rectnagular axes

OX, OY; OX′, OY′

in the plane of the motion of the particle P.

Let $\angle X'OX = \theta,$

so that, θ, determines the position of either system of axes relative to the other.

Let **i**, **j**; **i**′, **j**′ denote unit vectors along the axes and let $x, y; x', y'$ denote the co-ordinates of P relative to the two systems of axes. We have

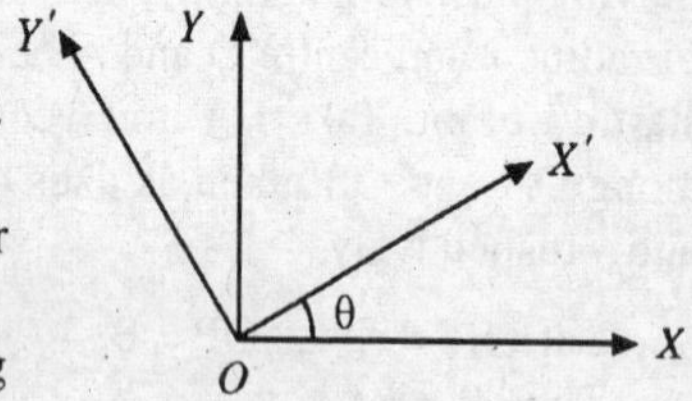

Fig. 4.2

$$\mathbf{r} = \overrightarrow{OP} = \mathbf{i}x + \mathbf{j}y = \mathbf{i}x' + \mathbf{j}'y'. \qquad ...(1)$$

We choose OX, OY as the reference frame and shall suppose that by the velocity and acceleration of P will be meant the same relative to OX, OY unless otherwise stated.

The vectors **i**′, **j**′ are changing with time relative to OX, OY.

From (1), we have

$$\mathbf{v} = \frac{d\mathbf{r}}{ds} = \mathbf{i}\frac{dx}{dt} + \mathbf{j}\frac{dy}{dt}$$

$$= \mathbf{i}'\frac{dx'}{dt} + \mathbf{j}'\frac{dy'}{dt} + x'\frac{d\mathbf{i}'}{ds} + y'\frac{d\mathbf{j}'}{dt} \qquad ...(2)$$

The vectors $d\mathbf{i}'/dt$, $d\mathbf{j}'/dt$ will now be expressed in therms of **i**′, **j**′ and $d\theta/dt$. We write

$$\frac{d\theta}{dt} = \omega.$$

As may be easily seen, we have

$$\mathbf{i}' = \mathbf{i}\cos\theta + \mathbf{j}\sin\theta,$$

$$\mathbf{j}' = -\,\mathbf{i}\sin\theta + \mathbf{j}\cos\theta.$$

$$\therefore \quad \frac{d\mathbf{i}'}{dt} = (-\,\mathbf{i}\sin\theta + \mathbf{j}\cos\theta)\,\frac{d\theta}{dt} = \mathbf{j}'\,\omega, \qquad ...(3)$$

$$\frac{d\mathbf{j}'}{dt} = (-\,\mathbf{i}\cos\theta - \mathbf{j}\sin\theta)\,\frac{d\theta}{dt} = -\,\mathbf{j}'\,\omega. \qquad ...(4)$$

From (2), (3) and (4), we have

$$\mathbf{v} = \frac{dx}{dt} + \mathbf{j}\frac{dy}{dt} = \left(\frac{dx}{dt} - y'\omega\right)\mathbf{i}' + \left(\frac{dy'}{dt} + x'\omega\right)\mathbf{j}'. \qquad ...(5)$$

Again taking time derivatives, we obtain

$$\mathbf{a} = \mathbf{i}\,\frac{d^2x}{dt^2} + \mathbf{j}\,\frac{d^2y}{dt^2}$$

$$= \left(\frac{d^2x}{dt^2} - \omega\,\frac{dy'}{dt} - y'\,\frac{d\omega}{dt}\right)\mathbf{i}'$$

$$+\left(\frac{d^2y}{dt^2}+\omega\,\frac{dx'}{dt}+x'\,\frac{d\omega}{dt}\right)\mathbf{j}'$$

$$+\left(\frac{dx'}{dt}-y'\omega\right)\frac{d\mathbf{i}'}{dt}+\left(\frac{dy'}{dt}+x'\omega\right)\frac{d\mathbf{j}'}{dt}$$

$$=\left(\frac{d^2x'}{dt^2}-x'\omega-2\omega\frac{dy'}{dt}-y'\,\frac{d\omega}{dt}\right)\mathbf{i}'$$

$$+\left(\frac{d^2y'}{dt^2}-y'\omega^2+2\omega\frac{dx'}{dt}+x'\,\frac{d\omega}{dt}\right)\mathbf{j}'. \quad ...(6)$$

From (5) and (6), we see that the components along OX′, OY′ [*i.e.*, along the positions of the moving axes at time *t*] of the velocity and acceleration at time *t* are

$$\frac{dx'}{dt}-y'\omega,\ \frac{dy'}{dt}+x'\omega. \quad ...(7)$$

$$\frac{d^2x'}{dt^2}-x'\omega^2-2\omega\frac{dy'}{dt}-y'\frac{d\omega}{dt},\ \frac{d^2y'}{dt^2}-y'\omega^2+2\omega\frac{dx'}{dt}+x'\frac{d\omega}{dt} \quad ...(8)$$

respectively.

The results obtained above may be compared to the statements that

$$\frac{dx'}{dt},\ \frac{dy'}{dt}$$

$$\frac{d^2x'}{dt^2},\ \frac{d^2y'}{dt^2}$$

are the components along OX′, OY′ of the velocity and acceleration relative to the same axes.

Ex. 1. Explain why the components (8) of acceleration are not the time derivative of the components (7) of velocity.

Ex. 2. Show directly that the components (7) and (8) are equal to

$$\frac{dx}{dt}\cos\theta+\frac{dy}{dt}\sin\theta,\ -\frac{dx}{dt}\sin\theta+\frac{dy}{dt}\cos\theta \quad ...(9)$$

$$\frac{d^2x}{dt^2}\cos\theta+\frac{d^2y}{dt^2}\sin\theta,\ -\frac{d^2x}{dt^2}\sin\theta+\frac{d^2y}{dt^2}\cos\theta \quad ...(10)$$

and explain why?

Note. The truth of the relations

$$\frac{d\mathbf{i}'}{dt}=\omega\mathbf{j}',\ \frac{d\mathbf{j}'}{dt}=-\,\omega\mathbf{i}',$$

may also be seen in another manner as follows:

The end point of the unit vector **i′** describes a circle of unit radius with O as centre. Thus, if, *s*, denotes the are of this circle upto P measured from the point where it meets OX, we have

$$s = \theta \text{ so that } ds/d\theta = 1.$$

Now $$\frac{d\mathbf{i}'}{dt} = \frac{d\mathbf{i}'}{dt}\frac{ds}{d\theta}\frac{d\theta}{dt} = \omega + \frac{d\mathbf{i}'}{ds}.$$

The vector $d\mathbf{i}'/ds$ which denotes unit tangent vector to the circle is perpendicular to $\mathbf{i}'$ and possesses sense corresponding to that of increasing s.

Thus $$\frac{d\mathbf{i}'}{ds} = \mathbf{j}'.$$

$\therefore$ $$\frac{d\mathbf{i}'}{dt} = \omega\mathbf{j}'.$$

Similarly $$\frac{d\mathbf{j}'}{dt} = -\omega\mathbf{i}.$$

4.4. SPECIAL CASES OF MOVING AXES FOR MOTION IN A PLANE

4.1.1. Radial and Transversal Components

Let OX, OY be the reference system of rectangular axes in the plane of motion of the moving particle P.

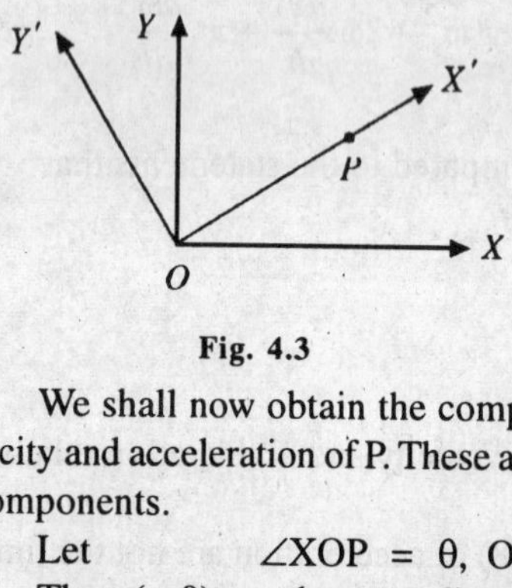

Fig. 4.3

Let another system of rectangular co-ordinate axes be formed by the line OP and the line through O perpendicular to OP. This latter system of axes is moving relative to the system OX, OY and its position at any time t is given by that of the moving point P at the time.

We shall now obtain the components along the moving axes of the velocity and acceleration of P. These are usually known as *Radial and Transveral* components.

Let $\angle XOP = \theta$, $OP = r$.

Then $(r, 0)$ are the co-ordinates of P relative to the moving axes OX′, OY′. Replacing x' by r and y' by 0 in the expressions (7) and (8) of § 4.3, we see that

$$\frac{d\mathbf{r}}{dt}, r\frac{d\theta}{dt},$$

$$\frac{d^2r}{dt^2} - r\left(\frac{d\theta}{dt}\right)^2, 2\frac{dr}{dt}\frac{d\theta}{dt} + r\frac{d^2\theta}{dt^2} = \frac{1}{r}\frac{d}{dt}\left(r^2\frac{d\theta}{dt}\right);$$

are the required radial and transversal components of the velocity and acceleration.

We may also obtain the same result *directly* as follows: If $\mathbf{i}'$, $\mathbf{j}'$, denote unit vectors along OP and perpendicular to it, we have

$$\overrightarrow{OP} = \mathbf{r} = r\,\mathbf{i}'.$$

Here, r, denotes the absolute value of the position vector $\mathbf{r}$ of the point P.

$$\therefore \qquad \mathbf{v} = \frac{d\mathbf{r}}{dt} = \frac{dr}{dt}\mathbf{i}' + r\frac{d\mathbf{i}'}{dt}$$

$$= \frac{dr}{dt}\mathbf{i}' + r\frac{d\mathbf{i}'}{d\theta}\frac{d\theta}{dt}$$

$$= \frac{dr}{dt}\mathbf{i}' + r\frac{d\theta}{dt}\mathbf{j}'. \qquad \text{(Note after § 4.3)}$$

Again
$$\mathbf{a} = \frac{d^2\mathbf{r}}{dt^2} = \frac{d^2r}{dt^2}\mathbf{i}' + \left(\frac{dr}{dt}\frac{d\theta}{dt} + r\frac{d^2\theta}{dt^2}\right)\mathbf{j}'$$

$$+ \frac{dr}{dt}\frac{d\mathbf{i}'}{dt} + r\frac{d\theta}{dt}\frac{d\mathbf{j}'}{dt}$$

$$= \left[\frac{d^2r}{dt^2} - r\left(\frac{d\theta}{dt}\right)^2\right]\mathbf{i}' + \frac{1}{r}\frac{d}{dt}\left(r^2\frac{d\theta}{dt}\right)\mathbf{j}'.$$

Hence the results.

4.4.2. Tangential and Normal Components

The investigation in this case will be applicable to the case of motion in *space* and we need not make the restriction to motion in plane.

We suppose that the moving axes at any time are parallel to the unit tangent, unit principal normal and unit binormal **t, n, b** to the trajectory of the particle at the position of the particle at the time.

We have
$$\mathbf{v} = \frac{ds}{dt}\mathbf{t}. \qquad (\text{§ } 4.1.1)$$

$$\mathbf{a} = \frac{d\mathbf{v}}{dt}$$

$$= \frac{d^2s}{dt^2}\mathbf{t} + \frac{ds}{dt}\frac{d\mathbf{t}}{dt}$$

$$= \frac{d^2s}{dt^2}\mathbf{t} + \frac{ds}{dt}\frac{d\mathbf{t}}{ds}\frac{ds}{dt}$$

$$= \frac{d^2s}{dt^2}\mathbf{t} + \left(\frac{ds}{dt}\right)^2\kappa\mathbf{n} + 0\mathbf{b}.$$

Then
$$\frac{d^2s}{dt^2},\ \left(\frac{ds}{dt}\right)^2\kappa\ 0.$$

are the components of acceleration along the vectors **t, n, b** respectively.

We can also state that if, v, denotes the speed of a particle, then the components of its acceleration along the tangent, principal normal and binormal are

$$\frac{dv}{dt}, \frac{v^2}{\rho}, 0.$$

Because of the component of acceleration along binormal being zero, we see that the *acceleration vector at any point of a trajectory lies in the osculating plane at the point.*

4.5. MOMENTUM AND MOMENT OF MOMENTUM. KINETIC ENERGY

By the *momentum* **M** of a particle at any time is mean.

$$\mathbf{M} = m\mathbf{v}.$$

where m is the mass and **v** the velocity of the particle.

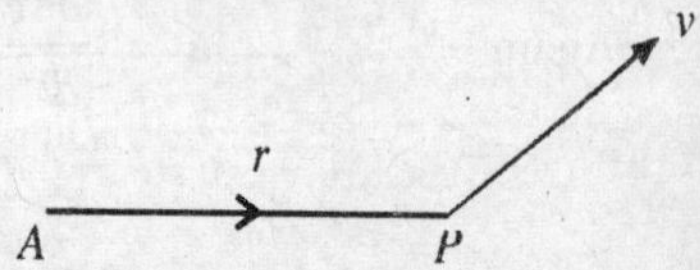

Fig. 4.4

If **M** be the momentum of a particle P at any time and if, **A**, be any point, then

$$\mathbf{H} = \overrightarrow{\mathrm{AP}} \times \mathbf{M}$$

where $\overrightarrow{\mathrm{AP}}$ is the position vector of P relative to A, is called the Moment of momentum or Angular momentum of the particle about A.

If, A, be the origin of the reference system, we have

$$\mathbf{H} = \mathbf{r} \times m\frac{d\mathbf{r}}{dt} = m\,\mathbf{r} \times \frac{d\mathbf{r}}{dt}.$$

The moment of momentum is in general different for different points A.

The scalar $\frac{1}{2} m\ \mathbf{v}^2$ is, by def. the kinetic energy of a particle of mass m and velocity **v**.

Note. Like velocity and acceleration, *momentum* is a relative concept having meaning with reference to some assigned Reference system only.

The moment of momentum is relative in *two* ways in as much as it is relative to the reference frame with respect to which momentum is considered and the point with respect to which the moment is taken.

4.5.1. Areal Velocity

Let P, Q be the positions of a moving particle at times t, $t + \delta t$ and let **r**, $\mathbf{r} + \delta\mathbf{r}$ be their position vector relative to the origin A of the Reference system. If $\delta\mathbf{A}$ denote the vector area of the triangle APQ, we have

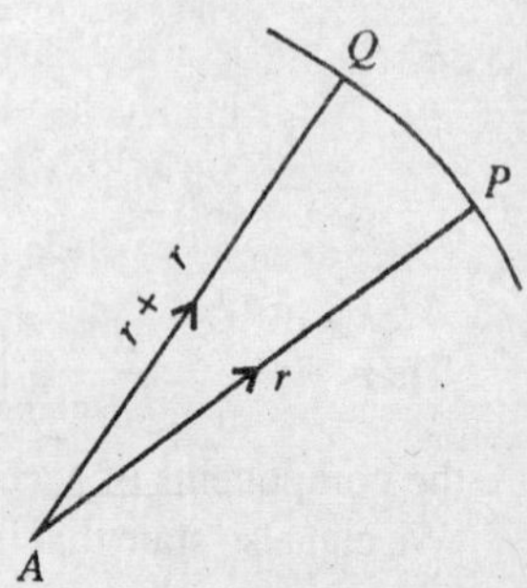

Fig. 4.5

$$\delta\mathbf{A} = \frac{1}{2}\mathbf{r} \times (\mathbf{r} + \delta\mathbf{r}) = \frac{1}{2}\mathbf{r} \times \delta\mathbf{r}.$$

$$\therefore \frac{\delta\mathbf{A}}{\delta t} = \frac{1}{2}\mathbf{r} \times \frac{\delta\mathbf{r}}{\delta t},$$

so that in the limit,

$$\frac{d\mathbf{A}}{dt} = \frac{1}{2}\mathbf{r}\times\frac{d\mathbf{r}}{dt} = \frac{1}{2}\mathbf{r}\times\mathbf{v}.$$

The time rate of change $d\mathbf{A}/dt$ of the area is called the *Areal velocity* of the particle.

Moment of force about a point. Lf **P** be the force acting at any point A, then the moment of the force about any point O is

$$\overrightarrow{\text{OA}}\times\mathbf{P}.$$

4.6. DYNAMICS OF A PARTICLE

The basis of Dynamics is the **Newton's Second Law of Motion** which states that the *time rate of change of momentum of a particle is proportional to the impressed force and takes place in the direction in which the force acts.*

The law states that if **P** be the force, then

$$\mathbf{P} \propto \frac{d}{dt}(m\mathbf{v}),$$

or

$$\mathbf{P} = k\frac{d}{dt}(m\mathbf{v}), \quad \text{...(1)}$$

where, k, is a scalar constant of proportionality.

Assuming that mass remains constant, we can rewrite (1) as

$$\mathbf{P} = km\frac{d\mathbf{v}}{dt} = k\,m\,\mathbf{a}. \quad \text{...(2)}$$

If, now, we choose the unit of force as one which produces in a unit mass a unit acceleration, we have, from (2)

$$k = 1.$$

Thus

$$\mathbf{P} = m\,\mathbf{a}. \quad \text{...(3)}$$

It is clear that the relation (3) cannot be true when the acceleration, **a**, is determined relative to any *arbitrary* frame of reference, for, otherwise, as may be easily seen, the acceleration caused by a certain given exertion could be thought of as having *any* value whatsoever relative to a suitably choosen reference frame. Thus the law can be true for accelerations measured relatively to some special class of Reference Frames only.

Inertial Frames. Now it has been found experimentally that one of the reference frames for which the relation (3) is true to a good degree of approximation is the Reference frame attached to the so called '*Fixed Stars*'.

In any case we shall say that any Reference frame for which the law is true is called an *Inertial Frame* and shall assume that the underlying reference frame is always as Inertial frame.

In the following we shall consider the applications of the principles developed so far to some important problems of *Particle Dynamics.*

4.7. MOTION IN THE GRAVITATIONAL FIELD

The equation of motion of a particle of mass m, subjected to the action of gravity only, is

$$m \frac{d^2\mathbf{r}}{dt^2} = -m\,g\,\mathbf{k}.$$

where **k** denotes the unit vector drawn vertically upwards.

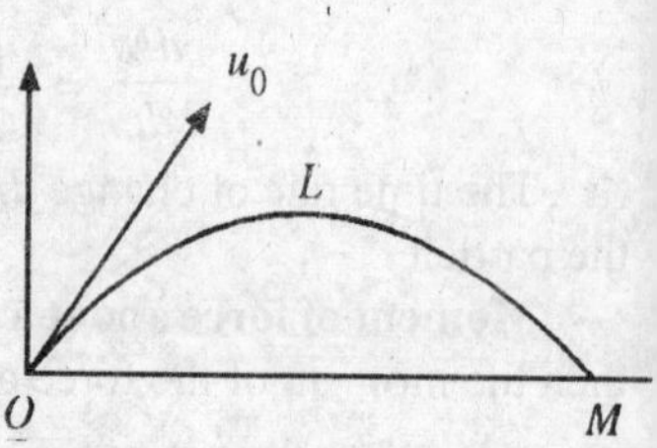

FIG. 4.6

We have

$$\frac{d^2\mathbf{r}}{dt^2} = -g\,\mathbf{k}. \qquad \text{...(1)}$$

Integrating, we get

$$\mathbf{v} = \frac{d\mathbf{r}}{dt} = -g\,t\,\mathbf{k} + \mathbf{b}, \qquad \text{...(2)}$$

where, **b**, is a constant vector.

Suppose that $\mathbf{v} = \mathbf{u}_0$ when $t = 0$. Therefore, from (2),

$$\mathbf{b} = \mathbf{u}_0$$

so that
$$\mathbf{v} = \frac{d\mathbf{r}}{dt} = \mathbf{u}_0 - gt\mathbf{k}. \qquad \text{...(3)}$$

Integrating again, we get

$$\mathbf{r} = \mathbf{u}_0 t - \frac{1}{2}\,gt^2\,\mathbf{k} + \mathbf{c}.$$

Supposing that $\mathbf{r} = \mathbf{0}$ when $t = 0$, we obtain $\mathbf{c} = \mathbf{0}$.

$$\therefore \quad \mathbf{r} = \mathbf{u}_0 t - \frac{1}{2}\,gt^2\,\mathbf{k}. \qquad \text{...(4)}$$

The equation (4) shows that the locus is a plane curve lying in the plane determined by the origin and the two vectors $\mathbf{u}_0$, **k**.

In case $\mathbf{u}_0$ has the direction of **k**, *i.e.*, the velocity at time **t** = 0 is vertical, we see from (4) that the particle constantly remains in the vertical line through the position of the particle at time $t = 0$.

In the general case when $\mathbf{u}_0$ and **k** have different directions, we take *x*-axis and *y*-axis along the vectors $\mathbf{u}_0$ and **k** respectively. Then the co-ordinates *x, y* at time *l* are given by

$$x = |\mathbf{u}_0|\,t, \quad y = -\frac{1}{2}\,gt^2,$$

so that, on eliminating *t*, we obtain

$$y = -\frac{1}{2}\,\frac{g}{|\mathbf{u}_0|}\,\tau^2 = -\frac{1}{2}\,\frac{g}{\mathbf{u}_0^2}\,x^2,$$

and accordingly the locus is a parabola. Thus the *locus of a particle moving freely in a constant gravitational field is a parabola.*

Time to reach the highest point. At the highest point, the velocity of the particle is horizontal so that it has no vertical component, *i.e.*,

$$\mathbf{v} \cdot \mathbf{k} = 0$$

Thus, from (3)

$$(\mathbf{u}_0 - g\mathbf{k}) \cdot \mathbf{k} = 0$$

or $$t = (\mathbf{u}_0 \cdot \mathbf{k})/g,$$

which gives the time to reach the highest point.

Time to reach the horizontal plane through the point of projection is given by

$$\mathbf{r} \cdot \mathbf{k} = 0.$$

i.e., $$\left(\mathbf{u}_0 t - \frac{1}{2} g t^2 \mathbf{k}\right) \cdot \mathbf{k} = 0,$$

$$\therefore \quad t = 0,\ 2\mathbf{u}_0 \cdot \mathbf{k}/g.$$

The position vectors of the highest point and the point of the trajectory on the horizontal plane can be obtained by putting

$$t = \mathbf{u}_0 \cdot \mathbf{k}/g \text{ and } t = 2\mathbf{u}_0 \cdot \mathbf{k}/g$$

respectively in (4).

Putting $t = \mathbf{u}_0 \cdot \mathbf{k}/g$ in (4), we see that the height of the highest point of the trajectory

$$= \left[\mathbf{u}_0 \left(\frac{\mathbf{u}_0 \cdot \mathbf{k}}{g}\right) - \frac{1}{2} g \frac{(\mathbf{u}_0 \cdot \mathbf{k})^2}{g^2} \mathbf{k}\right] \cdot \mathbf{k} = \frac{1}{2} \frac{\mathbf{u}_0 \cdot \mathbf{k}^2}{g}.$$

4.8. MOTION OF A PARTICLE UNDER TIME DEPENDENT APPLIED FORCE

Let the force be the function of time. Then

$$\mathbf{F} = \frac{d\mathbf{p}}{dt}$$

$$\Rightarrow \quad \mathbf{F} dt = d(m\mathbf{v})$$

$$\Rightarrow \quad m d\mathbf{v} = \mathbf{F}\, dt$$

If v_0 is the initial velocity of the particle at time to and **v** the velocity at any other instant t, then we have

$$m \int_{\mathbf{v}_0}^{\mathbf{v}} d\mathbf{v} = \int_{t_0}^{t} \mathbf{F}(t)\, dt$$

$$\Rightarrow \quad m\mathbf{v} - m\mathbf{v}_0 = \int_{t_0}^{t} \mathbf{F}(t)\, dt \qquad \text{...(1)}$$

But $$\mathbf{v} = \frac{d\mathbf{r}}{dt}, \text{ therefore}$$

$$\mathbf{v} = \frac{d\mathbf{r}}{dt} = \mathbf{v}_0 + \frac{1}{m} \int_{t_0}^{t} \mathbf{F}(t)\, dt \qquad \text{...(2)}$$

This equation gives the velocity of the particle at any instant t.

Equation (2) may also be written as

$$d\mathbf{r} = v_0\, dt + \frac{1}{m}\left[\int_{t_0}^{t} \mathbf{F}(t)\, dt\right] dt.$$

If $\mathbf{r}_0$ is the position of the particle at initial instant t_0 and **r** the position of the particle at instant t, then we have

$$\int_{\mathbf{r}_0}^{\mathbf{r}} d\mathbf{r} = \int_{t_0}^{t} \mathbf{v}_0\, dt + \frac{1}{m}\left[\int_{t_0}^{t} \mathbf{F}(t)\, dt\right] dt$$

$$\Rightarrow \quad \mathbf{r} - \mathbf{r}_0 = \mathbf{v}_0 (t - t_0) + \frac{1}{m}\int_{t_0}^{t}\left[\int_{t_0}^{t} \mathbf{F}(t)\, dt\right] dt$$

$$\Rightarrow \quad \mathbf{r} = \mathbf{r}_0 + \mathbf{v}_0 (t - t_0) + \frac{1}{m}\int_{t_0}^{t}\left[\int_{t_0}^{t} \mathbf{F}(t)\, dt\right] dt. \qquad ...(3)$$

This gives the position of the particle at any instant t.

4.9. MOTION IN THE GRAVITATIONAL FIELD SUBJECTED TO RESISTANCE PROPORTIONAL TO VELOCITY

The equation of motion of a particle moving in the gravitational field and subjected to resistance proportional to velocity is given by

$$m\frac{d^2\mathbf{r}}{dt^2} = -m\, g\, \mathbf{k} - m\, \mu\, \frac{d\mathbf{r}}{dt};$$

k being the unit vector in the direction of the upward drawn vertical. We have

$$\frac{d^2\mathbf{r}}{dt^2} + \mu\frac{d\mathbf{r}}{dt} = -g\,\mathbf{k}. \qquad ...(1)$$

Integrating, we get

$$\frac{d\mathbf{r}}{dt} + \mu\mathbf{r} = -gt\mathbf{k} + \mathbf{b},$$

where, **b**, is a constant vector. If $\mathbf{r} = \mathbf{0}$, $d\mathbf{r}/dt = \mathbf{u}_0$ when $t = 0$, we see $\mathbf{b} = \mathbf{u}_0$.

Thus $$\frac{d\mathbf{r}}{dt} + \mu\mathbf{r} = \mathbf{u}_0 - g\, t\, \mathbf{k}. \qquad ...(2)$$

To integrate (2), we multiply both sides by the scalar $e^{\mu t}$,

$$\therefore \quad e^{\mu t}\left[\frac{d\mathbf{r}}{dt} + \mu\mathbf{r}\right] = (\mathbf{u}_0 - gt\,\mathbf{k})\, e^{\mu t}.$$

Now integrating, we get

$$\mathbf{r}e^{\mu t} = \mathbf{u}_0 \int e^{\mu t}\, dt - g\mathbf{k}\int te^{\mu t}\, dt + \mathbf{c}$$

$$= \frac{\mathbf{u}_0}{\mu} e^{\mu t} - g\mathbf{k}\left[\frac{te^{\mu t}}{\mu} - \frac{e^{\mu t}}{\mu^2}\right] + \mathbf{c}.$$

As $\mathbf{r} = 0$ when $t = 0$, we get

$$\mathbf{0} = \frac{\mathbf{u}_0}{\mu} + \frac{g\mathbf{k}}{\mu^2} + \mathbf{c}.$$

$$\therefore \quad \mathbf{r} = \frac{\mathbf{u}_0}{\mu} - g\mathbf{k}\left[\frac{t}{\mu} - \frac{1}{\mu^2}\right] - \left[\frac{\mathbf{u}_0}{\mu} + \frac{g\mathbf{k}}{\mu^2}\right] e^{-\mu t}$$

$$= \frac{1}{\mu}\left(1-e^{-\mu t}\right)\mathbf{u}_0 + \left[\frac{9}{\mu^2}(1-e^{-\mu t}) - \frac{gt}{\mu}\right]\mathbf{k} \qquad ...(3)$$

Thus the locus is a plane curve whose plane passes through the origin and is parallel to the vectors $\mathbf{u}_0$ and $\mathbf{k}$.

Differentiating (3) with respect to t, we obtain

$$\mathbf{v} = \frac{d\mathbf{r}}{dt} = e^{-\mu t}\,\mathbf{u}_0 + \frac{g}{\mu}(e^{-\mu t} - 1)\,\mathbf{k}, \qquad ...(4)$$

so that when $t \to \infty$, $\mathbf{v} \to -\dfrac{g}{\mu}\mathbf{k}$ which is, therefore, the limiting velocity of the particle.

The expression (4) for velocity can also be obtained directly by re-writing (1) as

$$\frac{d\mathbf{v}}{dt} + \mu\mathbf{v} = -g\mathbf{k}$$

and integrating which can be done on multiplying by e^{μ}.

The particle will be moving horizontally at time t given by

$$\mathbf{v}\,.\,\mathbf{k} = 0,$$

i.e.,

$$e^{-\mu t}\,\mathbf{u}_0\,.\,\mathbf{k} + \frac{g}{\mu}(e^{-\mu t} - 1)\,\mathbf{k}\,.\,\mathbf{k} = 0.$$

This gives

$$t = \frac{1}{\mu}\log\left(1 + \frac{\mu}{g}\mathbf{u}_0\,.\,\mathbf{k}\right) = \frac{1}{\mu}\log\left(1 + \frac{\mu}{g}\mathbf{v}\right),$$

where v is the vertical component of the initial velocity.

4.10. MOTION OF A PARTICLE UNDER CENTRAL FORCE

A central force is one which is always directed towards or away from a fixed centre 0 and whose magnitude is always a function of distance from the point 0. If we choose the point 0 as origin, the central force may be expressed as

$$\mathbf{F} = f(r)\,\hat{\mathbf{r}} \qquad ...(1)$$

where $\hat{\mathbf{r}}$ is the unit vector along the vector $\mathbf{r}$.

The equation of motion of the particle under central force is given by

$$m\frac{d^2\mathbf{r}}{dt^2} = F(t)\,\hat{\mathbf{r}}$$

$$\Rightarrow \qquad m\frac{d^2\mathbf{r}}{dt^2} - F(t)\,\hat{\mathbf{r}} = 0.$$

This is the equation of motion of a particle under a central force and may be solved if $F(t)$ is given.

4.10.1. Harmonic Motion

The equation of motion of a particle subjected to force towards a fixed point and with magnitude proportional to its distance from the fixed point is given by

$$\frac{d^2\mathbf{r}}{dt^2} = -\mu\mathbf{r}; \qquad (\mu > 0) \quad ...(1)$$

the origin of the Inertial frame being taken at the fixed point.

The general solution of (1) is

$$\mathbf{r} = \mathbf{a}\cos\sqrt{\mu t} + \mathbf{b}\sin\sqrt{\mu t}, \qquad ...(2)$$

where **a, b**, are arbitrary constant vectors.

Differentiating (2) with respect to t, we obtain

$$\mathbf{v} = \frac{d\mathbf{r}}{dt} = -\sqrt{\mu}\ \mathbf{a}\sin\sqrt{\mu t} + \sqrt{\mu}\mathbf{b}\cos\sqrt{\mu t}. \quad ...(3)$$

From (2) and (3), we see that at time $t = 0$, we have

$$\mathbf{r} = \mathbf{r},\ \mathbf{v} = \sqrt{\mu}\mathbf{b}$$

so that **a** is the position vector of the particle and **b** is the product by the scalar $1/\sqrt{\mu}$ of the velocity at time $t = 0$.

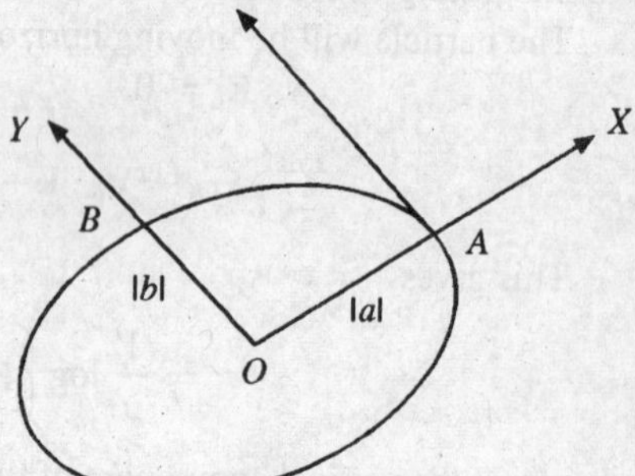

Fig. 4.7

If we take x-axis and y-axis parallel to the vectors **a, b**, then we see, from (2), that for any point x, y on the curve, we have

$$x = |\mathbf{a}|\cos\sqrt{\mu t},$$

$$y = |\mathbf{b}|\sin\sqrt{\mu t}$$

so that the path is the ellipse

$$\frac{x^2}{|\mathbf{a}|^2} + \frac{y^2}{|\mathbf{b}|^2} = 1,$$

i.e.,

$$\frac{x^2}{\mathbf{a}^2} + \frac{y^2}{\mathbf{b}^2} = 1$$

such that the diameters of the ellipse parallel to the vectors **a, b**, (*i.e.,* along the co-ordinate axes) are conjugate.

From (2) and (3), we see that **r** as well as **v** are periodic functions of t; the period being $2\pi/\sqrt{\mu}$. Thus the whole ellipse will be described once in time $2\pi/\sqrt{\mu}$.

Note. If $\mathbf{b} = k\mathbf{a}$ so that the velocity at the position A of the particle at time $t = 0$ is along OA, we obtain

$$\mathbf{r} = \mathbf{a}\left(\cos\sqrt{\mu t} + k\sin\sqrt{\mu t}\right),$$

so that the particle constantly lies along the straight line OA.

4.10.2. Planetary Motion. Motion of a particle moving under inverse square law

The equation of motion of a particle subjected to a force towards a fixed point, following Newton's inverse square law, is

$$\frac{d^2\mathbf{r}}{dt^2} = -\frac{u}{r^3}\mathbf{r} \qquad \text{where } |\mathbf{r}| = r, \qquad ...(1)$$

the fixed point being taken as the origin.

If is not found possible to investigate the motion by directly integrating the differential equation.

Multiplying (1) vectorially by **r**, we get

$$\mathbf{r} \times \frac{d^2\mathbf{r}}{dt^2} = \mathbf{0}. \qquad ...(2)$$

Integrating (2), we have

$$\mathbf{r} \times \frac{d\mathbf{r}}{dt} = \text{constant} = \mathbf{h}, \text{ say.} \qquad ...(3)$$

The equation (3) shows that the moment of momentum of the particle about the fixed point is constant. Also it shows that Areal velocity relative to the fixed point is constant.

Multiplying (3) scalarly by **r**, we see that

$$\mathbf{r} \cdot \mathbf{h} = 0, \qquad ...(4)$$

so that we prove that the locus of the point **r** is a plane curve on the plane through the fixed point perpendicular to the vector **h**.

We shall now proceed to show that the path is a conic. From (1) and (3), we have

$$\begin{aligned}
\mathbf{h} \times \frac{d^2\mathbf{r}}{dt^2} &= \left(\mathbf{r} \times \frac{d\mathbf{r}}{dt}\right) \times \left(-\frac{\mu}{r^3}\mathbf{r}\right) \\
&= -\frac{\mu}{r^3}\left[\mathbf{r}^2 \frac{d\mathbf{r}}{dt} - \mathbf{r} \cdot \frac{d\mathbf{r}}{dt}\mathbf{r}\right]. \\
&= -\frac{\mu}{r^3}\left[\mathbf{r}^2 \frac{d\mathbf{r}}{dt} - \mathbf{r}\frac{dr}{dt}\mathbf{r}\right] \\
&= -\mu\left[\frac{\mathbf{1}}{r}\frac{d\mathbf{r}}{dt} - \frac{\mathbf{1}}{r^2}\frac{dr}{dt}\mathbf{r}\right] \\
&= -\mu \frac{d}{dt}\left(\frac{\mathbf{r}}{r}\right). \qquad ...(5)
\end{aligned}$$

Integrating (5), we get

$$\mathbf{h} \times \frac{d\mathbf{r}}{dt} = -\mu \frac{\mathbf{r}}{r} + \mathbf{a}, \qquad ...(6)$$

where **a** is an arbitrary constant vector.

Multiplying (6) by **h**, we obtain

$$\mathbf{a}.\mathbf{h} = 0 \qquad \text{[Refer (4) above]}$$

so that the vector **a** necessarily lies in the plane of the motion; Now multiplying (6) scalarly by **r**, we obtain

$$\mathbf{h} \times \frac{d\mathbf{r}}{dt} . \mathbf{r} = -\mu \frac{\mathbf{r}.\mathbf{r}}{r} + \mathbf{a}.\mathbf{r},$$

or

$$-\mathbf{r} \times \frac{d\mathbf{r}}{dt} . \mathbf{h} = -\mu \frac{r^2}{r} + \mathbf{a}.\mathbf{r},$$

or

$$-\mathbf{h}^2 = -\mu r + \mathbf{a}.\mathbf{r},$$

or

$$\frac{\mathbf{h}^2}{\mu r} = 1 - \frac{\mathbf{a}}{\mu} - \frac{\mathbf{r}}{r}$$

$$= 1 - \left|\frac{\mathbf{a}}{\mu}\right| \cos\theta, \qquad \text{for } \left|\frac{\mathbf{r}}{r}\right| = 1$$

where θ is the angle between the variable vector **r** and the constant vector **a**. We write

$$\frac{\mathbf{h}^2}{\mu} = l, \quad \left|\frac{\mathbf{a}}{\mu}\right| = e,$$

and obtain

$$\frac{l}{r} = 1 - e\cos\theta \qquad ...(7)$$

which is the equation of a conic having the origin as a focus.

It will be seen that the vector **a** is parallel to an axis of the conic.

Hence *the path is a conic having the origin as a focus.* This conic is a parabola, ellipse or hyperbola according as

$$e = \left|\frac{\mathbf{a}}{\mu}\right|$$

is equal to, less than or greater than one,

i.e., according as $|\mathbf{a}| \begin{cases} = \mu \\ < \mu \\ > \mu. \end{cases}$

Velocity at any point of the path. Multiplying (1) scalarly by $d\mathbf{r}/dt$, we obtain

$$\frac{d\mathbf{r}}{dt} \cdot \frac{d^2\mathbf{r}}{dt^2} = -\frac{\mu}{r^3}\frac{d\mathbf{r}}{dt}\mathbf{r} = -\frac{\mu}{r^3} r \frac{dr}{dt} = -\frac{\mu}{r^2}\frac{dt}{dt}.$$

Integrating, we get

$$\frac{1}{2}\mathbf{v}^2 = \frac{1}{2}\left(\frac{d\mathbf{r}}{dt}\right)^2 = \frac{\mu}{r} + c, \quad ...(8)$$

where, c, is a constant scalar.

If $\mathbf{v} = \mathbf{v}_0$ when $\mathbf{r} = \mathbf{r}_0$, we have from (8),

$$\frac{1}{2}\mathbf{v}_0^2 = \frac{\mu}{r_0} + c. \quad ...(9)$$

From (8) and (9), we obtain

$$\mathbf{v}^2 = \mathbf{v}_0^2 + 2\mu\left(\frac{1}{r} - \frac{1}{r_0}\right). \quad ...(10)$$

This gives the speed at any point of the path in terms of the speed at any point of the same.

4.11. ACCELERATED FRAMES OF REFERENCE

The frames relative to which the body, not acted upon by external forces, is accelerated are called *accelerated* or *non-inertial* frames.

The motion relative to accelerated frames can be considered in the following two cases:

(*i*) *Frames of reference with translational accelerated.* Let S and S' be two frames, the latter moving with a constant translational acceleration a_0 relative to the former. If a particle of mass m has an acceleration a_i as observed from system S. Then, the force applied on the particle is given by

$$\mathbf{F} = m\mathbf{a}_i \quad ...(1)$$

If this force is observed from system S′, then it may be assumed that the system S is moving with an acceleration $-a_0$ relative to S′ and hence the particle will seem to have an additional force $-m\mathbf{a}_0$. Thus the force acting on the particle as observed from system S' is given by

$$\mathbf{F}' = m\,\mathbf{a}_i - m\,\mathbf{a}_0$$
$$= m\,(\mathbf{a}_i - \mathbf{a}_0) \quad ...(2)$$

i.e., the observer in system S' will observe the acceleration of the particle to the $(\mathbf{a}_i - \mathbf{a}_0)$.

The additional force $-m\,\mathbf{a}_0$ arising only due to the acceleration of the frame of reference is called the *Fictitious* or *apparent* or *prendo* or *inertial force.*

If the system S′ is moving with uniform acceleration $\mathbf{a}_0$ in the direction of negative z-axis then the impressed force in system S′ is given by

$$\mathbf{F}' = m\,\mathbf{a}_i + m\,a_0\,\mathbf{k}$$
$$= \mathbf{F} + m\,a_0\,\mathbf{k}, \quad ...(3)$$

where $\mathbf{k}$ is the unit vector along + ve z-axis.

Note. If we assume that $\mathbf{a}_0$, the acceleration of moving system, is identical with the acceleration due to gravity the fictitious or inertial force exactly balances the force of gravity $-mg\,\mathbf{k}$. Therefore the gravitational force of earth may be produced or neutralised by a proper choice of frame of reference. In other words, an accelerated frame of reference may be maintained equivalent to the gravitational field. Thus there is an equation between an accelerated frame of reference and a gravitational field which is called the *principal of equivalence.*

(ii) *Frames of reference rotating with constant angular velocity.*

Case I. *When the origins of the two frames coincide*

Consider two systems S and S′, the latter being in uniform rotation relative to the former which is at rest in absolute space. Let the origins of both the systems coincide. Let **i, j, k** be the unit vectors associated with axes X, Y, Z of system S and **i′**, **j′**, **k′** the unit vectors with axes X′, Y′, Z′ of system S′. Then

$$\mathbf{r} = \mathbf{i}x + \mathbf{j}y + \mathbf{k}z \qquad ...(1)$$

$$\mathbf{r} = \mathbf{i}'x' + \mathbf{j}'y' + \mathbf{k}'z' \qquad ...(2)$$

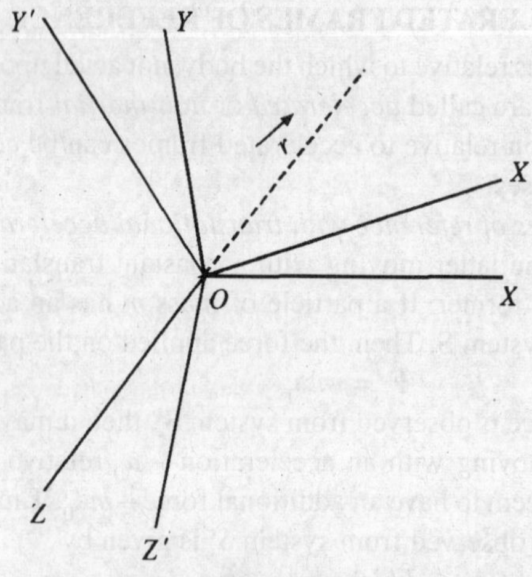

Fig. 4.8

An observer in system S′ situated at the origin 0 will observe the time derivative of **r** to be

$$\frac{d\mathbf{r}}{dt} = \frac{dx'}{dt}.\mathbf{i}' + \frac{dy'}{dt}.\mathbf{j}' + \frac{dz'}{dt}.\mathbf{k}' \qquad ...(3)$$

For the observer in system S will be

$$\frac{d\mathbf{r}}{dt} = \frac{dx'}{dt}\mathbf{i}' + \frac{dy'}{dt}\mathbf{j}' + \frac{dz'}{dt}\mathbf{k}'$$

$$+ x'\frac{d\mathbf{i}'}{dt} + y'\frac{d\mathbf{j}'}{dt} + z'\frac{d\mathbf{k}'}{dt} \qquad ...(4)$$

since the unit vector **i′**, **j′**, **k′** actually change with relative to the observer fixed in system S′.

Since **i′** is a unit vector, $\frac{d\mathbf{i}'}{dt}$ is perpendicular to **i′** and must therefore lie in the plane of **j′** and **k′**, then

$$\frac{d\mathbf{i}'}{dt} = a_1\,\mathbf{j}' + a_2\,\mathbf{k}' \qquad ...(5)$$

Similarly $$\frac{d\mathbf{j}'}{dt} = a_3\,\mathbf{k}' + a_4\,\mathbf{i}' \qquad ...(6)$$

$$\frac{d\mathbf{k}'}{dt} = a_5\,\mathbf{i}' + a_6\,\mathbf{j}' \qquad ...(7)$$

We have $$\mathbf{i}'\,.\,\mathbf{j}' = 0 \Rightarrow \mathbf{i}'\,.\,\frac{d\mathbf{j}'}{dt} + \frac{d\mathbf{i}'}{dt}\,.\,\mathbf{j}' = 0 \qquad ...(8)$$

Multiplying scalarly equations (5) with $\mathbf{j}'$ and (6) with $\mathbf{i}'$ on either side, we get

$$\mathbf{j}'\,.\,\frac{d\mathbf{i}'}{dt} = a_1 \text{ and } \mathbf{i}'\,.\,\frac{d\mathbf{j}'}{dt} = a_4$$

Substituting these values in (8),

$$a_4 + a_1 = 0 \Rightarrow a_4 = -a_1 \qquad ...(9)$$

Similarly, by $\mathbf{i}'\,.\,\mathbf{k}' = 0$ and $\mathbf{j}'\,.\,\mathbf{k}' = 0$, we have

$$a_2 + a_5 = 0 \qquad ...(10)$$

and $$a_3 + a_6 = 0 \qquad ...(11)$$

Then, we get

$$\frac{d\mathbf{i}'}{dt} = a_1\,\mathbf{j}' + a_2\,\mathbf{k}'$$

$$\frac{d\mathbf{j}'}{dt} = a_3\,\mathbf{k}' - a_1\,\mathbf{i}'$$

$$\frac{d\mathbf{k}'}{dt} = -a_2\,\mathbf{i}' - a_3\,\mathbf{j}'$$

Then we have

$$x'\frac{d\mathbf{i}'}{dt} + y'\frac{d\mathbf{j}'}{dt} + z'\frac{d\mathbf{k}'}{dt} = (-a_1y' - a_2z')\,\mathbf{i}' + (a_1x' - a_3z')\,\mathbf{j}' + (a_2y' - a_3x')\,\mathbf{k}'$$

$$= \begin{vmatrix} \mathbf{i}' & \mathbf{j}' & \mathbf{k}' \\ a_3 & -a_2 & a_1 \\ x' & y' & z' \end{vmatrix} \qquad ...(12)$$

Choosing $a_3 = \omega_x$, $-a_2 = \omega_g$ and $a_1 = \omega_z$. Where ω_x, ω_y and ω_z are the components of angular velocity vector $\vec{\omega}$ of system S′, *i.e.*,

$$\vec{\omega} = \mathbf{i}'\,\omega_x + \mathbf{j}'\,\omega_y + \mathbf{k}'\,\omega_z.$$

Then eqn. (12) may be written as

$$x'\frac{d\mathbf{i}'}{dt} + y'\frac{d\mathbf{j}'}{dt} + z'\frac{d\mathbf{k}'}{dt} = \begin{vmatrix} \mathbf{i}' & \mathbf{j}' & \mathbf{k}' \\ \omega_x & \omega_y & \omega_z \\ x' & y' & z' \end{vmatrix}$$

$$= \vec{\omega} \times \mathbf{r} \qquad ...(13)$$

From equations (4) and (13), we get

$$\frac{d\mathbf{r}}{dt} = \frac{dx'}{dt}\mathbf{i}' + \frac{dy'}{dt}\mathbf{j}' + \frac{dz'}{dt}\mathbf{k}' + \vec{\omega} \times \mathbf{r}$$

$$\therefore \quad \frac{d\mathbf{r}}{dt} = \frac{d'\mathbf{r}}{dt} + \vec{\omega} \times \mathbf{r} \qquad ...(14)$$

where we have denoted the differential referred to system S by $\frac{d'}{dt}$, *i.e.*,

$$\frac{d'\mathbf{r}}{dt} = \frac{dx'}{dt}\mathbf{i}' + \frac{dy'}{dt}\mathbf{j}' + \frac{dz'}{dt}\mathbf{k}'$$

operators $\frac{d}{dt}$ and $\frac{d'}{dt}$ are related by

$$\frac{d}{dt} = \frac{d'}{dt}\vec{\omega} \times$$

or

$$\frac{d}{dt} = \frac{d'}{dt} + [\vec{\omega}] \qquad ...(15)$$

Case II. *When the origins of the two frames do not coincide.*

Let there be two systems S and S′, the latter being in uniform rotation relative to former which is at rest in absolute space. Let **i**, **j**, **k** be the unit vector associated with axes X, Y, Z of the system S′ and **i**′, **j**′, **k**′, the unit vectors associated with axes X′, Y′, Z′ of the system S′. Let O and O′ be the origins of the systems S and S′ which do not coincide.

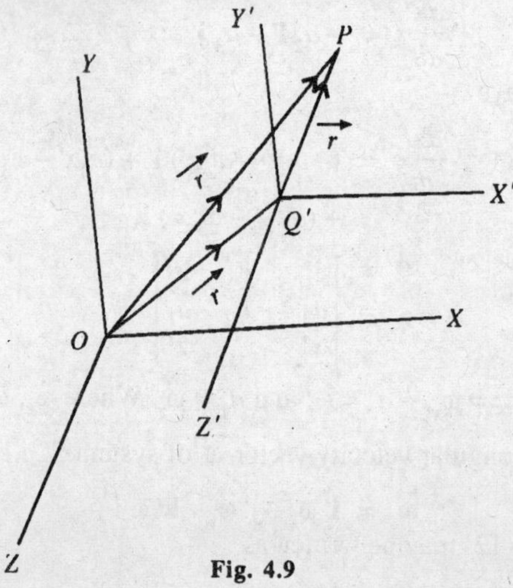

Fig. 4.9

Let $\mathbf{r}_0$ the position vector of origin 0 of system S′ relative to the origin 0 of the system S.

If **r** and **R** are the position vectors of the particle P relative to the origins O′ and relative to system S is given by

$$\frac{d\mathbf{R}}{dt} = \frac{d}{dt}(\mathbf{r}_0 + \mathbf{r}) = \frac{d\mathbf{r}_0}{dt} + \frac{d\mathbf{r}}{dt}$$

$$\Rightarrow \qquad \frac{d\mathbf{R}}{dt} = \frac{d\mathbf{r}_0}{dt} + \frac{d'\mathbf{r}}{dt} + \vec{\omega} \times \mathbf{r},$$

4.11.1. Coriols and centripetal forces

Differentiating equation (14) of § 4.11 × w.r.t.t,

$$\frac{d^2\mathbf{r}}{dt^2} = \frac{d}{dt}\left(\frac{d\mathbf{r}}{dt}\right) = \left(\frac{d'}{dt} + \omega \times\right)\left(\frac{d'\mathbf{r}}{dt} + \vec{\omega} \times \mathbf{r}\right)$$

$$= \frac{d'^2\mathbf{r}}{dt^2} + 2\left(\vec{\omega} \times \frac{d'\mathbf{r}}{dt}\right) + \vec{\omega} \times (\vec{\omega} \times \mathbf{r}) + \frac{d'\vec{\omega}}{dt} \times \mathbf{r} \quad ...(1)$$

The above formula is called *Coriolis Theorem.* In equation (1).

(*i*) $\frac{d'^2\mathbf{r}}{dt^2}$ represents the acceleration of the particle relative to system S.

(*ii*) $\frac{d'^2\mathbf{r}}{dt^2}$ represents the acceleration of the particle relative to system S′.

(*iii*) $2\left(\vec{\omega} \times \frac{d'\mathbf{r}}{dt}\right)$ is known as *coriolis acceleration* and appears only when the particle moves in the system S′.

(*iv*) $\frac{d'\vec{\omega}}{dt} \times \mathbf{r}$ vanishes for constant angular velocity of rotation about a fixed axis.

The impressed force on the particle of mass *m* in system S is given by

$$F = m\frac{d'^2\mathbf{r}}{dt^2} = m\frac{d'^2\mathbf{r}}{dt^2} + 2m\left(\vec{\omega} \times \frac{d'\mathbf{r}}{dt}\right)$$

$$+ m\,\vec{\omega} \times (\vec{\omega} \times \mathbf{r}) + m\frac{d'\vec{\omega}}{dt} \times \mathbf{r} \quad ...(2)$$

Then the forces acting on the particle of mass *m* in system S′ or the equation of motion of the particle in system S′ is given by

$$F' = m\frac{d'^2\mathbf{r}}{dt^2} = \mathbf{F} - 2m\left(\vec{\omega} \times \frac{d'\mathbf{r}}{dt}\right) + m\,\vec{\omega}$$

$$\times (\vec{\omega} \times \mathbf{r}) - \frac{d'\vec{\omega}}{dt} \times \mathbf{r} \quad ...(2)$$

Term $- m\,\vec{\omega} \times (\vec{\omega} \times \mathbf{r})$ is called the *centrifugal force.*

The term $-2m\left(\vec{\omega}\times\dfrac{d'\mathbf{r}}{dt}\right)$ represents the fictitions force acting on a particle in the non-inertial frame and is known as *coriolis force*. The term $-m\left(\dfrac{d'\vec{\omega}}{dt}\times\mathbf{r}\right)$ has no special name and appears only in the case of non-uniform rotation.

Note (*i*) Coriolis fo ce acts on a particle when it moves relative to the non-inertial frame and it is directly proportional to the velocity of the particle. Also the coriolis force is perpendicular to ω and $\dfrac{d'\mathbf{r}}{dt}$. –ve sign indicates that the coriolis force is in the direction opposite to that given by right-hand screw rule.

Again, $$Fa = -2m\,\omega\,\frac{d'\mathbf{r}}{dt}\sin\theta\,\hat{\mathbf{n}}$$

where $\hat{\mathbf{n}}$ is the unit vector perpendicular to both $\vec{\omega}$ and $\dfrac{d'\mathbf{r}}{dt}$ and θ is the angle between $\vec{\omega}$ and $\mathbf{v}'$, where $\mathbf{v}' = \dfrac{d'\mathbf{r}}{dt}$ the velocity of the particle in system S′.

(*ii*) If θ = 0, Fc = 0, *i.e.,* coriolis force is zero if the angular velocity of the frame S′ and the linear velocity of the particle relative to S′ are parallel.

The Coriolis force will be maximum when sin θ = 1, *i.e.,* when $\vec{\omega}$ and $\mathbf{v}'$ are perpendicular to each other and

$$Fc,\max = -2m\,\omega\mathbf{v}'$$

4.11.2. Deviation of freely falling bodies from the vertical

The earth is rotating with an angular velocity $\vec{\omega}$ with north-south as the axis of rotation. Let X′, Y′, Z′ be the axes along the east, north and vertical directions respectively and $\mathbf{i}'$, $\mathbf{j}'$, $\mathbf{k}'$ be the unit vectors along these axes. Let the body of mass *m* be falling freely on the surface of the earth at a place of latitude λ. Let $\mathbf{v}'$ be the velocity of the body at any instant *t*. Then

$$\vec{\omega} = \omega\cos\lambda\,\mathbf{j}' + \omega\sin\lambda\,\mathbf{j}'$$

and $$\mathbf{v}' = -v'\,\mathbf{k}'.$$

The Coriolis force is

$$\begin{aligned} \mathbf{Fc} &= -2m\,(\vec{\omega}\times\mathbf{v}') \\ &= -2m\,[(\omega\cos\lambda\,\text{‘}\mathbf{j}\text{’} + \omega\sin\lambda\,\mathbf{k}')\times(-v'\,\mathbf{k})] \\ &= 2mv\,\omega'\cos\lambda\,\mathbf{j}' \qquad \text{...(1)} \end{aligned}$$

Since the Coriolis force is acting along +ve X-axis, theretore the direction of the body will be towards east in the northern emisphese.

If $\dfrac{d^2\mathbf{x}}{dt^2}$ is the acceleration along X-axis, then

$$m\frac{d^2\mathbf{x}}{dt^2} = mv'\,\omega\cos\lambda\,\mathbf{i}'$$

$$\Rightarrow \quad \frac{d^2\mathbf{x}}{dt^2} = 2\mathbf{v}\,\omega\cos\lambda \qquad ...(2)$$

If g is the acceleration due to gravity, then

$$v' = gt$$

$$\Rightarrow \quad \frac{d^2\mathbf{x}}{dt^2} = 2g\,\omega\,t\cos\lambda \qquad ...(3)$$

$$\therefore \quad \frac{dx}{dt} = 0 \text{ at } t = 0, \text{ we have A} = 0$$

So,
$$\frac{dx}{dt} = g\omega\cos\lambda\,t^2$$

$$\Rightarrow \quad x = g\,\omega\cos\lambda\,\frac{t^3}{3} + \text{B}.$$

Initially deviation is zero yields B = 0.

$$\therefore \quad x = \frac{1}{3}g\omega\cos\lambda\,t^3.$$

If h is the height traversed by the body in time t, then $t = \sqrt{\left(\frac{2h}{g}\right)}$. Hence,

$$x = \frac{1}{3}g\,\omega\cos\lambda\left(\frac{2h}{g}\right)^{3/2} = \frac{2}{3}\,\omega\,h\cos\lambda\left(\frac{2h}{g}\right)^{1/2}$$

This is the deviation of the falling body from the vertical due to the rotation of the earth towards east in the northern hemisphere and towards south in the southern hemisphere.

EXAMPLES

1. *A particle of mass m is thrown vertically upward with speed v_0 from the surface of earth. Find (a) the position at any time, (b) the time taken to reach the highest point, (c) the maximum height reached, and (d) the speed of the particle in terms of its distance from the earth.*

Let the particle of mass m be located at the origin at $t = 0$. Take z-axis vertically upward having unit vector $\mathbf{k}$; then the equation of motion is given by

$$m\frac{d^2\mathbf{r}}{dt^2} = -m\,g\,\mathbf{k}, \qquad ...(1)$$

$$\Rightarrow \quad \frac{d^2\mathbf{r}}{dt^2} = -g\,\mathbf{k}, \qquad ...(2)$$

$$\Rightarrow \quad \frac{d\mathbf{r}}{dt} = -g\,\mathbf{k}$$

(*a*) $\Rightarrow$ $$\mathbf{v} = -g\,t\,\mathbf{k} + A$$

At $t = 0$, $v = v_0\mathbf{k}$, Þ $A = v_0\,\mathbf{k}$.

Then, $$\mathbf{v} = (v_0 - g\,t)\,\mathbf{k}$$

$$\Rightarrow \quad \frac{d\mathbf{r}}{dt} = (v_0 - gt)\,\mathbf{k} \qquad ...(3)$$

Integrating again, we get

$$\mathbf{r} = \left(v_0\,t - \frac{1}{2}g\,t^2\right)\mathbf{k} + B.$$

At $$t = 0,\ \mathbf{r} = 0 \ \Rightarrow\ B = 0.$$

Then the position vector **r** is given by

$$\mathbf{r} = \left(v_0\,t - \frac{1}{2}g\,t^2\right)\mathbf{k}, \qquad ...(4)$$

$$\Rightarrow \quad (\mathbf{i}x + \mathbf{j}y + \mathbf{k}z) = \left(v_0\,t - \frac{1}{2}g\,t^2\right)\mathbf{k},$$

$$\Rightarrow \quad x = 0,\ y = 0,\ z = vt - \frac{1}{2}\,gt^2 \qquad ...(5)$$

(*b*) At highest $\mathbf{v} = 0$, therefore equation (3) gives

$$v_0 - gt = 0$$

$$\Rightarrow \quad t = \frac{v_0}{g}.$$

Therefore highest point is reached after time $t = \dfrac{v_0}{g}$.

(*c*) Substituting $t = \dfrac{v_0}{g}$ in (5), the maximum height is given by

$$z = \frac{v_0^2}{2g}.$$

(*d*) Eliminating *t* from equations (3) and (5) we get

$$z = v_0\left(\frac{v_0 - v}{g}\right) - \frac{1}{2}\,g\left(\frac{v_0 - v}{g}\right)^2 = \frac{v_0^2 - v^2}{2g}$$

$$\Rightarrow \quad v^2 = v_0^{\,2} - gz.$$

2. *Obtain the equation of motion for a particle falling freely under the influence of gravity under a resistance proportional to the square of the velocity. Discuss the motion of the particle and find the maximum possible velocity.*

Let a particle of mass *m* be located at a distance *z* from the origin at time *t*. If **k** is the unit vector in the vertically downward direction and *b* the frictional constant then the equation of motion is given by

$$m\,\frac{d\mathbf{v}}{dt} = m\,g\,\mathbf{k} - bv^2\,\mathbf{k} \qquad ...(1)$$

$$\Rightarrow \quad m\frac{dv}{dt}\mathbf{k} = m\,g\,\mathbf{k} - bv^2\,\mathbf{k}$$

$$\Rightarrow \quad m\frac{dv}{dt} = mg - bv^2$$

$$\Rightarrow \quad \frac{dv}{dt} = g - \frac{b}{m}v = \frac{b}{m}\left(\frac{mg}{b} - v^2\right)$$

$$\Rightarrow \quad \frac{dv}{dt} = \frac{b}{m}(k^2 - v^2) \qquad \left[\because \frac{mg}{b} = k^2\right] \quad ...(2)$$

$$\Rightarrow \quad \frac{dv}{k^2 - v^2} = \frac{b}{m}dt$$

$$\Rightarrow \quad \frac{1}{2k}\log\frac{k+v}{k-v} = \frac{b}{m}t + \text{A} \quad ...(3)$$

At, $t = 0,\ v = 0 \Rightarrow \text{A} = 0$. Then

$$\Rightarrow \quad \frac{1}{2k}\log\frac{k+v}{k-v} = \frac{b}{m}t$$

$$\Rightarrow \quad \frac{k+v}{k-v} = e^{(2\,kb\,/\,m)\,t}$$

$$\Rightarrow \quad v = \frac{e^{(2kb/m)t} - 1}{e^{(2kb/m)t} + 1} = k\,\frac{e^{(kb/m)t} - e^{(-nb/m)t}}{e^{(kb/m)t} - e^{(-kb/m)t}}$$

$$= k\tan h\,\frac{k\,bt}{m}$$

$$\Rightarrow \quad v = \sqrt{\left(\frac{mg}{g}\right)}\tan h\sqrt{\left(\frac{bg}{m}\right)}t \quad ...(4)$$

This gives the velocity of the particle at any time t.

Expanding $\tan h\sqrt{\left(\frac{bg}{m}\right)}\,t$, we get

$$v = \begin{cases} -gt & \text{if} \quad t \le \sqrt{\left(\frac{m}{bg}\right)}, \\ -\sqrt{\left(\frac{mg}{b}\right)} & \text{if} \quad t \ge \sqrt{\left(\frac{m}{bg}\right)}, \end{cases}$$

Equation (2) may be rewritten as

$$v\frac{dv}{dz} = \frac{b}{m}(k^2 - v^2)$$

$$\Rightarrow \quad \frac{v}{k^2 - v^2}dv = \frac{b}{m}dz$$

$$\Rightarrow \quad -\frac{1}{2}\log(k^2 - v^2) = \frac{b}{m}z + B$$

at $t = 0,\ z = 0$ and $v = 0 \ \Rightarrow\ B = -\frac{1}{2}\log k^2.$

$$\therefore \quad -\frac{1}{2}\log(k^2 - v^2) = \frac{b}{m}z - \frac{1}{2}\log k^2$$

$$\Rightarrow \quad v^2 = \frac{mg}{b}[1 - e^{2bz/m}] \qquad ...(5)$$

This gives the velocity of the particle at a given distance from the origin, *i.e.*, at any position. Eliminating v from (4) and (5), we get

$$\frac{mg}{b}\tan h^2 \sqrt{\left(\frac{bg}{m}\right)}t = \frac{mg}{b}[1 - e^{2bz/m}]$$

$$\Rightarrow \quad e^{2bz/m} = \cos h^2 \sqrt{\left(\frac{bg}{m}\right)}\, t$$

$$\Rightarrow \quad \frac{2bz}{m} = \log\cos h^2 \sqrt{\left(\frac{bg}{m}\right)}\, t$$

$$\Rightarrow \quad z = \frac{m}{b}\log\cos h \sqrt{\left(\frac{bg}{m}\right)}\, t \qquad ...(6)$$

From equation (1), velocity v will be maximum if $\frac{dv}{dt} = 0$, *i.e.*, when

$$g - \frac{b}{m}v^2 = 0 \ \Rightarrow\ v_{max} = \sqrt{\left(\frac{mg}{b}\right)}.$$

This is maximum possible velocity or *terminal velocity.*

3. *Two points Q and R are taken on the tangent at a point P to the path of the moving point P such that*

$$\overrightarrow{QP} = \overrightarrow{PR} = k\mathbf{v},$$

where $\mathbf{v}$ *is the velocity of P and k is a constant. Show that if T be the point of intersection of the tangents to the paths of Q and R, then the vector* $k^{-2}\,\overrightarrow{TP}$ denotes the acceleration of P.

Let $\qquad \mathbf{r} = \mathbf{f}(t)$

e the path of P, The position vectors $\mathbf{R}_1$, $\mathbf{R}_2$ of Q and R are given by

$$\mathbf{R}_1 = \mathbf{r} - k\frac{d\mathbf{r}}{dt},\ \mathbf{R}_2 = \mathbf{r} + k\frac{d\mathbf{r}}{dt}.$$

Here $\mathbf{R}_1$, $\mathbf{R}_2$ are functions of t.

The velocities of Q and R being

$$\frac{d\mathbf{r}}{dt} - k\frac{d^2\mathbf{r}}{dt^2},\ \frac{d\mathbf{r}}{dt} + k\frac{d^2\mathbf{r}}{dt^2},$$

the equations of the tangents to the paths of Q and R at time t are

$$\mathbf{R} = \left(\mathbf{r} - k\,\frac{d\mathbf{r}}{dt}\right) + p\left(\frac{d\mathbf{r}}{dt} - k\,\frac{d^2\mathbf{r}}{dt^2}\right),$$

$$\mathbf{R} = \left(\mathbf{r} + k\,\frac{d\mathbf{r}}{dt}\right) + p\left(\frac{d\mathbf{r}}{dt} + k\,\frac{d^2\mathbf{r}}{dt^2}\right),$$

p, p' being parameters. At their point of intersection, we have

$$\left(\mathbf{r} - k\,\frac{d\mathbf{r}}{dt}\right) + p\left(\frac{d\mathbf{r}}{dt} - k\,\frac{d^2\mathbf{r}}{dt^2}\right) = \left(\mathbf{r} + k\,\frac{d\mathbf{r}}{dt}\right) + p\left(\frac{d\mathbf{r}}{dt} + k\,\frac{d^2\mathbf{r}}{dt^2}\right)$$

or $$(-k+p)\,\frac{d\mathbf{r}}{dt} - pk\,\frac{d^2\mathbf{r}}{dt^2} = (p'+k)\,\frac{d\mathbf{r}}{dt} + p' + k\,\frac{d^2\mathbf{r}}{dt^2}.$$

$$\therefore \quad \left.\begin{aligned} -k+p &= p'+k \\ -pk &= p'k. \end{aligned}\right\}.$$

These give

$$p = k;\; p' = -k,$$

$\therefore$ the point T is

$$\mathbf{r} - k\,\frac{d\mathbf{r}}{dt} + k\left(\frac{d\mathbf{r}}{dt} - k\,\frac{d^2\mathbf{r}}{dt^2}\right) = \mathbf{r} - k^2\,\frac{d^2\mathbf{r}}{dt^2}.$$

$$\therefore \quad \overrightarrow{\mathrm{TP}} = \mathbf{r} - \left(\mathbf{r} - k^2\,\frac{d^2\mathbf{r}}{dt^2}\right) = k^2\,\frac{d^2\mathbf{r}}{dt^2}.$$

Hence the result.

4. *A particle moves such that its acceleration at any instant is equal to* $\mu\mathbf{r} \times \mathbf{v}/\mathbf{r}^3$. *Show that it moves with constant speed and that* $r^2 = \alpha t^2 + 2\beta t + \gamma$ *where* α, β, γ *are constants. Show also that the angular momentum vector consists of two components, one constant in magnitude and direction and the other constant in magnitude in the direction of* $\mathbf{r}$.

Deduce that the particle always lies on a right circular cone with its vertex at the origin.

The equation of motion is

$$\frac{d^2\mathbf{r}}{dt^2} = \frac{\mu\mathbf{r}}{r^2} \times \frac{d\mathbf{r}}{dt}, \qquad \text{...(1)}$$

Multiplying scalarly with $d\mathbf{r}/dt$, we get

$$\frac{d\mathbf{r}}{dt} \cdot \frac{d^2\mathbf{r}}{dt^2} = 0.$$

Here $|\mathbf{r}| = \mathbf{r}.$

Integrating, we obtain

$$\left(\frac{d\mathbf{r}}{dt}\right)^2 = \text{constant} = \alpha;\ \text{say}. \qquad \text{...(2)}$$

Thus the speed is constant,

Multiplying (1) vectorially by **r**, we get

$$\mathbf{r} \times \frac{d^2\mathbf{r}}{dt^2} = \frac{\mu}{r^3}\mathbf{r} \times \left(\mathbf{r} \times \frac{d\mathbf{r}}{dt}\right)$$

$$= \frac{\mu}{r^3}\left[\left(\mathbf{r} \cdot \frac{d\mathbf{r}}{dt}\right)\mathbf{r} - (\mathbf{r}.\mathbf{r})\frac{d\mathbf{r}}{dt}\right]$$

$$= \frac{\mu}{r^3}\left[\mathbf{r}\frac{dr}{dt}\mathbf{r} - r^2\frac{d\mathbf{r}}{dt}\right]$$

$$= \mu \frac{d}{dt}\left(\frac{\mathbf{r}}{r}\right). \qquad ...(3)$$

Integrating (3), we obtain

$$\mathbf{r} \times \frac{d\mathbf{r}}{dt} = -\mu\frac{\mathbf{r}}{r} + \mathbf{a}, \qquad ...(4)$$

where **a** is a constant vector.

Thus the angular momentum, $\mathbf{r} \times m\, d\mathbf{r}/dt$, consists of two components, *viz.*, $m\,\mathbf{a}$ which is constant in magnitude and direction and $-\, m\mu\mathbf{r}/t$ whose magnitude is the constant $|-m\mu|$ and whose direction is that of **r**.

Now multiplying (1) scalarly with **r**, we get

$$\mathbf{r} . \frac{d^2\mathbf{r}}{dt^2} = 0, \qquad ...(5)$$

Adding (2) and (5), we obtain

$$\mathbf{r} . \frac{d^2\mathbf{r}}{dt^2} + \frac{d\mathbf{r}}{dt} . \frac{d\mathbf{r}}{dt} = \alpha,$$

or

$$\frac{d}{dt}\left(\mathbf{r} \cdot \frac{d\mathbf{r}}{dt}\right) = \alpha. \qquad ...(6)$$

Integrating, we have

$$\mathbf{r} \cdot \frac{d\mathbf{r}}{dt} = \alpha t + \beta,$$

or

$$r\frac{dr}{dt} = \alpha t + \beta.$$

Integrating again, we get

$$\frac{1}{2}r^2 = \frac{1}{2}\alpha t^2 + \beta t + \gamma'$$

or

$$r^2 = \alpha t^2 + 2\beta t + \gamma, \; (2\gamma' = \gamma)$$

where α, β, γ are constants.

Finally multiplying (4) scalarly with **r**, we get

$$0 = -\mu\frac{\mathbf{r}.\mathbf{r}}{r} + \mathbf{a}.\,\mathbf{r}$$

or $$\mathbf{a}\cdot\mathbf{r} = \mu\frac{r^2}{r} = \mu r$$

or $$\mathbf{a}\cdot\frac{\mathbf{r}}{r} = \mu. \qquad ...(7)$$

If P be the point with position vector $\mathbf{r}$ and $\overrightarrow{OA} = \mathbf{a}$, then (7) is equivalent to

$$OA \cos\alpha = \mu,$$

or $$\alpha = \cos^{-1}\left(\frac{\mu}{OA}\right).$$

Thus the angle α between OP and OA is constant. Hence P lies on a right circular cone with its vertex at the origin.

5. *A particle is fired horizontally with a velocity of 5×10^4 cm/sec at 30°N latitude. Calculate the horizontal component of coriolis acceleration and the consequent deflection of the projectile, as it hits a target 500 meters away. Also find the vertical displacement of the bullet due to gravity.*

The magnitude of horizontal component of Coriolis acceleration = $2\omega v'$ $\sin\lambda$.

Here, $$\omega = \frac{2\pi}{24\times60\times60} = 7.29 \times 10^{-5}\ \text{sec}^{-1},$$

$$v = 5 \times 10^4\ \text{cm/sec} \quad \text{and} \quad 1 = 30°$$

∴ The horizontal component of Coriolis acceleration

$$= 2 \times 5 \times 10^4 \times 7.29 \times 10^{-5} \times \frac{1}{2}$$
$$= 3.6\ \text{cm/sec}.$$

Time taken by the projectile to hit the target

$$= \frac{5\times10^4}{5\times10^4} = 1\ \text{sec}.$$

Consequent deflection of the projectile = $\frac{1}{2}ft^2$

$$= \frac{1}{2} \times 3.6 \times 1 = 1.8\ \text{cm}.$$

Vertical displacement of the bullet due to gravity

$$= \frac{1}{2}gt^2 = \frac{1}{2} \times 980 \times 1 = 490\ \text{cm}.$$

6. *The system S′ is rotating relative to a system S having the same origin and assumed to be fixed in space. The angular velocity of $-t^2\,\mathbf{j}' + 8\mathbf{k}'$, where t is the time. The position vector of the particle at the instant t relative to the system S′ is given by*

$$\mathbf{r} = t^2\,\mathbf{i}' - 6t\,\mathbf{j}' - 4t^3\,\mathbf{k}'$$

find (a) the apparent velocity, (b) the true velocity (c) the Coriolis acceleration, and (d) the centripetal acceleration at the time t = 1.

(*a*) The apparent velocity at the instant t is given by

$$\frac{d'\mathbf{r}}{dt} = 2t\,\mathbf{i}' - 6\mathbf{j}' - 12t^2\,\mathbf{k}'$$

(*b*) The true velocity at the instant t is given by

$$\begin{aligned}\frac{d'\mathbf{r}}{dt} &= \frac{d'\mathbf{r}}{dt} + \vec{\omega} \times \mathbf{r}\\ &= 2t\,\mathbf{i}' - 6\mathbf{j}' + 12t^2\,\mathbf{k}' + (2t\,\mathbf{i}' - t^2\,\mathbf{j}' + 8\mathbf{k}')\\ &\quad \times (t^2\,\mathbf{i}' - 6t\mathbf{j}' + 4t^3\,\mathbf{k}')\\ &= (50t - 4t^3)\,\mathbf{i}' + (6 - 8t^2 + 8t^4)\,\mathbf{j}' + t^4\,\mathbf{k}'\end{aligned}$$

At $t = 1$, the true velocity $= 46\mathbf{i}' - 6\mathbf{j}' + \mathbf{k}'$.

(*c*) Coriolis acceleration $\mathbf{F}c = 2\vec{\omega} \times \dfrac{d'\mathbf{r}}{dt}$

$$\begin{aligned}&= 2[(2t\,\mathbf{i}' - t^2\mathbf{j}' + 8\mathbf{k}') \times (2t\,\mathbf{i}' - 6\mathbf{j}' + 12t^2\,\mathbf{k}')]\\ &= 2\,[(48 - 12t^4)\,\mathbf{i}' + (16t - 24t^3)\,\mathbf{j}' - (12t - 2t^3)\,\mathbf{k}']\end{aligned}$$

At $t = 1$, the Coriolis acceleration $= 2[36\mathbf{i}' - 8\mathbf{j}' - 19\mathbf{k}']$

(*d*) The centripetal acceleration

$$\begin{aligned}&= \vec{\omega} \times (\vec{\omega} \times \mathbf{r})\\ &= (2t\,\mathbf{i}' - t^2\,\mathbf{j}' + 8\mathbf{k}') \times [(2t\,\mathbf{i}' - t^2\,\mathbf{j}' + 8\mathbf{k}')\\ &\quad \times (t^2\,\mathbf{i}' - 6t\,\mathbf{j}' + 4t^3\,\mathbf{k}')]\\ &= (-64\,t^2 + 76t^4 - t^6)\,\mathbf{i}' + (384t + 24t^3 - 34t^5)\,\mathbf{j}'\\ &\quad + (64\,t^3 - 16t^5 - 4t^7)\,\mathbf{k}'\end{aligned}$$

At $t = 1$, the centripetal acceleration

$$= 11\mathbf{i}' + 374\,\mathbf{j}' + 44\mathbf{k}'.$$

EXERCISES

1. A particle of mass 2 units has position vector $\mathbf{r}$ at time t, referred to fixed origin and axes given by

$$\mathbf{r} = (t^2 - 2t)\;\mathbf{i} + \left(\frac{1}{2}\,t^2 + 1\right)\mathbf{j} + \frac{1}{2}t^2\,\mathbf{k}.$$

Prove that the resultant force $\mathbf{F}$ acting on the particle is

$$\mathbf{F} = 4\mathbf{i} + 2\mathbf{j} + 2\mathbf{k}.$$

Also evaluate for the particle at time t

(*i*) its kinetic energy;

(*ii*) its angular momentum about the origin;

(*iii*) time rate of change of angular momentum about the origin;

(*iv*) the moment of the force $\mathbf{F}$ about the origin.

2. The position vector of particle of mass m at time t is

$$\mathbf{i}\;a\cos nt + \mathbf{j}\;a\sin nt + \mathbf{k}\;\frac{1}{2}\;at^2;$$

the origin being at 0.

Find expressions for the velocity and acceleration of the particle, the resultant force acting on the particle and the moment $\mathbf{L}$ of the force about 0.

If **H** is the angular momentum of the particle about, 0, show that

$$\frac{d\mathbf{H}}{dt} = \mathbf{L}.$$

3. The position vector of a particle of mass m at time t is

$$\mathbf{r} = \mathbf{i}\, cot \cos\frac{a}{t} + \mathbf{j}\, ct \sin\frac{a}{t};$$

where a, c are constants. Find the velocity; acceleration and angular momentum about the origin of the particle at time, t, and verify that

$$\frac{d^2\mathbf{r}}{dt^2} = -\frac{a^2c^4}{r^4}\mathbf{r}.$$

How does the force acting on the particle depend upon the position of the particle.

4. A particle of mass m has the position vector $\mathbf{r}$ at time t where

$$\mathbf{r} = \mathbf{i}\, at \cos nt + \mathbf{j}\, at \sin nt - \mathbf{k}\, bt;$$

find the velocity and the resultant force acting on the particle at time t. Also find the tangential component of the force.

Find the angular momentum **H** about the origin and the moment **L** of the force about the origin and verify that

$$\frac{d\mathbf{H}}{dt} = \mathbf{L}.$$

5. A particle of mass m is projected vertically under gravity, the resistance of air being proportional to velocity. Show that the greatest height attained by the particle is

$$\frac{v_c^2}{g}\,[\lambda - \text{long}\,(1+\lambda)]$$

where v_c is terminal velocity and $v_c\,\lambda$ is the initial velocity.

6. A projectile located at latitude λ is fired with velocity v_0 in a southward direction at an angle θ with horizontal. Find the position of the projectile after time t and prove that it is deflected toward the east of the original vertical plane by the amount

$$\frac{1}{3}\,\omega g \cos \lambda t^2 - \omega\, v_0 \sin(\theta + \lambda)\, t^2.$$

7. A constant force $\mathbf{F} = 6\mathbf{i} - 3\mathbf{j} + 3\mathbf{k}$, acts on a particle of mass 4 units having initially the velocity $\mathbf{u} = 4\mathbf{i} + 3\mathbf{j} - \mathbf{k}$.

Find the kinetic energy of the particle after 2 time units have elapsed.

8. A particle moves under the action of a force directed towards a fixed point and proportional to the cube of the distance, r, from the same; show that

$$r^2 = \alpha t^2 + \beta.$$

where α, β are constants.

9. Investigate the motion of the particle whose equation of motion is

$$\frac{d^2\mathbf{r}}{dt^2}+2\alpha\frac{d\mathbf{r}}{dt}+\beta\mathbf{r}=0,$$

where α, β are constants and α is positive.

10. A particle of mass m is subjected to a constant force $\mathbf{E}e$, and a force $\mathbf{H}\times\mathbf{v}$ where $\mathbf{v}$ is its velocity and $\mathbf{H}$ **a** constant vector. The particle starts from rest at the origin. Determine its displacement from the origin at time t. Show also that the motion is the superposition of a circular motion in a plane perpendicular to $\mathbf{H}$ and a parabolic motion in the plane of $\mathbf{H}$ and $\mathbf{E}\times\mathbf{H}$.

11. A particle moves along a curve whose intrinsic equation is

$$\mathbf{s} = 4a\sin\psi,$$

show that the tangential and normal components of the acceleration are

$$4a\,(\ddot{\psi}\cos\psi-\dot{\psi}^2\sin\psi),4a\,\dot{\psi}^2\cos\psi,$$

where

$$\dot{\psi} = d\psi\,/\,dt \text{ and } \ddot{\psi}-d^2\,\psi\,/\,dt^2.$$

4.12. MOMENTUM AND ANGULAR MOMENTUM OF A SYSTEM OF PARTICLES

If m be the mass and $\mathbf{v}$ the mass and $\mathbf{v}$ the velocity of any particle of system of particles, then the vector sum

$$\mathbf{M} = \Sigma m\mathbf{v},$$

of the momenta of the particles is called the momentum of the system; the summation extending to all the particles of the system.

Similarly if $\mathbf{r}$ be the position vector relative to a point O of a particle of mass m and velocity $\mathbf{v}$, then the vector sum

$$\mathbf{H} = \Sigma\mathbf{r}\times m\mathbf{v}=\Sigma m\mathbf{r}\times\mathbf{v},$$

of the angular momenta of the particles is called the angular momentum about 0 of the system; the summation again extending to all the particles of the system.

4.12.1. Comparison of Angular momenta about two points

If $\mathbf{H}$, $\mathbf{H}'$ *be the angular momenta of a system about two points, 0, 0′, respectively, then*

$$\mathbf{H}' = \mathbf{H}+\overrightarrow{0'0}\times\mathbf{M};$$

M, *being the momentum of the system.*

If m be the mass and $\mathbf{v}$ the velocity of a particle at the point P, we have

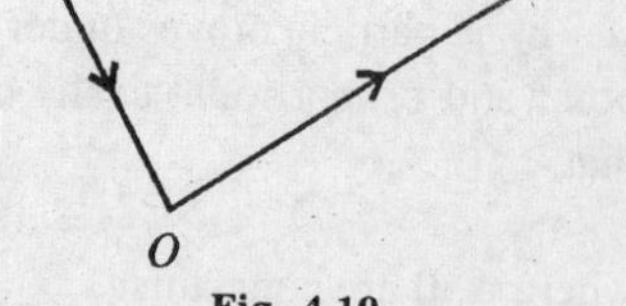

Fig. 4.10

$$\begin{aligned}\mathbf{H} &= \Sigma\overrightarrow{OP}\times m\mathbf{v}\\ \mathbf{H}' &= \Sigma\overrightarrow{O'P}\times m\mathbf{v}\\ &= \Sigma\,(\overrightarrow{O'O}+\overrightarrow{OP})\times m\mathbf{v}\\ &= \Sigma\,\overrightarrow{O'O}\times m\mathbf{v}+\Sigma\,\overrightarrow{OP}\times m\mathbf{v}\end{aligned}$$

$$= \overrightarrow{O'O} \times \Sigma m\mathbf{v} + \mathbf{H}$$

$$= \overrightarrow{O'O} \times \mathbf{M} + \mathbf{H} = \mathbf{H} + \overrightarrow{O'O} \times \mathbf{M}.$$

Cor. Angular momenta are the same about two points 0, 0′ if their join, 0′ 0 is parallel to the momentum vector **M**, for in this case

$$\overrightarrow{O'O} \times \mathbf{M} = 0.$$

4.13. CENTROID OF A SYSTEM

The discussion of the motion of a system is greatly facilitated by the concept of '*Centroid*' of the system, also known as *Centre of mass.* The point with position vector.

$$\mathbf{R} = \frac{\Sigma m\mathbf{r}}{\Sigma m},$$

is defined as the centroid; m being the mass and **r** the position vector of any particle of the system.

It can be easily shown that the position of the centroid of a system is independent of the choice of the origin of reference.

The centroid of a moving system of particles moves with the system and the velocity **V** of a particle moving with the centroid of a moving system is given by

$$\mathbf{V} = \frac{d\mathbf{r}}{dt} = \frac{\Sigma m \frac{d\mathbf{r}}{dt}}{\Sigma m} = \frac{\Sigma m\mathbf{v}}{\Sigma m}, \quad \text{...(1)}$$

4.13.1. Some properties of Centroid.

Theorem on Momentum. *The momentum of a system is the same as that of a particle whose mass is equal to the mass of the system and which is moving with the centroid of the same.*

Re-writing (1) as

$$(\Sigma m)\,\mathbf{V} = \Sigma m\mathbf{v} = \mathbf{M},$$

we have the result as stated.

Theorem on Angular momentum. *The angular momentum of a system of particles about the centroid G is equal to the angular momentum about G of the system in its motion relative to G.*

Let **R** be the position vector of the centroid G relative to a fixed point O and let **r**, ρ be respectively the position vectors of a particle P of mass m relative to O and G. The velocity **V** of the centorid is given by

$$\mathbf{V} = \frac{\Sigma m\mathbf{v}}{\Sigma m}.$$

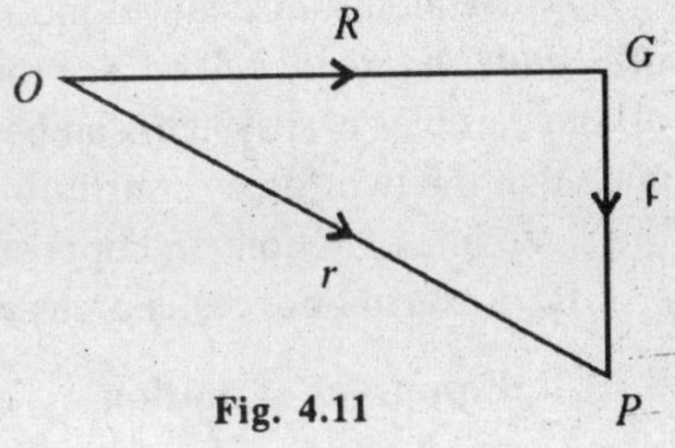

Fig. 4.11

The velocity of the particle at P relative to G

$$= \mathbf{v} - \mathbf{V}.$$

The angular momentum of the system about G in its motion relative to G

$$\begin{aligned}
&= \Sigma m\rho \times (\mathbf{v} - \mathbf{V}) \\
&= \Sigma m\rho \times \mathbf{v} - \Sigma m\rho \times \mathbf{V} \\
&= \Sigma m\rho \times \mathbf{v} - (\Sigma m\rho) \times \mathbf{V} \\
&= \Sigma m\rho \times \mathbf{v}, \text{ for } \Sigma m\rho = \mathbf{0} \\
&= \text{Angular momentum about G.}
\end{aligned}$$

4.14. Kinetic Energy of a system *is the sum of the kinetic energies of the different particles of the system.*

Theorem. *The kinetic energy of a system is equal to the kinetic energy in its motion relative to the centroid together with the kinetic energy of a particle whose mass is equal to that of the system and which is moving with the centroid.*

In terms of the notation of the preceding section, the kinetic energy of the system.

$$\begin{aligned}
&= \frac{1}{3}\,\Sigma m\mathbf{v}^2 \\
&= \frac{1}{2}\,\Sigma m\,[\mathbf{v} - \mathbf{V}) + \mathbf{V}]^2 \\
&= \frac{1}{2}\,\Sigma m(\mathbf{v} - \mathbf{V})^2 + \frac{1}{2}\,\Sigma m\mathbf{V}^2 + \Sigma m\mathbf{V}.(\mathbf{v} - \mathbf{V}).
\end{aligned}$$

$$\begin{aligned}
\text{Now, } \Sigma m\mathbf{V}.(\mathbf{v} - \mathbf{V}) &= \mathbf{V}.\Sigma m(\mathbf{v} - \mathbf{V}) \\
&= \mathbf{V}.[\Sigma m\mathbf{v} - (\Sigma m)\mathbf{V}] \\
&= \mathbf{V}.[(\Sigma m)\,\mathbf{V} - (\Sigma m)\,\mathbf{V}] = 0.
\end{aligned}$$

$$\therefore \quad \frac{1}{2}\,\Sigma m\mathbf{v}^2 = \frac{1}{2}\,\Sigma m\,(\mathbf{v} - \mathbf{V})^2 + \frac{1}{2}\,(\Sigma m)\,\mathbf{V}^2.$$

Hence the result.

4.15. DYNAMICS OF A SYSTEM OF PARTICLES

Forces acting on any particle of a system are of two types, Internal forces, and External forces.

I. Internal Forces. The forces which arise as a result of the connection of any particle with other particles of the system are called Internal forces.

By Newton's third law of motion which states that '*Action and Reaction are equal and opposite*', we see that the totality of Internal forces acting on all the particles of a system can be thought of as a system of pairs of forces such that the two forces constituting any pair are equal in magnitude, have the same line of action, and opposite senses.

II. External Forces. Forces which are not Internal are called External.

4.15.1. Equations of motion

Suppose that **P** and **Q** are the external and internal forces respectively acting on a particle of the system with mass m and position vector $\mathbf{r}$ relative to some Inertial Frame. By Newton's second law,

$$m\frac{d^2\mathbf{r}}{dt^2} = \mathbf{P}+\mathbf{Q}. \quad ...(1)$$

If we consider the totality of internal forces acting on all the particles of the system, we have

$$\Sigma\mathbf{Q} = \mathbf{0}, \quad ...(2)$$

and
$$\Sigma\mathbf{r}\times\mathbf{Q} = \mathbf{0}, \quad ...(3)$$

the summations extending to all the particles.

The equation (3) follows from the fact that the sum of the moments about any point of two equal and opposite forces acting along the same line is zero.

Adding equations of the type (1) for all the particles of the system and using (2), we get

$$\Sigma m\frac{d^2\mathbf{r}}{dt^2} = \Sigma\mathbf{P}. \quad ...(4)$$

Multiplying (1) vectorially by $\mathbf{r}$, we get

$$\mathbf{r}\times m\frac{d^2\mathbf{r}}{dt^2} = \mathbf{r}\times\mathbf{P}+\mathbf{r}\times\mathbf{Q} \quad ...(5)$$

Adding equations similar to (5) and using (3), we have

$$\Sigma m\mathbf{r}\frac{d^2\mathbf{r}}{dt^2} = \Sigma\mathbf{r}\times\mathbf{P}. \quad ...(6)$$

The equation (4) and (6), which we re-write below,

$$\Sigma m\frac{d^2\mathbf{r}}{dt^2} = \Sigma\mathbf{P},\ \Sigma m\mathbf{r}\times\frac{d^2\mathbf{r}}{dt^2} = \Sigma\mathbf{r}\times\mathbf{P},$$

are fundamental for the study of a system and will be referred to as the *Equations of motion.* They are free from Internal forces and involve External forces only.

These two vector equations are equivalent to six scalar equations.

In case we have n particles in motion, then there are $3n$ unknowns, *viz.*, the $3n$ coordinates of the n particles and accordingly the equations above will not, in general, be sufficient to determine the motion of a given system. Some further information about the *Internal configuration* of the system must necessarily be had for the determination of the motion is rigid so that the Internal configuration remains invariant, the equations do suffice to solve the problem of the determination of motion completely.

It is possible, however, to draw some important general conclusions from the Equations of motion about any material system and we will now proceed to do this. In this connection, the following two results are fundamental. They concern the *time rates of changes of Momentum and of Angular momentum of the system.*

4.15.2. Time Rates of change of Momentum and of Angular momentum

I. *The time rate of change of momentum of a system is equal to the sum of the external forces of the system.*

II. *The time rate of change of Angular momentum about a fixed point is equal to the sum of the moments of the external forces about the point.*

Take an inertial frame with its origin at the given fixed point. Let **M** denote the momentum and **H** the angular momentum about the origin. We have

$$\mathbf{M} = \Sigma m\mathbf{v} = \Sigma m \frac{d\mathbf{r}}{dt}. \qquad ...(1)$$

$$\mathbf{H} = \Sigma \mathbf{r} \times m\mathbf{v} = \Sigma m\mathbf{r} \times \frac{d\mathbf{r}}{dt}. \qquad ...(2)$$

Differentiating (1) and (2), we obtain

$$\frac{d\mathbf{M}}{dt} = \Sigma m \frac{d^2\mathbf{r}}{dt^2} = \Sigma \mathbf{P},$$

and

$$\frac{d\mathbf{H}}{dt} = \Sigma m\mathbf{r} \times \frac{d^2\mathbf{r}}{dt^2} = \Sigma \mathbf{r} \times \mathbf{P}.$$

Hence the results.

We shall usually denote $\Sigma\mathbf{P}$ by **F** and $\Sigma\mathbf{r} \times \mathbf{P}$ by **L** so that we have

$$\frac{d\mathbf{M}}{dt} = \mathbf{F}, \quad \frac{d\mathbf{H}}{dt} = \mathbf{L}.$$

Note. It should be remembered that the result $d\mathbf{H}/dt = \mathbf{L}$ is not true when the point about which we take moments is *not fixed.* The modified form of results when the point is not fixed will be obtained in the following section.

Cor. Principles of Conservation of Momentum and Angular Momentum.

I. *If the sum of the resolved part of the external forces along any fixed direction remains zero during a certain stage of motion, then the resolved part of the momentum of the system along the direction remains constant during the stage.*

II. *If the sum of the resolved parts along any fixed direction of the sum of the moments about a fixed point of the external forces remains zero during a certain stage of motion, then the resolved part along the direction of the angular momentum about the fixed point remains constant during the stage.*

Let, **a**, be the unit vector along the fixed direction. We have

$$\frac{d\mathbf{M}}{dt} = \mathbf{F}.$$

$$\therefore \quad \mathbf{a} \cdot \frac{d\mathbf{M}}{dt} = \mathbf{a} \cdot \mathbf{F} = 0, \text{ as given,}$$

or

$$\frac{d(\mathbf{a}.\mathbf{M})}{dt} = 0.$$

$\therefore$ **a . M** is constant. Hence I., Again we have

$$\frac{d\mathbf{H}}{dt} = \mathbf{L}.$$

$$\therefore \quad \mathbf{a} \cdot \frac{d\mathbf{H}}{dt} = \mathbf{a} \cdot \mathbf{L} = 0,$$

or
$$\frac{d(\mathbf{a} \cdot \mathbf{H})}{dt} = 0.$$

$\therefore$ **a . H** is constant. Hence H.

4.15.3. Time rate of change of angular momentum about a moving origin

We shall now examine the modification needed in the result $d\mathbf{H}/dt = \mathbf{L}$ when the moments are taken about a *moving* point instead of a *fixed* point. The investigation will also enable us to determine conditions under which the statement remains true without any modification even in case moments are taken about a moving point.

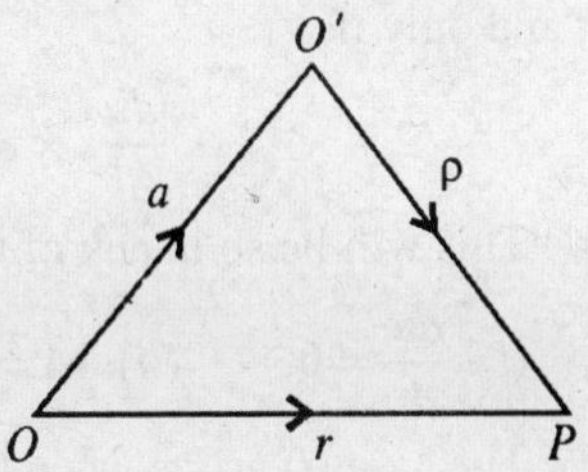

Fig. 4.12

Let O be a fixed point and O′ a point moving relatively to O and let P be any point of the system.

Let $\overrightarrow{OP} = \mathbf{r}, \quad \overrightarrow{O'P} = \mathrm{r}, \quad \overrightarrow{OO'} = \mathbf{a},$

so that $\rho = \mathbf{r} - \mathbf{a}.$

Let **H**, **H′** denote the angular moments about O and O′ respectively. We have

$$\mathbf{H} = \Sigma \mathbf{r} \times m\mathbf{v},$$
$$\mathbf{H}' = \Sigma \rho \times m\mathbf{v},$$

where **v** is the velocity of the particle at P. Since O is a fixed point, we have

$$\mathbf{v} = \frac{d\mathbf{r}}{dt}.$$

Thus

$$\mathbf{H}' = \Sigma (\mathbf{r} - \mathbf{a}) \times m \frac{d\mathbf{r}}{dt}$$
$$= \Sigma \mathbf{r} \times m \frac{d\mathbf{r}}{dt} - \mathbf{a} \times \Sigma m \frac{d\mathbf{r}}{dt}$$
$$= \mathbf{H} - \mathbf{a} \times \mathbf{M}.$$

$$\frac{d\mathbf{H}'}{dt} = \frac{d\mathbf{H}}{dt} - \frac{d\mathbf{a}}{dt} \times \mathbf{M} - \mathbf{a} \times \frac{d\mathbf{M}}{dt}\frac{d\mathbf{r}}{dt}$$
$$= \Sigma \mathbf{r} \times \mathbf{P} - \frac{d\mathbf{a}}{dt} \times \mathbf{M} - \mathbf{a} \times \Sigma \mathbf{P} \qquad [\S\ 4.15.2]$$
$$= \Sigma (\mathbf{r} - \mathbf{a}) \times \mathbf{P} - \frac{d\mathbf{a}}{dt} \times \mathbf{M}$$

$$= \Sigma \rho \times \mathbf{P} - \frac{d\mathbf{a}}{dt} \times \mathbf{M}$$

$$= \mathbf{L}' - \frac{d\mathbf{a}}{dt} \times \mathbf{M},$$

where $\mathbf{L}'$ denotes the sum of the moments of the external forces about the moving point O′.

Thus we see that, in general, the rate of change of angular momentum $d\mathbf{H}'/dt$ about a moving point is not equal to the sum $\mathbf{L}'$ of the moments of the external forces about the same the two differing by

$$\frac{d\mathbf{a}}{dt} \times \mathbf{M}.$$

The equality $d\mathbf{H}'/dt = \mathbf{L}'$ will be valid even in the case of moving point if, and only if,

$$\frac{d\mathbf{a}}{dt} \times \mathbf{M} = \mathbf{0}.$$

This will be so in any of the following three cases:

(*i*) $\frac{d\mathbf{a}}{dt} = \mathbf{0}$. (*ii*) $\mathbf{M} = \mathbf{0}$. (*iii*) $\frac{d\mathbf{a}}{dt}$ is parallel to $\mathbf{M}$.

The case (*i*) arises when O′ is itself a fixed point.

The case (*ii*) will arise when the centroid is a fixed point, for $\mathbf{M}$ is the momentum of a particle moving with the centroid of the system. (§ 4.13.1)

The case (*iii*) will arise when the velocity $d\mathbf{a}/dt$ of O′ is parallel to that of the centroid.

Thus we see that the equation

$$\frac{d\mathbf{H}'}{dt} = \mathbf{L}'$$

is valid if and only if

(*i*) the point about which moments are taken is fixed, or (*ii*) the centroid is fixed, or (*iii*) the velocity of the point about which moments are taken is parallel to that of the centroid.

The alternative (*iii*) contains the important and useful fact *that the time rate of change of angular momentum about the centroid is equal to the moment of the external forces about the same,* irrespective of the nature of motion of the centroid.

4.16. TIME RATE OF CHANGE OF MOMENT ABOUT A MOVING POINT OF THE MOMENTUM RELATIVE TO THE POINT

We shall now consider the case when the momentum considered is also relative to the moving point.

We adopt the notation of the preceding section.

Let $\mathbf{H}''$ denote the moment about the moving point O″ of the momentum relative to O′. Thus we have

$$\mathbf{H} = \Sigma \mathbf{r} \times m \frac{d\mathbf{r}}{dt}$$

$$\mathbf{H}'' = \Sigma \rho \times m \frac{d\rho}{dt}$$

$$= \Sigma (\mathbf{r} - \mathbf{a}) \times m \left(\frac{d\mathbf{r}}{dt} - \frac{d\mathbf{a}}{dt}\right).$$

$$\therefore \quad \frac{d\mathbf{H}''}{dt} = \Sigma(\mathbf{r} - \mathbf{a}) \times m \left(\frac{d^2\mathbf{r}}{dt^2} - \frac{d^2\mathbf{a}}{dt^2}\right)$$

$$= \Sigma m\mathbf{r} \times \frac{d^2\mathbf{r}}{dt^2} - \Sigma \mathbf{a} \times m \frac{d^2\mathbf{r}}{dt^2} - \Sigma m\mathbf{r} \times \frac{d^2\mathbf{a}}{dt^2} + \Sigma \mathbf{a} \times m \frac{d^2\mathbf{a}}{dt^2}$$

$$= \Sigma m\mathbf{r} \times \frac{d^2\mathbf{r}}{dt^2} - \mathbf{a} \times \Sigma m \frac{d^2\mathbf{r}}{dt^2} - (\Sigma m\mathbf{r}) \times \frac{d^2\mathbf{a}}{dt^2}$$

$$+ (\Sigma m\mathbf{a}) \times \frac{d^2\mathbf{a}}{dt^2}$$

$$= \Sigma \mathbf{r} \times \mathbf{P} - \mathbf{a} \times \Sigma \mathbf{P} - (\Sigma m\mathbf{r} - \Sigma m\mathbf{a}) \times \frac{d^2\mathbf{a}}{dt^2}$$

$$= \Sigma(\mathbf{r} - \mathbf{a}) \times \mathbf{P} - (\Sigma m) (\mathbf{R} - \mathbf{a}) \times \frac{d^2\mathbf{a}}{dt^2};$$

R denoting the position vector of the centroid relative to **O**.

Here $\quad \Sigma(\mathbf{r} - \mathbf{a}) \times \mathbf{P} = \times \mathbf{P}$

denotes the moment of the external forces about **O′**.

Thus we see that the time rate of change of moment about a moving point of the momentum relative to the point will be equal to the moment of the external forces about the point;

i.e., $\quad d\mathbf{H}''/dt = \mathbf{L}''$

even in the case of the momentum being considered relative to the moving point, if and only if

$$(\mathbf{R} - \mathbf{a}) \times \frac{d^2\mathbf{a}}{dt^2} = \mathbf{0}.$$

i.e., if

(*i*) the moving point is the centroid.

or (*ii*) the acceleration of the moving point is zero,

or (*iii*) the line joining the centroid to the moving point is parallel to the acceleration of the moving point.

The case (*i*) is specially important. It states that *the time rate of change of moment about the centroid of the momentum relative to the same is equal to the sum of the moments of the external forces about the centroid.*

4.17. SYSTEM OF VARIABLE MASS

Let a system of mass M be moving at a velocity $\mathbf{v}$ at any instant t in a particular frame of reference. Let, at a later instant $t + \delta t$, a part δM separated from the system be moving with a velocity $\mathbf{u}$ and the remaining system M – δM with new velocity $\mathbf{v} + \delta\mathbf{v}$. If we still treat both parts as forming one and the same system, then for the finite time-interval δt, we can write

$$\mathbf{F}_{ext} = \frac{[(M-\delta M)(\mathbf{v}+\delta\mathbf{v})+\delta M\,\mathbf{u}]-M\,\mathbf{v}}{\delta t}$$

$$= M\frac{\delta\mathbf{v}}{\delta t} - \mathbf{v}\frac{\delta M}{\delta t} - \delta\mathbf{v}\frac{\delta M}{\delta t} + \mathbf{u}\frac{\delta M}{\delta t}$$

Now if $\delta t \to 0$, *i.e.*, the mass of the system is continuously decreasing, then

$$\frac{\delta\mathbf{v}}{\delta t} = \frac{\delta\mathbf{v}}{dt};\quad \frac{\delta M}{dt} = -\frac{\delta M}{dt};\quad \delta\mathbf{v} = 0,$$

so that
$$\mathbf{F}_{ext} = M\frac{\delta\mathbf{v}}{dt} + \mathbf{v}\frac{\delta M}{dt} - \mathbf{u}\frac{\delta M}{dt}$$

$$= \frac{d}{dt}(M\mathbf{v}) - \mathbf{u}\frac{dM}{dt}.$$

This expresses Newton's second law as applied to a body of *variable mass*.

About equation can be written as

$$M\frac{d\mathbf{v}}{dt} = \mathbf{F}_{ext} + (\mathbf{u}-\mathbf{v})\frac{dM}{dt},$$

where $\mathbf{v}_{rel}$ is the velocity of the ejected mass *relative to the main body.* The last term in the above equation is the rate of change of momentum of the system due to the mass leaving it. It can be taken as the reaction force exerted on the system by the leaving mass. Thus we can write

$$M\frac{d\mathbf{v}}{dt} = \mathbf{F}_{ext} + \mathbf{F}_{reaction}.$$

4.18. IMPULSE

When a constant force $\mathbf{F}$ acts upon a particle for a time-interval Δt, then the product $\mathbf{F}\,\Delta t$ is called the *impulse* of the force.

If the force $\mathbf{F}$ varies in magnitude, and acts from a time t, to t_2, then the integral $\int_{t_1}^{t_2} \mathbf{F}\,dt$ is the impulse of the force.

If $\mathbf{p}$ is the momentum of a particle, then

$$\mathbf{F} = \frac{d\mathbf{p}}{dt}$$

$$\Rightarrow \quad d\mathbf{p} = \mathbf{F}\,dt.$$

If the force acts from a time t_1 to a time t_2, then the change in momentum during this time-interval is

$$\mathbf{p}_2 - \mathbf{p}_1 = \int_{p_1}^{p_2} d\mathbf{p} = \int_{t_1}^{t_2} \mathbf{F}\, dt.$$

Hence the *change in momentum is equal to the impulse.*

4.19. CONSERVATION OF MOMENTUM DURING COLLISION

Let us consider a collision between two bodies of masses m_1 and m_2. During the collision they exert large forces on one another which are equal to opposite at all instants. Let $\mathbf{F}_{12}(t)$ and $\mathbf{F}_{21}(t)$ be the forces exerted on body 1 by body 2 and on body 2 by body 1 respectively at any instant t.

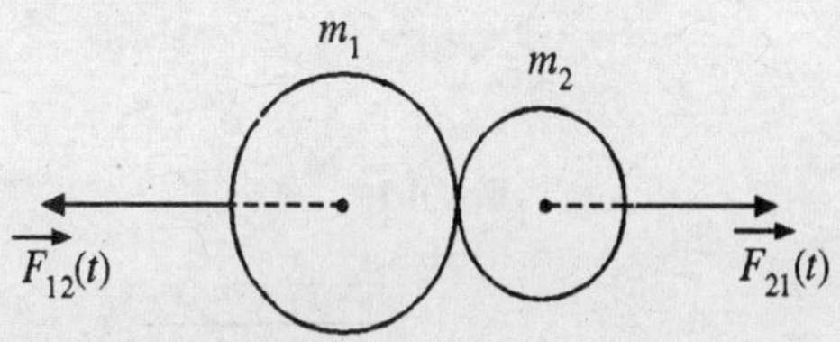

Fig. 4.13

The change in momentum of body 1 resulting from the collision is

$$\Delta\mathbf{p}_1 = \int_{t_i}^{t_f} \mathbf{F}_{12}\, dt = \mathbf{F}_{12}\, \Delta t,$$

where $\mathbf{F}_{12}$ is the average value of $\mathbf{F}_{12}$ during the collision time interval $\Delta t = t_f - t_i$.

Similarly, change in momentum of body 2 is

$$\Delta\mathbf{p}_2 = \int_{t_i}^{t_f} \mathbf{F}_{12}\, dt = \mathbf{F}_{12}\, \Delta t$$

By $\qquad \mathbf{F}_{12} = -\mathbf{F}_{21} \quad$ (Newton's law)

$\therefore \qquad \Delta\mathbf{p}_1 = -\Delta\mathbf{p}_2$

$\Rightarrow \qquad \Delta\mathbf{p}_1 + \Delta\mathbf{p}_2 = 0,$

i.e., if there are no external forces, the total change in momentum of the system as a result of collision is zero.

EXAMPLES

1. *Show that the centre of mass of two particles must lie on the line joining them, and the ratio of the distances of the two particles from the centre of mass is the inverse ratio of their masses.*

Let c be the centre of mass whose position vectors w.r.t. m_1 and m_2 are $\mathbf{d}_1$ and $\mathbf{d}_2$.

The position vector of c with respect to the origin O is defined by

$$\mathbf{r}_{cm} = \frac{m_1\mathbf{r}_1 + m_2\mathbf{r}_2}{m_1 + m_2}$$

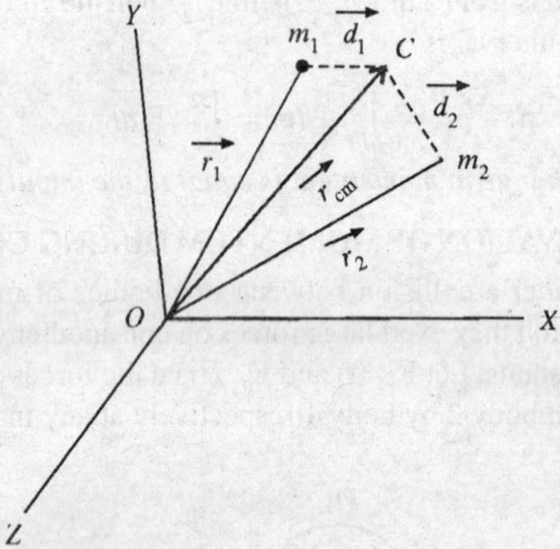

Fig. 4.14

we have,
$$\mathbf{d}_1 = \mathbf{r}_{cm} - \mathbf{r}_1 = \frac{m_1\mathbf{r}_1 + m_2\mathbf{r}_2}{m_1 + m_2}\ \mathbf{r}_1$$

$$= \frac{m_2}{m_1 + m_2}(\mathbf{r}_2 - \mathbf{r}_1)$$

and
$$\mathbf{d}_2 = \mathbf{r}_2 - \mathbf{r}_{cm} = \mathbf{r}_2 - \frac{m_1\mathbf{r}_1 + m_2\mathbf{r}_2}{m_1 + m_2}$$

$$= \frac{m_2}{m_1 + m_2}(\mathbf{r}_2 - \mathbf{r}_1)$$

$$\Rightarrow \qquad d_1 = \frac{m_2}{m_1}\,\mathbf{d}_2.$$

Thus the vectors $\mathbf{d}_1$ and $\mathbf{d}_2$ are collinear, *i.e.*, the centre of mass C lies on the line joining m_1 and m_2.

Further, we have

$$\frac{\mathbf{d}_1}{\mathbf{d}_2} = \frac{m_2}{m_1}.$$

2. *Two particles P and Q of different masses are initially at rest. They attract each other with a constant force. No external forces act on the system. Describe the motion of the centre of mass. At what distance from the original position of P do the particles collide?*

Initially the system is at rest so that its centre of mass is also at rest. If M be the total mass of the system and $\mathbf{a}_{cm}$ the acceleration of the centre of mass, then

$$M\,\mathbf{a}_{cm} = \mathbf{F}_{ext}$$

Since no external forces act on the system, we have

$$\mathbf{a}_{cm} = \frac{d\mathbf{v}_{cm}}{dt} = 0$$

$$\Rightarrow \qquad \mathbf{v}_{cm} = \text{constant.}$$

Hence the *centre of mass remains at rest.*

Let the particle m_1 lie at the origin of coordinates, and the particle m_2 at a distance r from it along the x-axis. The distance of the centre of mass of the system from the origin would be

$$x_{cm} = \frac{(m_1 \times 0) + (m_2 \times r)}{m_1 + m_2} = \frac{m_2}{m_1 + m_2} r.$$

Each particle is acted upon by a constant force F. Therefore the accelerations of these particles would be

$$a_1 = \frac{f}{m_1} \text{ and } a_2 = \frac{f}{m_2}.$$

Suppose the particles collide after time t at a distance x from the origin. This means that in time t, the particle m_1 travels a distance x and the particle m_2 travels a distance $r - x$. Thus,

$$x = \frac{1}{2} a_1 t^2 \quad \text{and} \quad r - x = \frac{1}{2} a_2 t^2$$

$$\Rightarrow \qquad \frac{r-x}{x} = \frac{a_2}{a_1} = \frac{m_1}{m_2}$$

$$\Rightarrow \qquad \frac{r}{x} = \frac{m_1}{m_2} + 1 = \frac{m_1 + m_2}{m_2}$$

$$\Rightarrow \qquad x = \frac{m_2}{m_1 + m_2} r.$$

3. *Show that if a particle moves only in the xy-plane, its angular momentum vector has only a z-component.*

The angular momentum of a particle

$$\begin{aligned} \mathbf{H} &= \mathbf{r} \times \mathbf{p} = \mathbf{r} \times m\mathbf{v} = m(\mathbf{r} \times \mathbf{v}) \\ &= m\,[(x\mathbf{i} + y\mathbf{j}) \times v_x\,\mathbf{i} + v_y\,\mathbf{j})] \\ &= m\,(xv_y - yv_x)\,\mathbf{k}. \end{aligned}$$

4. *A man of mass m is standing on a trolly of mass M which is moving with a velocity* **v** *on frictionless horizontal rails. If the man starts running opposite to the direction of motion of the trolly and lies velocity relative to the trolly is* $\mathbf{v}_{rel}$ *just before he jumps off, find the change in velocity of the trolly.*

Let us now assume that there are n men, each of mass m, on the trolly. Should they all run and jump off together, or should they do so one by one in order to give a greater velocity to the trolly.

The initial mass of the (trolly + man) system is (M + n) which is moving with a velocity **v** in a fixed frame of reference. As the man jumps, he acquires a backward momentum and the trolly acquires forward momentum. Then, in the absence of external forces,

$$(\mathrm{M}+n)\frac{d\mathbf{v}}{dt} = \mathbf{v}_{\text{rel}}\frac{d(\mathrm{M}+n)}{dt}$$

$$\Rightarrow \quad (\mathrm{M}+n)\,d\mathbf{v} = \mathbf{v}_{\text{rel}}\,d(\mathrm{M}+n) \qquad \text{...(1)}$$

$$\Rightarrow \quad (\mathrm{M}+n)\,\Delta\mathbf{v} = \mathbf{v}_{\text{rel}}\,\Delta(\mathrm{M}+n)$$

Here $\Delta(\mathrm{M}+n) = -m$ (the man of mass m jumps off).

$$\therefore \quad (\mathrm{M}+n)\,\Delta\mathbf{v} = -\mathbf{v}_{\text{rel}}\,m$$

$$\Rightarrow \quad \Delta\mathbf{v} = -\mathbf{v}_{\text{rel}}\frac{m}{\mathrm{M}+m}$$

So, the speed of the trolly will increase by $\dfrac{m\,v_{\text{rel}}}{(\mathrm{M}+m)}$ in its initial direction of motion.

Let us now suppose that n men are standing on that trolly. If they all jump off together, the change in velocity would be

$$\Delta\mathbf{v} = -\mathbf{v}_{\text{rel}}\frac{mn}{\mathrm{M}+mn} \qquad \text{...(2)}$$

But, if they jump one by one, the mass of the system will go on changing till the last man jumps. In this case eqn. (1) will be

$$d\mathbf{v} = \mathbf{v}_{\text{rel}}\frac{d(\mathrm{M}+mn)}{(\mathrm{M}+mn)}$$

$$\Rightarrow \quad \Delta\mathbf{v} = \mathbf{v}_{\text{real}}\int_{\mathrm{M}+mn}^{\mathrm{M}}\frac{d(\mathrm{M}+mn)}{(\mathrm{M}+mn)}$$

$$= \mathbf{v}_{\text{rel}}\left[\log_e(\mathrm{M}+mn)\right]_{\mathrm{M}+mn}^{\mathrm{M}}$$

$$= \mathbf{v}_{\text{rel}}\log_e\left(\frac{\mathrm{M}}{\mathrm{M}+mn}\right) = -\mathbf{v}_{\text{rel}}\log_e\left(1+\frac{mn}{\mathrm{M}}\right)$$

$$= -\mathbf{v}_{\text{rel}}\frac{mn}{\mathrm{M}}\ \text{(approx)} \qquad \text{...(3)}$$

A comparison of equations (2) and (3) shows that the men would give a greater velocity to the trolly by running and jumping off one by one.

5. *Five particles of sand are being dropped continuously from a fixed container on to a moving belt. Find out the force necessary to have the belt moving at a constant speed, and show that the power supplied by this force is twice the rate of increase of the kinetic energy of the system.*

Suppose the belt is moving with a constant velocity $\mathbf{v}$ in a certain reference frame. The container is fixed in this reference frame. As sand particles are falling on the belt, the mass of the system is continuously increasing. Let M be the mass of the system at any instant, $d\mathrm{M}/dt$ the rate at which material is falling on the belt. Then

$$\mathrm{M}\frac{d\mathbf{v}}{dt} = \mathbf{F}_{\text{ext}} + \mathbf{v}_{\text{rel}}\frac{d\mathrm{M}}{dt}, \qquad \text{...(1)}$$

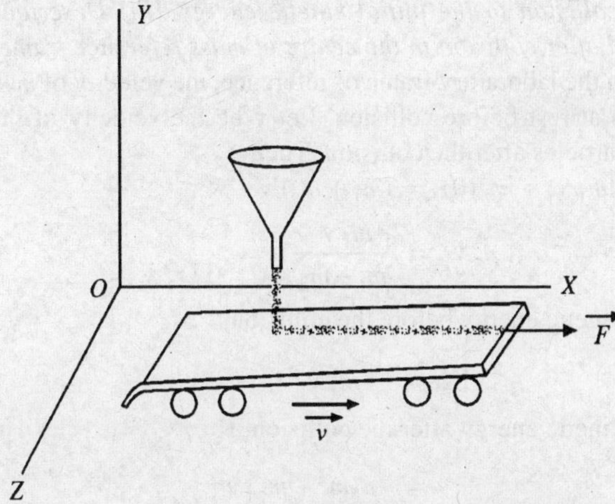

Fig. 4.15

where $\mathbf{F}_{ext}$ is the external force acting on the system and $\mathbf{v}_{rel}$ is the relative horizontal velocity of the falling mass relative to the belt.

$$\frac{d\mathbf{v}}{dt} = 0, \quad \text{(as } \mathbf{v} \text{ is constant)}$$

$$\Rightarrow \qquad v_{rel} = -\mathbf{v}$$

and $\dfrac{d\mathrm{M}}{dt}$ is positive.

Therefore the equation (1) becomes

$$0 = \mathbf{F}_{ext} - \mathbf{v}\frac{d\mathrm{M}}{dt}$$

$$\Rightarrow \qquad \mathbf{F}_{ext} = \mathbf{v}\frac{d\mathrm{M}}{dt} \qquad ...(2)$$

This is the external force acting on the system under which it is moving at a constant rate.

Now, the power P supplied by the force $\mathbf{F}_{ext}$ is the rate of doing work. Then

$$\mathbf{P} = \mathbf{F}_{ext}\,.\,\mathbf{v} = \mathbf{v}\frac{d\mathrm{M}}{dt}\,.\,\mathbf{v}$$

$$= \mathbf{v}^2\frac{d\mathrm{M}}{dt} = \frac{d}{dt}(\mathrm{M}v^2)$$

$$= 2\frac{d}{dt}\left(\frac{1}{2}\mathrm{M}v^2\right) = 2\frac{d\mathrm{K}}{dt}.$$

where K is the kinetic energy.

6. *A particle of mass m_1 moving with velocity $\mathbf{v}_1$ collides head-on with another particle of mass m_2, initially at rest, and sticks to it. Find (i) the velocity of the system after the collision, and (ii) ratio of the kinetic energy* *after the collision to the initial kinetic energy, (iii) Describe the motion before and after collision in the centre of mass reference frame.*

(*i*) In the laboratory frame of reference, the velocity of m_1 is $\mathbf{v}_1$ and m_2 is at rest before collision. Let $\mathbf{v}$ be the velocity of the combined particles after the collision. Then

$$m_1 \mathbf{v}_1 + m_2 (0) = (m_1 + m_2) \mathbf{v}$$

$$\Rightarrow \quad \mathbf{v}_1 = \frac{m_1 \mathbf{v}_1}{m_1 + m_2} \qquad ...(1)$$

(*ii*) Kinetic energy before the collision,

$$\text{K}i = \frac{1}{2} m_1 v_1^2$$

Kinetic energy after the collision, K_f

$$= \frac{1}{2} (m_1 + m_2) v^2$$

$$\therefore \quad \frac{\text{K}_f}{\text{K}_i} = \frac{(m_1 + m_2) v^2}{m_1 v_1^2}$$

$$= \frac{(m_1 + m_2)}{m_1 v_1^2} \cdot \frac{m_1 v_1^2}{(m_1 + m_2)^2} = \frac{m_1}{m_1 + m_2}.$$

Thus $\text{K}_f < \text{K}_i$. Thus the energy is lost and the lost fraction is

$$1 - \frac{\text{K}_f}{\text{K}_i} = \frac{m_2}{m_1 + m_2}.$$

(*iii*) Before collision, the velocity of the centre of mass of the system is

$$\mathbf{v}_{\text{cm}} = \frac{m_1 \mathbf{v}_1}{m_1 + m_2}.$$

In the centre of mass frame the velocity of the centre of mass is zero. Therefore, in this frame the velocity of m_1 is

$$\mathbf{v}'_1 = \mathbf{v}_1 - \mathbf{v}_{\text{cm}} = \mathbf{v}_1 - \frac{m_1 \mathbf{v}_1}{m_1 + m_2} = \frac{m_2 \mathbf{v}_1}{m_1 + m_2}$$

and that of m_2 is

$$\mathbf{v}'_2 = 0 - \mathbf{v}_{\text{cm}} = -\mathbf{v}_{\text{cm}} = -\frac{m_1 \mathbf{v}_1}{m_1 + m_2}.$$

Thus, before collision, both particles seem to approach the observer sitting at the centre of mass.

After collision, the velocity $\mathbf{v}'$ the velocity $\mathbf{v}'$ of the combined particle in the centre-of-mass frame is given by

$$(m_1 + m_2) \mathbf{v}' = m_1 \mathbf{v}'_1 + m_2 \mathbf{v}_2'$$

$$= \frac{m_1 m_2 \mathbf{v}_1}{m_1 + m_2} - \frac{m_1 m_2 \mathbf{v}_1}{m_1 + m_2} = 0.$$

That is in the centre-of-mass frame the system is at rest.

EXERCISES

1. Three particles, of masses 1, 2, 3 units, have position vectors at time t given respectively by

$$\mathbf{r}_1 = 4a \cos t\mathbf{i} + 2b \sin t\mathbf{j} + 5t^2\mathbf{k},$$
$$\mathbf{r}_2 = -2a \cos t\mathbf{i} + b \sin t\mathbf{j} - (t^2 + 3t)\,\mathbf{k},$$
$$\mathbf{r}_3 = 6a \cos t\mathbf{i} - 3b \sin t\mathbf{j} - (t^2 + 2t)\,\mathbf{k},$$

referred to a fixed origin O and fixed axes, a and b being constants. Find the position vector of the centroid of the system and verify that the momentum of the system for motion relative to the centroid is zero.

Also find the magnitude and direction of the sum of the external forces acting on the system and the sum of the moments of the external forces about (*i*) the origin, (*ii*) the centroid.

2. Two particles of masses m, M moving with velocities $\mathbf{u}$ and $\mathbf{v}$ respectively collide and coalesce. Show that the loss of kinetic energy is

$$\frac{1}{2}\frac{mM}{m+M}(\mathbf{u}-\mathbf{v})^2.$$

3. The velocity of two particles of masses m_1 and m_2 relative to an inertial observer are $\mathbf{v}_1$ and $\mathbf{v}_2$. Determine the velocity of the centre of mass relative to the observer and the velocity of each particle relative to the centre of mass.

4. Two masses, one n times heavier than the other, are dropped from the same height. How do their momenta compare just before they hit the ground? If these masses have same kinetic energy then how do their momenta compare?

5. A machine gun is mounted on a trolly which is moving in a horizontal frictionless fixed reference frame. The mass of the system (trolly + gun) is M at a particular instant. At that instant the gun is firing bullets, each of mass m, at the rate of n per second, in a direction opposite to the motion of the trolly. The velocity of the bullets in the same reference frame is $\mathbf{u}$. Compute the acceleration of the trolly, and the average thrust of the ejecting bullets on the system.

6. A wagon filled with sand has a hold so that sand leaks through the bottom at a constant rate $-\dfrac{dm}{dt} = \lambda$. A force $\mathbf{F}$ acts on the wagon in the direction of its motion. If its instantaneous velocity is $\mathbf{v}$, write the equation of motion.

7. A ball of mass m makes head-on elastic collision with a ball of mass nm which is initially at rest. Show that the fractional transfer of energy by the first ball is $4n/(1+n)^2$. Deduce the value of n for which the transfer is maximum.

8. Two balls A and B, having different but unknown masses collide. A is initially at rest while b has a speed v. After collision B has a speed $v/2$ and moves at right angles to its original motion. Find the direction in which ball

A moves after collision. Can you determine the speed of A from the given information?

OBJECTIVE QUESTIONS

For each of the following questions, four alternatives are given for the answer. Only one of them is correct. Choose the correct alternative.

1. Component of acceleration along binormal is

(*a*) $\frac{dv}{dt}$ (*b*) v^2/ρ (*c*) 0 (*d*) ∞

2. Acceleration vector at any point of a trajectory lies in

(*a*) Normal plane (*b*) Osculating plane

(*c*) Tangent plane (*d*) In any plane

3. Time rate of change of momentum of a particle is proportional to the

(*a*) impressed force (*b*) angular momentum

(*c*) velocity (*d*) Displacement

4. The locus of aparticle moving freedly in a constant gravitational field is a

(*a*) Straight line (*b*) Circle

(*c*) Parabola (*d*) Hyperbola

5. The force arising only due to the acceleration of the frame of reference is

(*a*) Fictitious force (*b*) Impressed force

(*c*) Central force (*d*) None of these

6. A man is standing in an accelerated frame of acceleration $\mathbf{a}_0$. If $\mathbf{a}_0$ is identical with the acceleration due to gravity then he feels

(*a*) Loss in weight (*b*) Increase in weight

(*c*) Weight less ness (*d*) No change in weight

7. If the angular velocity of the frame S′ and the linear velocity of the particle relative to S′ are parallel, Coriolis force is

(*a*) Maximum (*b*) Negative

(*c*) Positive (*d*) Zero

8. Angular momenta are the same about two points O, O′ if O′ O is

(*a*) Parallel to the momentum vector

(*b*) Perpendicular to the momentum vector

(*c*) Inclined to the momentum vector

(*d*) None of these

9. The time rate of change of momentum of a system is equal to

(*a*) Impressed force (*b*) Sum of the external forces

(*c*) Angular momentum (*d*) None of these

ANSWERS

1. (*c*) **2.** (*b*) **3.** (*a*) **4.** (*c*) **5.** (*a*)

6. (*c*) **7.** (*d*) **8.** (*a*) **9.** (*b*)

5

Mechanics II

RIGID BODY

5.1. KINEMATICS OF A RIGID BODY

At any instant of time, each particle of a body has a certain velocity and as such a complete description of the motion of a body requires that we state the velocity of each particle of the body. It will now be seen that this description is possible in a very convenient manner when the body is rigid.

Firstly we see that in the case of a rigid body, the velocities of different particles cannot entirely be independent of one another. Thus if $\mathbf{r}_1$, $\mathbf{r}_2$ be the position vectors of any two particles P_1, P_2 of a rigid system, we have

$$P_1P_2^2 = (\mathbf{r}_1 - \mathbf{r}_2)^2 = \text{constant},$$

so that taking time derivatives, we obtain

$$(\mathbf{r}_1 - \mathbf{r}_2) \cdot \left(\frac{d\mathbf{r}_1}{dt} - \frac{d\mathbf{r}_2}{dt}\right) = 0.$$

This latter equation means that the velocity

$$\frac{d\mathbf{r}_1}{dt} - \frac{d\mathbf{r}_2}{dt}$$

of P_1 relative to P_2 at any instant is perpendicular to the line P_1P_2. This holds good for any pair of particles of a rigid body.

As a result of the limitation of rigidity imposed on a system, it will be possible to show that the velocity of every particle of a rigid body can be expressed in terms of two vectors.

Before taking up the proper investigation, we define and introduce the important concept of *Angular Velocity* in the following sub-section.

5.1.1. Angular Velocity of a rigid body moving about a fixed axis

Let a rigid body be set rotating about a fixed line OA. Let ϕ be the angle at any time t between two planes through OA, one of which is fixed in the body and the other fixed in space. Then $d\phi/dt$ is called the *Angular Speed* of the body about OA. Clearly $d\phi/dt$ is independent of the particular plane fixed in the body or in space.

If $\mathbf{k}$ denote the unit vector along the axis of rotation, then the vector

$$\omega = \mathbf{k}\frac{d\phi}{dt},$$

will be called the Angular velocity of the body.

Velocity of any point of the body in terms of Angular velocity.

Take any point P of the rigid body. Let $\overrightarrow{OP} = \mathbf{r}$.

Draw $PM \perp OA$. Then the point P is describing a circle whose centre is M, radius PM and plane perpendicular to the axis OM. Let $\mathbf{v}$ denote the velocity of P. Then

(*i*) the magnitude of $\mathbf{v}$ is

$$\mathbf{MP}\frac{d\phi}{dt},$$

(*ii*) the direction of $\mathbf{v}$ is normal to the plane OMP,

(*iii*) the sense of $\mathbf{v}$ is such that

$\omega, \mathbf{r}, \mathbf{v}$

form a right handed system.

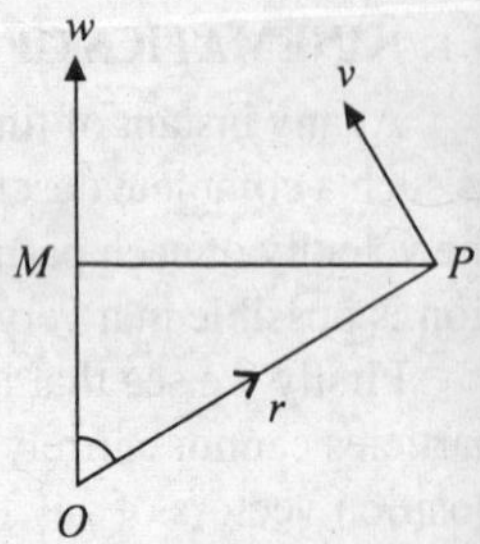

Fig. 5.1

The facts stated above above, $\mathbf{v}$, by (*i*), (*ii*), (*iii*) can be restated by the single vector equation

$$\mathbf{v} = \omega \times \mathbf{r}.$$

We may notice that

$$|\mathbf{v}| = |\omega|\ |\mathbf{r}|\ \sin\theta$$

$$= \frac{d\phi}{dt} OP \sin\theta = MP\frac{d\phi}{dt}.$$

5.1.2. Rigid body moving with one point fixed Theorem

If a rigid body is moving with one point O fixed, then, at any time, there exists a vector, ω, such that the velocity $\mathbf{v}$, *at the time, of any point P of the body is given by*

$$\mathbf{v} = \omega \times \mathbf{r},$$

where, $\mathbf{r}$, *is the position vector $\overrightarrow{OP}$ of P relative to O.*

Let P_1, P_2 be any two particles of the rigid body and let

$$\overrightarrow{OP_1} = \mathbf{r}_1 \quad \overrightarrow{OP_2} = \mathbf{r}_2.$$

Denoting time rates of change by a *dot*, the velocities of the particles are given by

$$\dot{\mathbf{r}}_1,\ \dot{\mathbf{r}}_2.$$

Firstly we shall show that there exists one and only one vector, ω, such that

$$\dot{\mathbf{r}}_1 = \omega \times \mathbf{r}_1,\ \dot{\mathbf{r}}_2 = \omega \times \mathbf{r}_2. \quad ...(1)$$

As O, P_1, P_2 are three particles of a rigid body,

$$OP_1^2 = \mathbf{r}_1^2, \; OP_2^2, = \mathbf{r}_2^2, \; P_1P_2^2 = (\mathbf{r}_2 - \mathbf{r}_1)^2,$$

are constants so that their time derivatives are zero. Thus

$$\mathbf{r}_1 . \dot{\mathbf{r}}_1 = 0, \; \mathbf{r}_2 . \dot{\mathbf{r}}_2 = 0, \; (\mathbf{r}_2 - \mathbf{r}_1) . (\dot{\mathbf{r}}_2 - \dot{\mathbf{r}}_1) = 0.$$

These conditions of rigidity are equivalent to

$$\mathbf{r}_1 . \dot{\mathbf{r}}_1 = 0, \; \mathbf{r}_2 . \dot{\mathbf{r}}_2 = 0, \; \mathbf{r}_1 . \dot{\mathbf{r}}_2 + \mathbf{r}_2 . \dot{\mathbf{r}}_1 = 0. \qquad ...(2)$$

We see from (1) that, ω, if it exists, must be perpendicular to both $\dot{\mathbf{r}}_1$ and $\dot{\mathbf{r}}_2$ and accordingly it must be of the form

$$k\dot{\mathbf{r}}_1 \times \dot{\mathbf{r}}_2, \qquad ...(3)$$

where, k, is some scalar. Suppose that

$$\omega = k\dot{\mathbf{r}}_1 \times \dot{\mathbf{r}}_2,$$

satisfies first of the two equations in (1). Therefore

$$\begin{aligned} \dot{\mathbf{r}}_1 &= k(\dot{\mathbf{r}}_1 \times \dot{\mathbf{r}}_2) \times \mathbf{r}_1 \\ &= k[(\dot{\mathbf{r}}_1 . \dot{\mathbf{r}}_1)\, \dot{\mathbf{r}}_2 - (\mathbf{r}_1 . \dot{\mathbf{r}}_2)\dot{\mathbf{r}}_1] \\ &= -k\,(\mathbf{r}_1 . \dot{\mathbf{r}}_2)\, \dot{\mathbf{r}}_1, \qquad \text{[By (2)]} \end{aligned}$$

or

$$k = -\frac{1}{\mathbf{r}_1 . \dot{\mathbf{r}}_2}. \qquad ...(4)$$

Similarly we may see that $\omega = k\, \dot{\mathbf{r}}_1 \times \dot{\mathbf{r}}_2$ will satisfy second of the two equations in (1) if

$$k = \frac{1}{\mathbf{r}_2 . \dot{\mathbf{r}}_1}. \qquad ...(5)$$

But the third of the conditions of rigidity in (2), we see that

$$-\frac{1}{\mathbf{r}_1 . \dot{\mathbf{r}}_2} = \frac{1}{\mathbf{r}_2 . \dot{\mathbf{r}}_1}.$$

Thus we see that for

$$\omega = k\dot{\mathbf{r}}_1 \times \dot{\mathbf{r}}_2 \qquad ...(6)$$

where

$$k = -\frac{1}{\mathbf{r}_1 . \dot{\mathbf{r}}_2} = \frac{1}{\mathbf{r}_2 . \dot{\mathbf{r}}_1},$$

we have

$$\dot{\mathbf{r}}_1 = \omega \times \mathbf{r}_1 \text{ and } \dot{\mathbf{r}}_2 = \omega \times \mathbf{r}_2.$$

As *second* step, we shall show that ω, as determined above, is such that

$$\dot{\mathbf{r}} = \omega \times \mathbf{r}, \qquad ...(7)$$

where $\mathbf{r}$ is the position vector of any particle P of the system. Consider P, we have the following additional conditions of rigidity

$$\mathbf{r}.\dot{\mathbf{r}} = 0, \mathbf{r}.\dot{\mathbf{r}}_1 + \dot{\mathbf{r}}_1.\mathbf{r} = 0, \mathbf{r}.\dot{\mathbf{r}}_2 + \dot{\mathbf{r}}.\mathbf{r}_2 = 0. \qquad ...(8)$$

Now we have

$$(\omega \times \mathbf{r} - \dot{\mathbf{r}}) . \dot{\mathbf{r}}_1 = 0, \qquad \text{By (8)}$$

$$(\omega \times \mathbf{r} - \dot{\mathbf{r}}) . \dot{\mathbf{r}}_1 = \omega \times \mathbf{r}.\dot{\mathbf{r}}_1 - \dot{\mathbf{r}}.\dot{\mathbf{r}}_1$$

$$= -\omega \times \mathbf{r}_1.\mathbf{r} - \dot{\mathbf{r}}.\mathbf{r}_1$$

$$= -\dot{\mathbf{r}}_1.\mathbf{r} - \mathbf{r}.\dot{\mathbf{r}}_1 \qquad \text{By (8)}$$

$$= 0.$$

Similarly

$$(\omega \times \mathbf{r} - \dot{\mathbf{r}}).\dot{\mathbf{r}}_2 = 0$$

Thus, unless **r** is coplanar with $\mathbf{r}_1$ and $\mathbf{r}_2$, we have

$$\omega \times \mathbf{r} - \dot{\mathbf{r}} = \mathbf{0}, \text{ i.e., } \omega \times \mathbf{r} = \dot{\mathbf{r}}.$$

Suppose now that **r** is coplanar with $\mathbf{r}_1$ and $\mathbf{r}_2$ so that we have a relation

$$\mathbf{r} = p_1\mathbf{r}_1 + p_2\mathbf{r}_2 \qquad ...(9)$$

where p_1, p_2 are two scalars. In view of the condition of rigidity, the scalars p_1 and p_2 do not change with time so that, we obtain from (9),

$$\dot{\mathbf{r}} = p_1\dot{\mathbf{r}}_2 + p_2\dot{\mathbf{r}}_2.$$

$$\therefore \quad \omega \times \mathbf{r} = \omega \times (p_1\mathbf{r}_1 + p_2\mathbf{r}_2)$$

$$= p_1\omega \times \mathbf{r}_1 + p_2\omega \times \mathbf{r}_2$$

$$= p_1\dot{\mathbf{r}}_1 + p_2\dot{\mathbf{r}}_2 = \dot{\mathbf{r}}.$$

Thus we have shown that there exists a vector ω such that

$$\dot{\mathbf{r}} = \omega \times \mathbf{r},$$

for every particle of the rigid body.

Def. *The vector, ω, is called the angular velocity. The line through O parallel to the vector, ω, is called the* **Axis of Rotation.**

It should be carefully understood that in the case of motion about a fixed point the vector, ω, describes only the *instantaneous* state of motion of the body and it may be different at different times. The velocity of every point on the axis of motion is zero at the instant under consideration.

If, ω, is constant then we have a *permanent* axis of rotation [Refer § 5,1] as against an *instantaneous* axis of rotation if ω, varies with time.

Cor. If $\mathbf{r}_1$, $\mathbf{r}_2$ be the position vectors of any two points of a rigid body with reference to the fixed point O thereof, then the angular velocity of the body may be written as

$$\frac{\dot{\mathbf{r}}_1 \times \dot{\mathbf{r}}_2}{\dot{\mathbf{r}}_1 . \dot{\mathbf{r}}_2}$$

5.1.3. Motion of a rigid body in the General Case.

Theorem. *If a rigid body is moving in any manner and A be any point of the body, then there exists, at any instant, a vector ω such that the velocity of any particle P of the body at the instant is*

$$\mathbf{u} = \omega \times \mathbf{r},$$

where **u** *is the velocity of A and* **r** *the position vector* $\overrightarrow{AP}$ *of* P *relative to A. Also ω is independent of the choice of the point A.*

Let, O, be the origin of the reference system. Consider any two points P_1, P_2, of the rigid body and let P be any arbitrary point.

Let $\overrightarrow{OA} = \mathbf{a}$, $\overrightarrow{OP_1} = \rho_1$,

$\overrightarrow{OP_2} = \rho_2$, $\overrightarrow{OP} = \rho$,

$\overrightarrow{AP_1} = \mathbf{r}_1$, $\overrightarrow{AP_2} = \mathbf{r}_2$,

$\overrightarrow{AP} = \mathbf{r}$.

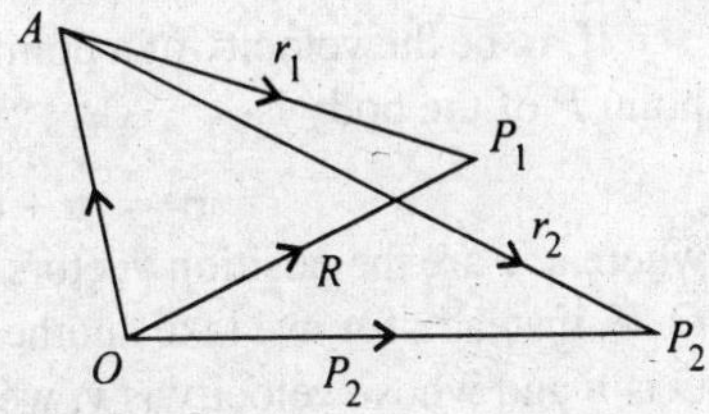

Fig. 5.2

Now, $AP_1^{\,2} = \mathbf{r}_1^{\,2}$, $AP_2^{\,2}$,

$P_1P_2^{\,2} = (\mathbf{r}_2 - \mathbf{r}_1)^2$

are constants so that, as in the preceding § 5·1·2,

$$\mathbf{r}_1.\dot{\mathbf{r}}_2 = 0, \quad \mathbf{r}_2.\dot{\mathbf{r}}_2 = 0, \ \mathbf{r}_1.\dot{\mathbf{r}}_2 + \mathbf{r}_2.\dot{\mathbf{r}}_1 = 0.$$

Thus, as in § 5.1.2, there exists a vector

$$\omega = -\frac{\dot{\mathbf{r}}_1 \times \dot{\mathbf{r}}_2}{\dot{\mathbf{r}}_1 . \mathbf{r}_2}$$

such that $\dot{\mathbf{r}}_1 = \omega \times \mathbf{r}_1, \ \dot{\mathbf{r}}_2 = \omega \times \mathbf{r}_2.$

Also the vector ω is such that

$$\dot{\mathbf{r}} = \omega \times \mathbf{r}. \qquad ...(1)$$

Now $\mathbf{r} = \overrightarrow{AP} = \overrightarrow{OP} - \overrightarrow{OA} = \rho - \mathbf{a}$.

$\therefore$ $\dot{\mathbf{r}} = \dot{\rho} - \dot{\mathbf{a}} = \dot{\rho} - \mathbf{u};$

$\mathbf{u}$ being the velocity of **A**. Substituting in (1), we obtain

$$\rho - \dot{\mathbf{u}} = \omega \times (\rho - \mathbf{a})$$

or $\dot{\rho} = \mathbf{u} + \omega \times \mathbf{r}.$

Hence the first part of the theorem.

Suppose that B is another particle of the rigid body with position vector $\mathbf{b}$ relative to the origin O. Let, if possible, ω, ω' be two different vectors such that

$$\dot{\rho} = \dot{\mathbf{a}} + \omega \times \overrightarrow{AP}, \qquad ...(2)$$

$$\dot{\rho} = \dot{\mathbf{b}} + \omega' \times \overrightarrow{BP}. \qquad ...(3)$$

Applying (3) to the point A, we have

$$\dot{\mathbf{a}} = \dot{\mathbf{b}} + \omega' \times \overrightarrow{BA}. \qquad ...(4)$$

From (2), (3), (4), eliminating $\dot{\mathbf{a}}$, $\dot{\mathbf{b}}$, $\dot{\rho}$, we obtain

$$\omega \times \overrightarrow{AP} = \omega' \times (\overrightarrow{BP} - \overrightarrow{BA}) = \omega' \times \overrightarrow{AP}$$

or $(\omega - \omega') \times \overrightarrow{AP} = \mathbf{0}.$

As P is any arbitrary point of the rigid body, we deduce that

$$\omega - \omega' = \mathbf{0}, \ \textit{i.e.}, \ \omega = \omega'.$$

Thus ω is independent of the choice of the point A.

Cor. If P, Q be any two points of a rigid body, then the velocity of Q relative to P

$$= \omega \times \overrightarrow{PQ}.$$

5.1.4. Reduction to Screw Motion

If, **u**, be the velocity of a point A of a rigid body, then the velocity of any point P of the body is

$$\dot{\mathbf{r}} = \mathbf{u} + \omega \times \overrightarrow{AP} = \mathbf{u} + \omega \times (\mathbf{r} - \mathbf{a}), \qquad ...(1)$$

where **a**, **r** are the position vectors of the points A, P relative to a fixed origin O. If, instead of A, we take another point B whose position vector relative to O is **b** and whose velocity is **v**, we have also

$$\dot{\mathbf{r}} = \mathbf{v} + \omega \times (\mathbf{r} - \mathbf{b}). \qquad ...(2)$$

If will now be shown that it is possible to choose points B such that *the velocity* **v** *of B is parallel to* ω, *i.e.*,

$$\mathbf{v} \times \omega = \mathbf{0}. \qquad ...(3)$$

Suppose that B is such a point.

Applying (1) to the particle B, we have

$$\mathbf{v} = \mathbf{u} + \omega \times (\mathbf{b} - \mathbf{a}). \qquad ...(4)$$

Multiplying (4) vectorially with ω, we get

$$\mathbf{0} = \mathbf{u} \times \omega + \omega^2 (\mathbf{b} - \mathbf{a}) - [(\mathbf{b} - \mathbf{a}).\, \omega]\, \omega$$

or

$$\mathbf{b} = \mathbf{a} - \frac{\mathbf{u} \times \omega}{\omega^2} + k\omega, \qquad ...(5)$$

where $k = (\mathbf{b} - \mathbf{a}).\, \omega/\omega^2$ is a scalar.

Substituting the value of **b** from (5) in (4), we obtain

$$\mathbf{v} = \mathbf{u} + \omega \times \left[-\frac{\mathbf{u} \times \omega}{\omega^2} + k\omega \right] = \frac{\omega\,.\,\mathbf{u}}{\omega^2}\, \omega$$

so that **v** is parallel to ω.

The locus of points, **b**, given by (5) for different values of k is the straight line

$$\mathbf{r} = \left[\mathbf{a} - \frac{\mathbf{u} \times \omega}{\omega^2} \right] + k\omega;\ k, \text{ a parameter}$$

where is parallel to ω. The velocity **v** of any point on this straight line is

$$\frac{\omega\,.\,\mathbf{u}}{\omega^2}\, \omega.$$

Thus we have shown that there exists at any instant a line parallel to the vector ω such that the velocity of every point of the line is parallel to ω; *every point of this line is moving along the line while the body is rotating about the line.*

We say that the *instantaneous motion of a rigid body is equivalent to a Screw Motion.*

5.2. KINETIC ENERGY OF A RIGID BODY MOVING WITH ONE POINT FIXED

Let O be the fixed point and ω the angular velocity at any time. Then the velocity of a particle P

$$= \omega\, \mathbf{r} \text{ where } \mathbf{r} = \overrightarrow{OP}.$$

If m be the mass of this particle, the kinetic energy T of the system is given by

$$T = \Sigma \frac{1}{2} m\,(\omega \times \mathbf{r})^2,$$

where the summation extends to all the particles of the system. Now

$$\begin{aligned}(\omega \times \mathbf{r}) \,.\, (\omega \times \mathbf{r}) &= (\omega \times \mathbf{r}) \times \omega \,.\, \mathbf{r}\\ &= [\omega^2\mathbf{r} - (\omega \,.\, \mathbf{r})\,\omega] \,.\, \mathbf{r}\\ &= \omega^2\mathbf{r}^2 - (\omega \,.\, \mathbf{r})^2.\end{aligned}$$

Thus $$T = \Sigma \frac{1}{2} m\,[\omega^2\,\mathbf{r}^2 - (\omega \,.\, \mathbf{r})^2].$$

We write

$$\mathbf{r} = \mathbf{i}x + \mathbf{j}y + \mathbf{k}z,$$
$$\omega = \mathbf{i}\omega_1 + \mathbf{j}\omega_2 + \mathbf{k}\omega_3.$$

$$\begin{aligned}\therefore \quad T &= \Sigma \frac{1}{2} m\,[(\omega_1^{\,2} + \omega_2^{\,2} + \omega_3^{\,2})\,(x^2 + y^2 + z^2) - (\omega_1 x + \omega_2 y + \omega_2 z)^2]\\ &= A\omega_1^{\,2} + B\omega_2^{\,2} + C\omega_2^{\,2} - 2\omega_1\omega_2 F - 2\omega_2\omega_3\,D - 2\omega_3\omega_1\,E,\end{aligned}$$

where
$$A = \Sigma m(y^2 + z^2),\ B = \Sigma m(z^2 + x^2),\ C = \Sigma m(x^2 + y^2)$$
$$D = \Sigma myz,\ E = \Sigma mzx,\ F = \Sigma mxy.$$

5.3. ANGULAR MOMENTUM ABOUT THE FIXED POINT OF A RIGID BODY

If $\mathbf{H}$ denotes the angular momentum about the fixed point O, we have

$$\begin{aligned}\mathbf{H} &= \Sigma m\mathbf{r} \times (\omega \times \mathbf{r})\\ &= \Sigma m\,[\mathbf{r}^2\omega - (\mathbf{r} \,.\, \omega)\mathbf{r}]\\ &= (\Sigma m\mathbf{r}^2)\omega - \Sigma m\,(\mathbf{r} \,.\, \omega)\,\mathbf{r}\\ &= (A\omega_1 - B\omega_2 - C\omega_3)\mathbf{i} + (-D\omega_1 + E\omega_2 - F\omega_2)\mathbf{j} +\\ &\qquad (-A\omega_1 + B\omega_2 + C\omega_3)\mathbf{k},\end{aligned}$$

in terms of the notation of the preceding section.

5.4. COMPARISON OF THE RATES OF CHANGE OF A VECTOR RELATIVE TO TWO RELATIVELY MOVING SYSTEMS OF AXES

We shall now take up the problem referred to in Note 2 after § 4.1.2.

Let $OX, OY, OZ;\ O'X', O'Y', O'Z'$ be two systems of axes thought of as rigid systems and moving relative to each other.

We shall call (or regard) the system

$$OX, OY, OZ$$

as fixed. The motion of the system

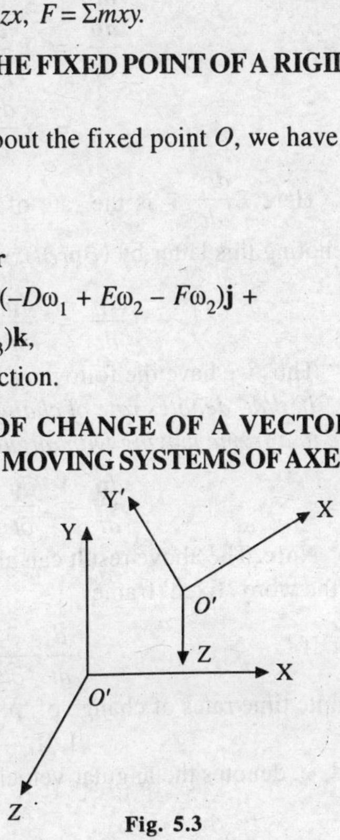

Fig. 5.3

$O'X', O'Y', O'Z'$

[to be called moving system] is given by the velocity $\mathbf{u}$ of O' and an angular velocity ω about an axis through O'.

Let $\mathbf{i}', \mathbf{j}', \mathbf{k}'$ denote unit vectors along $O'X', O'Y', O'Z'$ respectively so that they are functions of time. Then if p', q', r' be the components of a variable vector $\mathbf{p}$ at time t relative to the position of the moving axes at the time, we have

$$\mathbf{p} = p'\mathbf{i}' + q'\mathbf{j}' + r'\mathbf{k}'.$$

$\therefore$ the time rate of change of $\mathbf{p}$ (relative to the fixed axes OX, OY, OZ) is given by

$$\frac{d\mathbf{p}}{dt} = \Sigma \frac{dp'}{dt}\mathbf{i}' + \Sigma p' \frac{d\mathbf{i}'}{dt}.$$

As $\mathbf{i}'$ is the position vector of some point A' on $O'X'$ at unit distance from O', $d\mathbf{i}'/dt$ is the velocity of A' relatively to O' and accordingly, by cor. to § 5.1.3.

$$\frac{d\mathbf{i}'}{dt} = \omega \times \mathbf{i}'.$$

Similarly $$\frac{d\mathbf{j}'}{dt} = \omega \times \mathbf{j}', \quad \frac{d\mathbf{k}'}{dt} = \omega \times \mathbf{k}'.$$

$$\therefore \quad \frac{d\mathbf{p}}{dt} = \Sigma \frac{dp'}{dt}\mathbf{i}' + \Sigma p'\, \omega \times \mathbf{i}'$$

$$= \Sigma \frac{dp'}{dt}\mathbf{i}' + \omega \times \Sigma p'\,\mathbf{i}'. \quad \text{...(1)}$$

Here $\Sigma \frac{dp'}{dt}\mathbf{i}'$ is the rate of change of $\mathbf{p}$ relative to the moving axes. Denoting this latter by $(\partial \mathbf{p}/\partial t)$, we re-write (1) as

$$\frac{d\mathbf{p}}{dt} = \frac{d\mathbf{p}}{\partial t} + \omega \times \mathbf{p}.$$

Thus we have the following important results:

If, d/dt, denotes rate of change relative to a fixed frame and, $\partial/\partial t$, relative to a frame moving with angular velocity ω, then for any vector $\mathbf{p}$.

$$\frac{d\mathbf{p}}{dt} = \frac{\partial \mathbf{p}}{\partial t} + \omega \times \mathbf{p}.$$

Note. The above result can also be restated in another form without use of the word 'fixed' frame.

If $$\frac{d}{dt}, \frac{\partial}{\partial t}$$

denote time rates of change of, $\mathbf{p}$, relative to two reference frames

I, II,

and, ω, denotes the angular velocity of II relative to I, then

$$\frac{d\mathbf{p}}{dt} = \frac{\partial \mathbf{p}}{\partial t} + \omega \times \mathbf{p}.$$

We may also re-write it as

$$\frac{\partial \mathbf{p}}{\partial t} = \frac{d\mathbf{p}}{dt} + (-\omega) \times \mathbf{p},$$

where, now $-\omega$, denotes the angular velocity of I relative to II.

Cor. $$\frac{d\omega}{dt} = \frac{\partial \omega}{\partial t}.$$

In fact replacing **p** by ω, we have

$$\frac{d\omega}{dt} = \frac{\partial \omega}{\partial t} + \omega \times \omega = \frac{\partial \omega}{\partial t}.$$

5.5. COMPARISON OF VELOCITIES AND ACCELERATIONS OF A PARTICLE RELATIVE TO TWO RELATIVELY MOVING SYSTEMS OF AXES

We adopt the notation of the preceding section.

Lct P be a moving particle and let

$$\overrightarrow{OO'} = \mathbf{b}, \ \overrightarrow{OP} = \mathbf{r}, \ \overrightarrow{O'P} = \mathbf{r},$$

so that $$\mathbf{r} = \mathbf{b} + \mathbf{r}'.$$

Let x', y', z' be the co-ordinates of P relative to $O'x'$ $O'y'$, $O'z'$.

We then have

$$\mathbf{r}' = x'\mathbf{i}' + y'\mathbf{j}' + z'\mathbf{k}',$$

so that $$\mathbf{r} = \mathbf{b} + x'\mathbf{i}' + y'\mathbf{j}' + z'\mathbf{k}'.$$

$$\therefore \quad \frac{d\mathbf{r}}{dt} = \frac{d\mathbf{b}}{dt} + \Sigma\frac{dx'}{dt}\mathbf{i}' + \Sigma x'\frac{d\mathbf{i}'}{dt}$$

$$= \mathbf{u} + \Sigma\frac{dx'}{dt}\mathbf{i}' + \Sigma x'\,\omega \times \mathbf{i}'$$

$$= \mathbf{u} + \Sigma\frac{dx'}{dt}\mathbf{i}' + \omega \times \mathbf{r}'$$

or $$\frac{d\mathbf{r}}{dt} = \Sigma\frac{dx'}{dt}\mathbf{i}' + \mathbf{u} + \omega \times \mathbf{r}'. \qquad ...(1)$$

Here $$\Sigma\frac{dx'}{dt}\mathbf{i}'$$

is the velocity of P relative to the system $O'x'$, $O'y'$, $O'z'$. Denoting it by $\partial\mathbf{r}'/\partial t$, we re-write (1) as

$$\frac{d\mathbf{r}}{dt} = \frac{\partial \mathbf{r}'}{\partial t} + \mathbf{u} + \omega \times \mathbf{r}'. \qquad ...(2)$$

The equation (2) connects the velocities $d\mathbf{r}/dt$ and $\partial\mathbf{r}'/\partial t$ of P relative to the systems of axes.

From (2), we now have

$$\frac{d^2\mathbf{r}}{dt^2} = \frac{d}{dt}\left(\frac{\partial \mathbf{r}'}{\partial t}\right) + \frac{d\mathbf{u}}{dt} + \frac{d}{dt}(\omega \times \mathbf{r}'). \quad \text{...(3)}$$

By § 5·4,

$$\frac{d}{dt}\left(\frac{\partial \mathbf{r}'}{\partial t}\right) = \frac{\partial}{\partial t}\left(\frac{\partial \mathbf{r}'}{\partial t}\right) + \omega \times \frac{\partial \mathbf{r}'}{\partial t} = \frac{\partial^2 \mathbf{r}'}{\partial t^2} + \omega \times \frac{\partial \mathbf{r}'}{\partial t} \quad \text{...(4)}$$

$$\frac{d}{dt}(\omega \times \mathbf{r}') = \frac{\partial}{\partial t}(\omega \times \mathbf{r}') + \omega \times (\omega \times \mathbf{r}')$$

$$= \frac{\partial \omega}{\partial t} \times \mathbf{r}' + \omega \times \frac{\partial \mathbf{r}'}{\partial t} + \omega \times (\omega \times \mathbf{r}')$$

$$= \frac{d\omega}{dt} \times \mathbf{r}' + \omega \times \frac{\partial \mathbf{r}'}{\partial t} + \omega \times (\omega \times \mathbf{r}'). \quad \text{...(5)}$$

From (3), (4) and (5), we have

$$\frac{d^2\mathbf{r}}{dt^2} = \frac{\partial^2 \mathbf{r}'}{\partial t^2} + \frac{d\mathbf{u}}{dt} + 2\omega \times \frac{\partial \mathbf{r}'}{\partial t} + \frac{\partial \omega}{\partial t} \times \mathbf{r}' + \omega \times (\omega \times \mathbf{r}') \quad \text{...(6)}$$

Since $\dfrac{\partial \omega}{\partial t} = \dfrac{\partial \omega}{\partial t}$.

The equation (6) connects the accelerations $d^2\mathbf{r}/dt^2$ and $\partial^2\mathbf{r}'/\partial t^2$ relative to the two system of axes.

5.6. EQUATIONS OF MOTION OF A RIGID BODY

We have seen that for *any* system of particles,

$$\frac{d\mathbf{M}}{dt} = \mathbf{F}, \frac{d\mathbf{H}}{dt} = \mathbf{L} \quad \text{...(1)}$$

where **M** is the momentum and **H** angular momentum of the system relative to the origin of a fixed reference system and d/dt denotes time rate of change relative to the same. Also F is the sum of the external forces and **L** the sum of the moments of the same about the origin. We suppose that the external forces are known.

As a particular case the equations (1) hold for a rigid body also.

5.6.1. If, **R**, denotes the position vector of the centroid and **M** the mass of the body, we can re-write $d\mathbf{M}/dt = \mathbf{F}$ as

$$M\frac{d^2\mathbf{R}}{dt^2} = \mathbf{F}. \quad \text{...(2)}$$

If $\mathbf{R} = \mathbf{i}x + \mathbf{j}y + \mathbf{k}z, \ \mathbf{F} = \mathbf{i}F_1 + \mathbf{j}F_2 + \mathbf{k}F_3,$

then (2) is equivalent to the three scalar equations

$$M\frac{d^2 x}{dt^2} = F_1, M\frac{d^2 y}{dt^2} = F_2, M\frac{d^2 z}{dt^2} = F_3, \quad \text{...(3)}$$

which are ordinary differential equations of second order. The general solution of the differential equations (3) for x, y, z, will involve *six* arbitrary constants whose values can be determined if the position and the velocity of the centroid at any given time be known. Thus we determine the motion of centroid.

We have seen (§ 4.15.3) that the equation

$$\frac{d\mathbf{H}}{dt} = \mathbf{L},$$

remains true when **H** is the moment about the centroid of the momentum relative to the centroid and **L** the sum of the moments of the external forces about the centroid; d/dt denoting time rate of change with respect to axes through the centroid parallel to the fixed axes. Thus we have the same equation as when the centroid is a fixed point. Without any loss of generality we may, therefore, consider the case of motion of a rigid body about a fixed point.

We shall now obtain the well-known *Euler's equations* of motion for motion of a rigid body about a fixed point.

5.6.2. Euler's Equations of Motion

Consider a rigid body moving about a fixed point O thereof. Let OX, OY, OZ denote principal axes of the body through O. The system of co-ordinate axes formed by these is fixed relative to the body. Let A, B, C denote principal moments of inertia about OX, OY, OZ. These will not change with time. Also the products of inertia relative to these axes in pairs are also zero.

Let $\mathbf{i}', \mathbf{j}', \mathbf{k}'$, denote unit vectors along OX, OY, OZ. Let at any time t, ω denote the angular velocity about O.

We write $\qquad \omega = \omega_1 \mathbf{i}' + \omega_2\mathbf{j}' + \omega_3\mathbf{k}'.$

Also, $\qquad \mathbf{H} = A\omega_1\mathbf{i}' + B\omega_2\mathbf{j}' + C\omega_3\mathbf{k}'.$

We write, **L**, the sum of the moments of the external forces about O as

$$\mathbf{L} = L_1\mathbf{i}' + L_2\mathbf{j}' + L_3\mathbf{k}'.$$

By § 5.4,

$$\frac{d\mathbf{H}}{dt} = \frac{\partial \mathbf{H}}{dt} + \omega \times \mathbf{H},$$

where d/dt denotes time rate of change relative to axes fixed in space and $\partial/\partial t$ relative to axes OX, OY, OZ fixed in the body.

We have

$$\mathbf{L} = \frac{d\mathbf{H}}{dt} = \frac{\partial \mathbf{H}}{dt} + \omega \times \mathbf{H}.$$

$$\therefore \quad (L_1\mathbf{i}' + L_2\mathbf{j}' + L_3\mathbf{k}') = \frac{\partial}{\partial t}(A\omega_1\mathbf{i}' + B\omega_2\mathbf{j}' + C\omega_3\mathbf{k}') +$$

$$(\omega_1\mathbf{i}' + \omega_2\mathbf{j}' + \omega_3\mathbf{k}') \times (A\omega_1\mathbf{i}' + B\omega_2\mathbf{j}' + C\omega_3\mathbf{k}') +$$

$$= (A\omega_1\mathbf{i}' + B\omega_2\mathbf{J}' + C\omega_3\mathbf{k}') +$$

$$(C-B)\,\omega_2\omega_3\mathbf{i}' + (A-C)\,\omega_3\omega_1\mathbf{j}' + (B-A)\,\omega_1\omega_2\mathbf{k}',$$

where $\dot{\omega}_1 = \dfrac{\partial \omega_1}{\partial t}$, etc. Here *A, B, C*, are constants.

Thus we have

$$A\dot{\omega}_1 + (C-B)\,\omega_2\omega_3 = L_1,$$
$$B\dot{\omega}_2 + (A-C)\,\omega_3\omega_1 = L_2,$$
$$C\dot{\omega}_3 + (B-A)\,\omega_1\omega_2 = L_3,$$

which are the well known equations of motion due to *Eular.*

EXAMPLES

1. *Find the acceleration of a particle of a circular wheel rolling with constant speed along a straight level track.*

Taking A the centre of the wheel as the base point, let its velocity be **V** which will be a constant vector, lying horizontally in the plane of the wheel. Let $\vec{\omega}$ be the angular velocity of the wheel.

Consider a point P of the wheel distance **r** from A, i.e., $\overrightarrow{AP} = \mathbf{r}$. Then, the velocity of P is given by

$$\mathbf{v} = \mathbf{V} + \vec{\omega} \times \mathbf{r}$$

The acceleration **a** of the particle P is

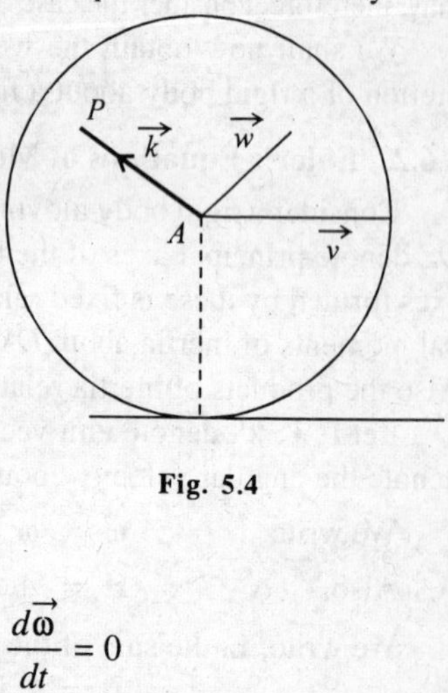

Fig. 5.4

$$\mathbf{a} = \frac{d\mathbf{v}}{dt} = \frac{d\mathbf{V}}{dt} + \frac{d\vec{\omega}}{dt} \times \mathbf{r} + \vec{\omega} \times \frac{d\mathbf{r}}{dt}$$

where **V** and $\vec{\omega}$ being constant, $\dfrac{d\mathbf{V}}{dt} = 0,\ \dfrac{d\vec{\omega}}{dt} = 0$

$$\therefore \quad \mathbf{a} = \vec{\omega} \times \frac{d\mathbf{r}}{dt}$$

$$= \vec{\omega} \times (\vec{\omega} \times \mathbf{r}) \qquad \left[\because \ \frac{d\mathbf{r}}{dt} = \vec{\omega} \times \mathbf{r}\right]$$

$$= (\vec{\omega}\,.\,\mathbf{r})\,\vec{\omega} - (\vec{\omega}\,.\,\vec{\omega})\,\mathbf{r}.$$

But **r** lies in the plane of the wheel, while $\vec{\omega}$ lies in a plane perpendicular to the wheel. Therefore, $\vec{\omega}$ and **r** are mutually at right angles, *i.e.,*

$$\vec{\omega}\,.\,\mathbf{r} = 0$$

$$\therefore \quad \mathbf{a} = -(\vec{\omega}\,.\,\vec{\omega})\,\mathbf{r}$$
$$= -\mathbf{r}\,\omega^2.$$

2. *A rigid body is moving with one point fixed under the action of no* *external forces; prove that*

$$A\omega_1^2 + B\omega_2^2 + C\omega_3^2 = \text{constant};$$
$$A^2\omega_1^2 + B^2\omega_2^2 + C^2\omega_3^2 = \text{constant};$$

where A, B, C are the principal moments of inertia at the fixed point and ω_1, ω_2, ω_3 *are the components of the angular velocity at any time along the directions of the principal axes at the fixed point at the time.*

The equations of motion are

$$A\frac{d\omega_1}{dt} - (B - C)\,\omega_2\omega_3 = 0, \quad ...(1)$$

$$B\frac{d\omega_2}{dt} - (C - B)\,\omega_3\omega_1 = 0, \quad ...(2)$$

$$C\frac{d\omega_3}{dt} - (A - B)\,\omega_1\omega_2 = 0. \quad ...(3)$$

Multiplying (1), (2) and (3) by ω_1, ω_2, ω_3 respectively and adding, we get

$$A\omega_1\frac{d\omega_1}{dt} + B\omega_2\frac{d\omega_2}{dt} + C\omega_3\frac{d\omega_3}{dt} = 0.$$

Integrating, we obtain

$$A\omega_1^2 + B\omega_2^2 + C\omega_3^2 = \text{constant}.$$

Again multiplying (1), (2), (3) by $A\omega_1$, $B\omega_2$, $C\omega_3$ respectively, adding and integrating, we shall obtain

$$A^2\omega_1^2 + B^2\omega_2^2 + C^2\omega_3^2 = \text{constant}.$$

Hence the results.

3. *Prove that for a rigid body moving about a fixed point.*

$$\frac{dT}{dt} = \omega.\,\mathbf{L},$$

(with usual notation) and deduce that the kinetic energy is constant if the body is subjected to forces passing through the fixed point only or if **L** *is perpendicular to* ω.

We have
$$T = \Sigma\frac{1}{2}m\,(\omega \times \mathbf{r})^2.$$

$$\therefore \quad \frac{dT}{dt} = \Sigma m(\omega \times \mathbf{r})\,.\,(\dot{\omega} \times \mathbf{r} + \omega \times \dot{\mathbf{r}})$$
$$= \Sigma m(\omega \times \mathbf{r})\,.\,(\omega \times \mathbf{r}) + \Sigma m(\omega \times \mathbf{r})\,.\,\omega \times (\omega \times \mathbf{r})$$
$$= \Sigma m(\omega \times \mathbf{r})\,.\,(\dot{\omega} \times \mathbf{r}),$$

for the scalar triple produce

$$(\omega \times \mathbf{r})\,.\,\omega \times (\omega \times \mathbf{r}) \text{ is zero.}$$

Again

$$\mathbf{H} = \Sigma m\mathbf{r} \times (\omega \times \mathbf{r}).$$

$$\therefore \quad L = \frac{d\mathbf{H}}{dt} = \Sigma m\mathbf{r} \times (\dot{\omega} \times \mathbf{r} + \omega \times \dot{\mathbf{r}}) + \Sigma m\dot{\mathbf{r}} \times (\omega \times \mathbf{r})$$

$$= \Sigma m\mathbf{r} \times (\dot{\omega} \times \mathbf{r} + \omega \times \dot{\mathbf{r}}) + \Sigma m(\omega \times \mathbf{r}) \times (\omega \times \mathbf{r})$$

$$= \Sigma m\mathbf{r} \times (\dot{\omega} \times \mathbf{r}) + \Sigma m\mathbf{r} \times (\omega \times \dot{\mathbf{r}})$$

$$\therefore \quad \mathbf{L} \,.\, \omega = \Sigma m\mathbf{r} \times (\dot{\omega} \times \mathbf{r}) \,.\, \omega + \Sigma m\mathbf{r} \times (\omega \times \dot{\mathbf{r}}) \,.\, \omega$$

Now $\quad \mathbf{r} \times (\dot{\omega} \times \dot{\mathbf{r}}) \,.\, \omega = (\omega \times \mathbf{r}) \,.\, (\omega \times \mathbf{r})$

$$= (\omega \times \mathbf{r}) \,.\, [\omega \times (\omega \times \mathbf{r})]$$

$$= 0.$$

$$\therefore \quad \mathbf{L} \,.\, \omega = \Sigma m(\omega \times \mathbf{r}) \,.\, (\dot{\omega} \times \mathbf{r})$$

$$= \frac{dT}{dt}.$$

The second part now follows at ones.

4. *A rigid body symmetric about an axis has one point fixed on this axis. Discuss the rotational motion of the body when there is no other force acting except the reaction force at the fixed point.*

Taking **i**, **j**, **k** as unit vectors along the principal axes, which may be taken coinciding with the axes of x, y, z. If axis **k** be the axis of symmetry, then A = B, so that Eular's equations of § 5.6.2 become

$$A\dot{\omega}_1 - (A - C)\omega_2\omega_3 = 0 \qquad \text{...(1)}$$

$$A\dot{\omega}_2 - (C - A)\omega_3\omega_1 = 0 \qquad \text{...(2)}$$

$$C\,\dot{\omega}_3 = 0 \qquad \text{...(3)}$$

Equation (3) gives ω_3 = constant = δ (say) and then (1) and (2) become

$$\dot{\omega}_1 + \left(\frac{C-A}{A}\right)\delta \,.\, \omega_2 = 0 \qquad \text{...(4)}$$

$$\dot{\omega}_2 + \left(\frac{A-C}{A}\right)\delta \,.\, \omega_1 = 0 \qquad \text{...(5)}$$

Differentiating (5) w.r.t. t, we get

$$\ddot{\omega}_2 = \frac{A-C}{A}\delta \,.\, \dot{\omega}_1 = 0$$

or

$$\ddot{\omega}_2 + \left(\frac{A-C}{A}\right)^2 \delta^2 \,.\, \dot{\omega}_2 = 0 \qquad \text{by (4)}$$

Putting $\left(\frac{A-C}{A}\right)^2 \delta^2 = \alpha$, we get

$$\ddot{\omega}_2 + \alpha^2\omega_2 = 0$$

Its solution is $\omega_2 = \beta \cos \alpha t + \gamma \sin \alpha t$ ...(7)

Applying the initial condition, such that $\omega_2 = 0$

When $t = 0$, so that $\beta = 0$.

$\therefore \quad \omega_2 = \gamma \sin \alpha t$

Putting this value in (5), we get

$$\omega_1 = \gamma \cos \alpha t.$$

$$\therefore \quad \vec{\omega} = \gamma \cos \alpha t\, \mathbf{i} + \gamma \sin \alpha t\, \mathbf{j} + \delta \mathbf{k}$$

$\Rightarrow \quad \omega = |\vec{\omega}| = \sqrt{(\gamma^2 + \delta^2)}$, which gives the magnitude of the angular velocity.

$$\text{Frequency of precession} = \frac{\alpha}{2\pi} = \left|\frac{A-C}{2\pi A}\right| \delta.$$

5. *O and P are vertices of a cube at opposite ends of a diagonal. If the point O of the cube is fixed, show that the angular momentum about O in any general motion can be expressed in the form*

$$\frac{1}{3} Ma^2 \left(11\mathbf{k} \times \frac{d\mathbf{k}}{dt} + 2n\mathbf{k}\right)$$

where **k** *is the unit vector along OP and n is the component of the angular velocity about OP and M is the mass and, a, an edge of the cube.*

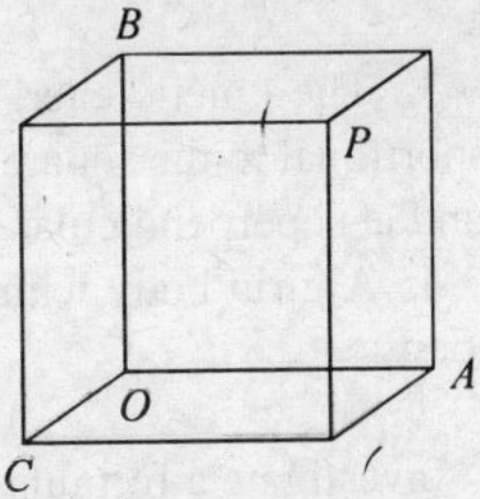

Fig. 5.5

If **i′**, **j′**, **k′** denote unit vectors along the edges OA, OB, OC and ω_1, ω_2, ω_3 the components of the angular velocity ω about A, OB, OC, the A.M., **H**, about O is given by

$$\mathbf{H} = a^2\left(\frac{8}{3}\omega_1 - \omega_2 - \omega_3\right)\mathbf{i}'$$

$$+ a^2\left(-\omega_1 + \frac{8}{3}\omega_2 - \omega_3\right)\mathbf{j}' + a^2\left(-\omega_1 - \omega_2 + \frac{8}{3}\omega_3\right)\mathbf{k}'$$

(Refer § 5.3)

Here **i′**, **j′**, **k′** are variable vectors.

We have $\quad \mathbf{k} = \dfrac{(\mathbf{i}' + \mathbf{j}' + \mathbf{k}')}{\sqrt{3}}$

$$\frac{d\mathbf{k}}{dt} = \frac{1}{\sqrt{3}}\left(\frac{d\mathbf{i}'}{dt} + \frac{d\mathbf{j}'}{dt} + \frac{d\mathbf{k}'}{dt}\right)$$

$$= \frac{1}{\sqrt{3}}\, \omega \times (\mathbf{i}' + \mathbf{j}' + \mathbf{k}')$$

$$\therefore \quad \mathbf{k} \times \frac{d\mathbf{k}}{dt} = \frac{1}{3}(\mathbf{i}' + \mathbf{j}' + \mathbf{k}') \times [\omega \times (\mathbf{i}' + \mathbf{j}' + \mathbf{k}')]$$

$$= \frac{1}{3}[(\mathbf{i}' + \mathbf{j}' + \mathbf{k}')^2 \omega - (\mathbf{i}' + \mathbf{j}' + \mathbf{k}') \cdot \omega\, (\mathbf{i}' + \mathbf{j}' + \mathbf{k}')]$$

$$= \frac{1}{3}[3(\omega_1\mathbf{i}' + \omega_2\mathbf{j}' + \omega_3\mathbf{k}') - (\omega_1 + \omega_2 + \omega_3) . (\mathbf{i}' + \mathbf{j}' + \mathbf{k}')]$$

Also
$$n\mathbf{k} = \left(\omega . \frac{\mathbf{i}'+\mathbf{j}'+\mathbf{k}'}{\sqrt{3}}\right)\frac{\mathbf{i}'+\mathbf{j}'+\mathbf{k}'}{\sqrt{3}}$$

$$= \frac{1}{3}(\omega_1 + \omega_2 + \omega_3)(\mathbf{i}' + \mathbf{j}' + \mathbf{k}').$$

We may now easily verify the truth of the result.

Exercises

1. At a certain instant a particle is at the point $3\mathbf{i} + 2\mathbf{j} + \mathbf{k}$, its angular velocity about the origin O is $\mathbf{i} + \mathbf{j} - 2\mathbf{k}$ and its acceleration is $c\mathbf{V}$ where $\mathbf{V}$ is the velocity and c is a scalar. Find the moment about O of the force acting on the particle.

2. Prove that for a rigid body moving with one point fixed

$$T = \frac{1}{2} H. \omega.$$

3. The kinetic energy of a rigid body moving with one point fixed is proportional to the square of the angular momentum; show that the plane of $\mathbf{H}$ and ω is perpendicular to that of $\mathbf{H}$ and $\mathbf{L}$.

4. A girid body with one of its points O fixed is rotating with angular velocity

$$\mathbf{i}\omega_1 + \mathbf{j}\omega_2 + \mathbf{k}\omega_3.$$

Investigate a formula for the angular momentum about O.

5. A rigid body has a point fixed but is otherwise free. Show that if throughout a certain motion, the instantaneous axis of rotation of the body is a line fixed i the body, then the motion is a rotation of the body about a line fixed in space.

[Note that $\partial\omega/\partial t = d\omega/dt$.]

6. A rigid body of mass M with centre of mass G has angular velocity ω while G has velocity $\mathbf{v}$. Equal particles each of mass $\frac{1}{2}M$ are attached at points A and B and move with the body. The angular momentum of the body about a given point P is the same as the angular momentum of the body about P and the particles have the same total momentum as the body.

If, C, the mid-point of AB does not coincide with G, show that

$$\overrightarrow{GC}$$

is parallel to ω and the angular momentum of the body about, C, is perpendicular to AB in the plane ABC.

7. Show that if, at a certain instant in the general motion of a rigid body, the direction of the axis of the equivalent screw is not stationary, then there is one and only one point of the body which has zero acceleration at the instant.

8. Prove that if no point of a rigid body is fixed and if the velocity of one point O is $\mathbf{u}$ and the angular velocity is $\omega\mathbf{n}$ where ω is a scalar and $\mathbf{n}$ is a unit vector, the body is twisting about a screw of pitch $\omega^{-1}\mathbf{n}.\ \mathbf{V}$, the axis of which is in the direction of $\mathbf{n}$ and passes through a point $\omega^{-1}\ \mathbf{n} \times \mathbf{V}$ relative to O.

9. The position vector of a particle relative to a point on the earth's surface is $\mathbf{r}$, the velocity and acceleration of the particle relative to the earth are $\mathbf{r}$ and $\ddot{\mathbf{r}}$. If the particle moves under the earth's attraction and a force $\mathbf{F}$ per unit mass, prove that

$$\ddot{\mathbf{r}} + 2\omega \times \dot{\mathbf{r}} = \mathbf{g}(\mathbf{r}) + \mathbf{F},$$

where ω is the earth's angular velocity and $\mathbf{g}(\mathbf{r})$ is the acceleration (relative to earth) of a freely falling particle instantaneously at rest relative to the earth at the point $\mathbf{r}$.

10. The motion of a dynamical system of mass M is referred to a rigid reference frame rotating with angular velocity ω about an origin whose acceleration is $\mathbf{a}$. The angular momentum about the origin is $\mathbf{H}$ and the angular momentum in the motion relative to the origin is $\mathbf{h}$. Show that the rate of change of $\mathbf{H}$ is

$$\mathbf{h} + \omega \times \mathbf{h} + M\mathbf{R} \times \mathbf{a},$$

where $\mathbf{R}$ is the position vector of the centroid.

11. One vertex O of a cube is moving with speed v in the direction of an edge OA and the cube is also spinning about the diagonal OB of a face $OABC$ with angular speed $v\sqrt{2}/a$ where, a, is the length of an edge. Find the axis of the equivalent screw.

12. A rigid lamina is freely movable about its centroid which is fixed. It is started rotating about a line in its plane about which its moment of inertia is I. Show that if $A > B$, the ratio of the greatest to least angular velocity in the subsequent motion is

$$\sqrt{\{(A+I)/(A+B)\}}$$

where A and B are the principal moments of inertia.

13. A uniform elliptic lamina free to move about its centre, but, under no forces, is set rotating so that the components of the initial angular velocity referred to axes along the major and minor axes of the ellipseand normal to it are all equal to k. Show that the components of the angular velocity can be expressed in terms of a parameter θ in the form

$$\omega_1 = k\sqrt{2}\ \cos\theta,\ \omega_2 = k\sqrt{2}\ \sin\theta,$$

$$\omega_2 = k\sqrt{\left[\frac{2(1-e^2\sin^2\theta)}{2-e^2}\right]} = \theta,$$

where e is the eccentricity of the ellipse.

14. A uniform rectangular plate whose sides are $2a$, $2b$ $(a > b)$ is free to move about its centre O under forces which have no moment about O. At some instant it is spinning about a diagonal with a given angular velocity. Find the components of the angular velocity about the principal axes when the axis about which it is spinning lies in the plane through O which is perpendicular to the plate and parallel to its longer sides.

15. The ends of a uniform rod of mass m have velocities $\mathbf{u}$ and $\mathbf{v}$. Show that its kinetic energy is

$$\frac{1}{6} m \{\mathbf{u}^2 + \mathbf{u} \cdot \mathbf{v} + \mathbf{v}^2\}.$$

OBJECTIVE QUESTIONS

For each of the following questions, four alternatives are given for the answer. Only one of them is correct. Choose the correct alternative.

1. Relation between velocity and angular velocity is

(*a*) $\mathbf{v} = \vec{\omega} \times \mathbf{r}$ (*b*) $\mathbf{v} = \mathbf{r} \times \vec{\omega}$

(*c*) $\vec{\omega} = \mathbf{r} \times \mathbf{v}$ (*d*) $\vec{\omega} = \mathbf{v} \times \mathbf{r}$

2. If $\mathbf{r}_1, \mathbf{r}_2$ be the position vectors of any two points of a rigid body with reference to the fixed point O, then the angular velocity of the body is

(*a*) $\dfrac{\mathbf{r}_1 \times \mathbf{r}_2}{\mathbf{r}_1 \cdot \mathbf{r}_2}$ (*b*) $\dfrac{\dot{\mathbf{r}}_1 \times \dot{\mathbf{r}}_2}{\mathbf{r}_1 \cdot \mathbf{r}_2}$

(*c*) $\dfrac{\dot{\mathbf{r}}_1 \times \dot{\mathbf{r}}_2}{\mathbf{r}_1\, \mathbf{r}_2}$ (*d*) $\dfrac{\dot{\mathbf{r}}_1 \cdot \dot{\mathbf{r}}_2}{\mathbf{r}_1 \times \mathbf{r}_2}$

3. If A, B be any two points of a rigid body, then the velocity of B relative to A is

(*a*) $\vec{\omega} \times \overrightarrow{AB}$ (*b*) $\overrightarrow{AB} \times \vec{\omega}$

(*c*) $\vec{\omega} + \overrightarrow{AB}$ (*d*) None of these

4. If d/dt denotes rate of change relative to a fixed frame and $\partial/\partial t$ relative to a frame moving with angular velocity ω, then for any vector $\mathbf{p}$

(*a*) $\dfrac{d\mathbf{p}}{dt} = \dfrac{\partial \mathbf{p}}{\partial t}$ (*b*) $\dfrac{d\mathbf{p}}{dt} = \dfrac{\partial \mathbf{p}}{\partial t} + \mathbf{p}$

(*c*) $\dfrac{d\mathbf{p}}{dt} = \dfrac{\partial \mathbf{p}}{\partial t} + \vec{\omega} \times \mathbf{p}$ (*d*) $\dfrac{d\mathbf{p}}{dt} = \dfrac{\partial \mathbf{p}}{\partial t} + \mathbf{p} \times \vec{\omega}$

5. If $\mathbf{H}$ be the angular momentum of a body rotating with angular velocity $\vec{\omega}$, then kinetic energy T is given by

(*a*) $T = \vec{\omega} \cdot \mathbf{H}$ (*b*) $\dfrac{1}{2} \vec{\omega} \cdot \mathbf{H}$

(*c*) $\dfrac{1}{2} \omega^2 H^2$ (*d*) $\omega^2 H^2$

ANSWERS

1. (*a*) **2.** (*b*) **3.** (*a*) **4.** (*c*) **5.** (*b*)

6

Differential Operators

Introduction. This chapter is devoted to a study of the Differential Operators — *Gradient, Divergence* and *Curl* — which, as a result of operation on point functions, leads to new point functions. These operators play a very important part in the study of Hydrodynamics and Electrodynamics and other branches of Applied Mathematics. The operators will be defined and the rules of combination of the same will be obtained in terms of rectangular cartesian axes and a section will also be devoted to the determination of the expressions of these operators in terms of general orthogonal curvilinear co-ordinates.

6.1. POINT FUNCTIONS

6.1.1. Scalar Valued Point Functions

Consider any region R of space and suppose that to each point P of the same, there corresponds, by any law whatsoever, a scalar denoted by $\phi(P)$. We then say that ϕ is a scalar point function defined for the region R.

Illustrations. Consider some material body occupying a region R and let $\phi(P)$ denote the density of the body at any point P thereof. Then ϕ is a scalar function defined for the region R.

The temperature at any point of a body occupying a certain region is a scalar point function defined for the region.

6.1.2. Vector Valued Point Functions

If to any point P of a region there corresponds a vector $\mathbf{f}(P)$, then we say that $\mathbf{f}$ is a vector function defined for the region.

Illustration. Consider any body in motion. At any given time t if $\mathbf{f}(P)$ denote the velocity of the particle at a point P, then $\mathbf{f}$ is a vector point function defined for the region occupied by the body at time t.

Notes. 1. If **r** denotes the position vector of a point P relative to any origin, then we may write

$$f(\mathbf{r}) \text{ and } \mathbf{f}(\mathbf{r}) \text{ instead of } f(P) \text{ and } \mathbf{f}(P)$$

respectively.

2. In the following, the regions of definition of point functions will not, in general, be mentioned.

6.2. LIMITS

6.2.1. A Scalar Point Function $\phi(P)$ *is said to tend to a limit l when P tends to a given point A, if to every pre-assigned positive number ε, there corresponds a positive number δ such that*

$$|\phi(P) - l| < \varepsilon,$$

for all points P within the sphere with centre A and radius δ, except possibly for the point A.

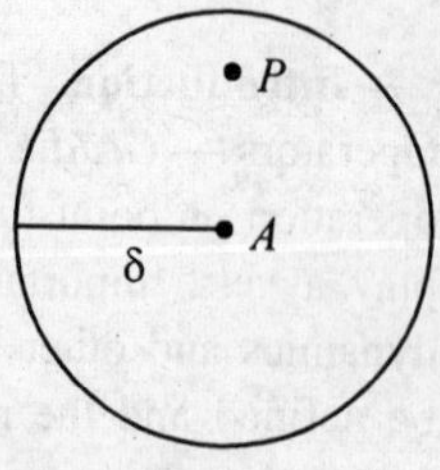

Fig. 6.1.

Also, we then write

$$\lim_{P \to A} \phi(P) = l.$$

This means that to every pre-assigned positive number ε, there corresponds a sphere with its centre at A such that for every point P within the sphere (except possibly A), $\phi(P)$ differs from l numerically by a number less than ε.

6.2.2. A Vector Point Function f (P), *is said to tend to a limit* **l** *when P tends to a given point A, if to every pre-assigned positive number ε, there corresponds a positive number δ such that*

$$|\mathbf{f}(P) - \mathbf{l}| < \varepsilon,$$

for all points P within the sphere with centre A and radius δ, except possibly for the point A.

We write

$$\lim_{P \to A} \mathbf{f}(P) = \mathbf{l}.$$

6.2.3. Continuity. A scalar point function $\phi(P)$ is said to be continuous at a point A, if

$$\lim_{P \to A} \phi(P) = \phi(A).$$

Also a scalar point function is said to be continuous, if it is continuous at every point of its region of definition.

The continuity of vector point functions may be similarly defined.

6.3. DIRECTIONAL DERIVATIVES

6.3.1. Scalar Point Functions

Let $\phi(P)$ be any scalar point function and A any given point of its region of direction. Also, let l be any given semi-line through A, *i.e.*, drawn on one side of A only and let P be any point on the semi-line.

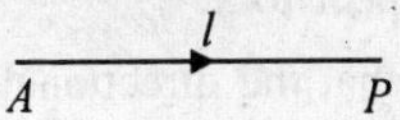

Fig. 6.2.

Then

$$\lim \frac{\phi(P)-\phi(A)}{AP}, \text{ when } P \to A,$$

if it exists, is called the *directional derivative* of the scalar function at A along the direction of the line l.

The length AP is to be regarded positive.

6.3.2. Vector Point Functions

In terms of the above notation

$$\lim \frac{\mathbf{f}(P)-\mathbf{f}(A)}{AP}, \text{ when } P \to A,$$

if it exists, is called the *directional derivative* of the vector function at A along the line l.

The directional derivative of a vector (scalar) point function at a point along any direction is a vector (scalar).

6.4. CARTESIAN REPRESENTATION OF POINT FUNCTIONS AND THEIR DIRECTIONAL DERIVATIVES

We introduce a system of orthogonal cartesian axes OX, OY, OZ.

Let (x, y, z) be the co-ordinates of a point P. We may now write

$$\phi(P) = \phi(x, y, z),$$

so that a scalar point function appears as a scalar function of three scalar variables x, y, z.

We similarly have

$$\mathbf{f}(P) = \mathbf{f}(x, y, z).$$

6.4.1. Directional Derivatives of Scalar Point Functions along Co-ordinate Axes

Take a point $P(x + \delta x, y, z)$ on a line through $A(x, y, z)$ drawn parallel to the positive direction of x-axis. We have AP $= \delta x$ and δx is positive.

We have

$$\frac{\phi(P)-\phi(A)}{AP}=\frac{\phi(x+\delta x, y, z)-\phi(x, y, z)}{\delta x}.$$

This shows that the directional derivative at any point A (x, y, z) along the positive direction of x-axis is the partial derivative $\partial\phi/\partial x$ of the function with respect to x.

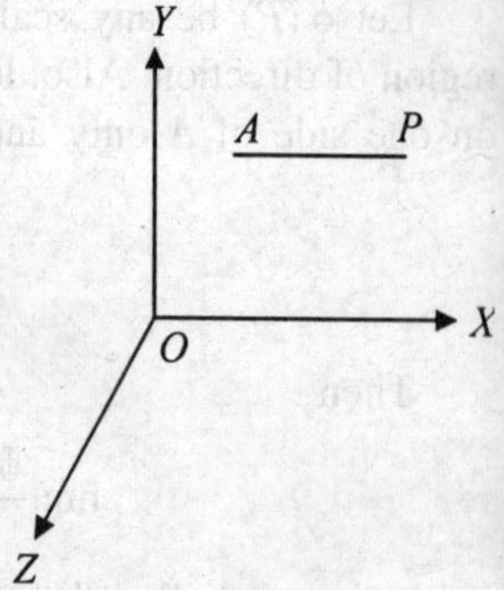

Fig. 6.3.

Similarly, we may see that the directional derivatives at A (x, y, z) along the positive directions of y-axis and z-axis are the partial derivatives $\partial\phi/\partial y$ and $\partial\phi/\partial z$ with respect to y and z respectively.

6.4.2. Directional Derivative along any Line

Consider now *any* semi-line through A (x, y, z) with direction cosines l, m, n.

The co-ordinates of the point P on this line at a distance r from A are

$$(x + lr, \ y + mr, \ z + nr).$$

We have

$$\frac{\phi(P)-\phi(A)}{AP}$$

$$=\frac{\phi(x+lr, \ y+mr, \ z+nr)-\phi(x, y, z)}{r}.$$

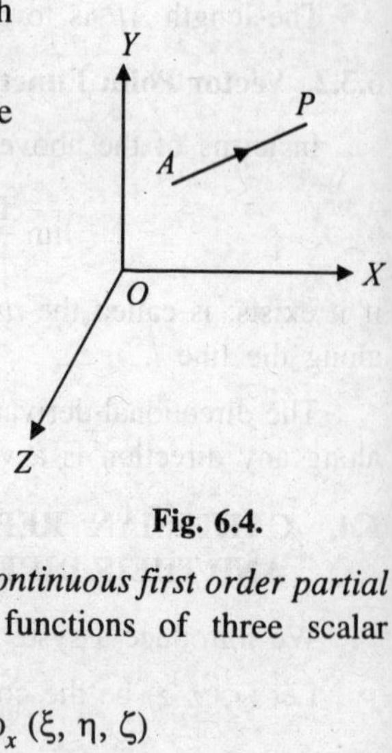

Fig. 6.4.

We now suppose that $\phi(x, y, z)$ *possesses continuous first order partial derivatives.* Then by Taylor's theorem for functions of three scalar variables, we have

$$\phi(x + lr, y + mr, z + nr) = \phi(x, y, z) + lr\phi_x(\xi, \eta, \zeta) + mr\phi_y(\xi, \eta, \zeta) + nr\phi_z(\xi, \eta, \zeta),$$

where (ξ, η, ζ) is a point of the line AP, joining A (x, y, z) to P $(x + lr, y + mr, z + nr)$. Thus,

$$\frac{\phi(P)-\phi(A)}{AP}=l\phi_x(\xi, \eta, \zeta)+m\phi_y(\xi, \eta, \zeta)+n\phi_z(\xi, \eta, \zeta)$$

so that when $P \to A$, we obtain

$$\frac{\phi(P)-\phi(A)}{AP}=l\phi_x(x, y, z)+m\phi_y(x, y, z)+n\phi_z(x, y, z)$$

$$= l\phi_x + m\phi_y + n\phi_z.$$

Thus, *the directional derivative at A (x, y, z) along a line with direction cosines l, m, n is*

$$l\phi_x + m\phi_y + n\phi_z,$$

so that we can write the directional derivative along any line in terms of those along the three co-ordinate axes.

We have seen that if $\phi(x, y, z)$ possesses continuous first order partial derivatives, then it possesses directional derivative along every direction and we say that $\phi(x, y, z)$ is *continuously differentiable.*

Note. The direction cosines of any semi-line through a point are equal and opposite to those of the semi-line produced on to the other side. Thus, if l, m, n are the direction cosines of AP, then $-l, -m, -n$ are those of AQ. Hence, the directional derivatives along the semi-lines AP, AQ are

$$l\phi_x + m\phi_y + n\phi_z, \quad -l\phi_x - m\phi_y - n\phi_z$$

respectively so that each is the negative of the other.

From above or otherwise directly, we may see that the directional derivatives along the negative directions of the coordinate axes are

$$-\phi_x - \phi_y - \phi_z.$$

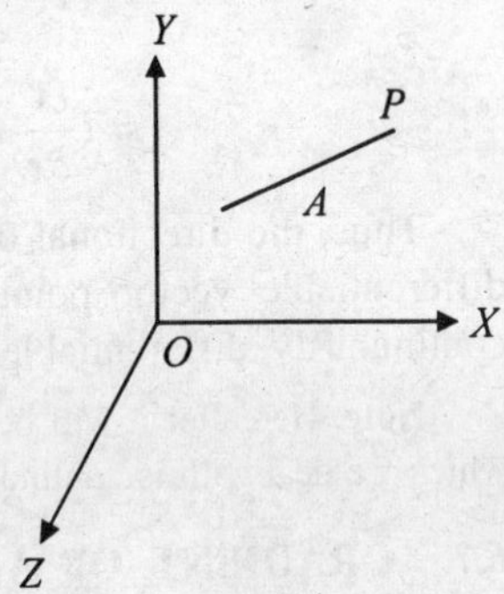

Fig. 6.5.

6.5. DIRECTIONAL DERIVATIVES OF VECTOR POINT FUNCTIONS ALONG CO-ORDINATE AXES

As in the case of scalar point functions, we may show that the directional derivatives of a vector point function $\mathbf{f}(\mathbf{r}) = \mathbf{f}(x, y, z)$ at any point (x, y, z) along the positive directions of co-ordinate axes are

$$\frac{\partial \mathbf{f}}{\partial x}, \frac{\partial \mathbf{f}}{\partial y}, \frac{\partial \mathbf{f}}{\partial z}.$$

6.6. DIRECTIONAL DERIVATIVES ALONG ANY LINE

Let l, m, n be the direction cosines of any semi-line through $A(x, y, z)$. We write

$$\mathbf{f}(x, y, z) = \mathbf{i} f_1(x, y, z) + \mathbf{j} f_2(x, y, z) + \mathbf{k} f_3(x, y, z)$$

and suppose that $\mathbf{f}(x, y, z)$ possesses continuous first order partial derivatives. Then, by § 9.2, 9.3 the scalar functions f_1, f_2 and f_3 also possesses continuous first order partial derivatives. In this case we say that the vector functions $\mathbf{f}$ is continuously differentiable.

We have

$$\frac{\mathbf{f}(P) - \mathbf{f}(A)}{AP} = \mathbf{i}\frac{f_1(P) - f_1(A)}{AP} + \mathbf{j}\frac{f_2(P) - f_2(A)}{AP} + \mathbf{k}\frac{f_3(P) - f_3(A)}{AP}$$

Let $P \to A$. Then by § 6.5.2,

$$\lim \frac{\mathbf{f}(P)-\mathbf{f}(A)}{AP} = \mathbf{i}\left[l\frac{\partial f_1}{\partial x}+m\frac{\partial f_1}{\partial y}+n\frac{\partial f_1}{\partial z}\right]+\mathbf{j}\left[l\frac{\partial f_2}{\partial x}+m\frac{\partial f_2}{\partial y}+n\frac{\partial f_2}{\partial z}\right]+$$
$$\mathbf{k}\left[l\frac{\partial f_3}{\partial x}+m\frac{\partial f_3}{\partial y}+n\frac{\partial f_3}{\partial z}\right]$$
$$= l\left[\mathbf{i}\frac{\partial f_1}{\partial x}+\mathbf{j}\frac{\partial f_2}{\partial x}+\mathbf{k}\frac{\partial f_3}{\partial x}\right]+m\left[\mathbf{i}\frac{\partial f_1}{\partial y}+\mathbf{j}\frac{\partial f_2}{\partial y}+\mathbf{k}\frac{\partial f_3}{\partial y}\right]+$$
$$n\left[\mathbf{i}\frac{\partial f_1}{\partial z}+\mathbf{j}\frac{\partial f_2}{\partial z}+\mathbf{k}\frac{\partial f_3}{\partial z}\right]$$
$$= l\frac{\partial \mathbf{f}}{\partial x}+m\frac{\partial \mathbf{f}}{\partial y}+n\frac{\partial \mathbf{f}}{\partial z}.$$

Thus, the directional derivative along any direction of a continuously differentiable vector point function is of the same form as that of a continuously differentiable scalar point function.

Note. Hereafter it will be assumed that every point function, scalar or vector, which we deal with is continuously differentiable.

6.7. GRADIENT OF A SCALAR POINT FUNCTION

Let, ϕ, be any continuously differentiable scalar point function.

We have

$$l\frac{\partial \phi}{\partial x}+m\frac{\partial \phi}{\partial y}+n\frac{\partial \phi}{\partial z} = (\mathbf{i}l+\mathbf{j}m+\mathbf{k}n)\cdot\left(\mathbf{i}\frac{\partial \phi}{\partial x}+\mathbf{j}\frac{\partial \phi}{\partial y}+\mathbf{k}\frac{\partial \phi}{\partial z}\right)$$

so that the directional derivative on the left appears as a scalar product of the two vectors

$$\mathbf{i}l+\mathbf{j}m+\mathbf{k}n \text{ and } \mathbf{i}\frac{\partial \phi}{\partial x}+\mathbf{j}\frac{\partial \phi}{\partial y}+\mathbf{k}\frac{\partial \phi}{\partial z}$$

of which the former is the unit vector along the line with direction cosines l, m, n and the latter is such that it depends only on the point (x, y, z) and not on any direction. Thus, we see that *the directional derivative along any direction can be obtained on multiplying the vector*

$$\mathbf{i}\frac{\partial \phi}{\partial x}+\mathbf{j}\frac{\partial \phi}{\partial y}+\mathbf{k}\frac{\partial \phi}{\partial z},$$

scalarly with the unit vector along the direction in question.

Thus, we are lead to introduce the notion of Gradient as follows :

Gradient. Def. *The vector function*

$$\mathbf{i}\frac{\partial \phi}{\partial x}+\mathbf{j}\frac{\partial \phi}{\partial y}+\mathbf{k}\frac{\partial \phi}{\partial z},$$

is called the gradient of the scalar point function ϕ *and is written as grad* ϕ. Thus,

$$\text{grad } \phi = \mathbf{i}\frac{\partial\phi}{\partial x} + \mathbf{j}\frac{\partial\phi}{\partial y} + \mathbf{k}\frac{\partial\phi}{\partial z}.$$

Here, we have assumed that ϕ is a continuously differentiable function.

Thus, the directional derivative of the scalar point function ϕ along the direction of the unit vector **a** can be written as

$$\mathbf{a} \cdot \text{grad } \phi.$$

6.7.1. Gradient of a Constant

If ϕ is constant then $\partial\phi/\partial x$, $\partial\phi/\partial y$, $\partial\phi/\partial z$ are all zero so that grad ϕ = 0. Conversely if, grad ϕ is zero, the partial derivatives are all zero and accordingly the function is constant. *Thus,*

$$\text{grad } \phi = 0,$$

if, and only if, the function ϕ *is constant.*

6.7.2. Derivative along a Curve

Let

$$\mathbf{r} = \mathbf{F}(s) = \mathbf{i}\phi_1(s) + \mathbf{j}\phi_2(s) + \mathbf{k}\phi_3(s)$$

be any curve. Along the curve $\phi(x, y, z)$ is a function of s. Then if A, P be two points on the curve, we have

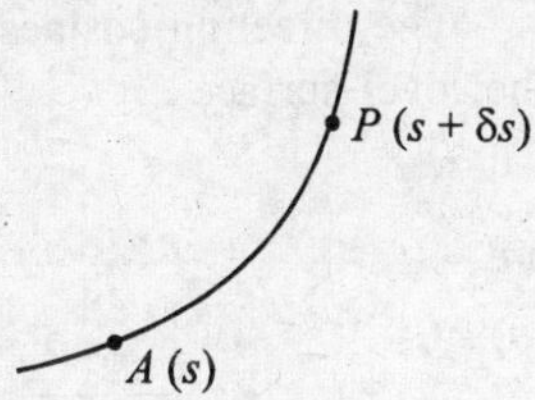

Fig. 6.6.

$$\lim_{P \to A} \frac{\phi(P) - \phi(A)}{AP} = \lim_{\delta s \to 0} \frac{\phi(s + \delta s) - \phi(s)}{\delta s}$$

$$= \frac{d\phi}{ds}$$

$$= \frac{\partial\phi}{\partial x}\frac{dx}{ds} + \frac{\partial\phi}{\partial y}\frac{dy}{ds} + \frac{\partial\phi}{\partial z}\frac{dz}{ds}$$

$$= \left(\mathbf{i}\frac{dx}{ds} + \mathbf{j}\frac{dy}{ds} + \mathbf{k}\frac{dz}{ds}\right) \cdot \left(\mathbf{i}\frac{\partial\phi}{\partial x} + \mathbf{j}\frac{\partial\phi}{\partial y} + \mathbf{k}\frac{\partial\phi}{\partial z}\right)$$

$$= \mathbf{t} \cdot \text{grad } \phi,$$

where **t** denotes unit tangent to the curve.

It can be shown *that the derivative*

$$\mathbf{t} \cdot \text{grad } \phi$$

along a curve is zero if, and only if, the function ϕ *is constant along the curve.*

6.7.3. Differential dϕ

We have

$$d\phi = \frac{\partial \phi}{\partial x}dx + \frac{\partial \phi}{\partial y}dy + \frac{\partial \phi}{\partial z}dz$$

$$= (\mathbf{i}dx + \mathbf{j}dy + \mathbf{k}dz) \cdot \left(\mathbf{i}\frac{\partial \phi}{\partial x} + \mathbf{j}\frac{\partial \phi}{\partial y} + \mathbf{k}\frac{\partial \phi}{\partial z} \right)$$

$$= d\mathbf{r} \,.\, \text{grad } \phi.$$

Thus,

$$d\phi = d\mathbf{r} \,.\, \text{grad } \phi.$$

6.8. GEOMETRICAL INTERPRETATION OF Grad ϕ

Level surfaces. The locus of point P such that

$$\phi\,(P) = \phi\,(A),$$

where A is any given point, is called the level surface of the scalar point function ϕ through the point A. This surface is such that the value of the function at any point thereof is the same as that at A. The level surface for any point A is also the level surface for any point B of the surface.

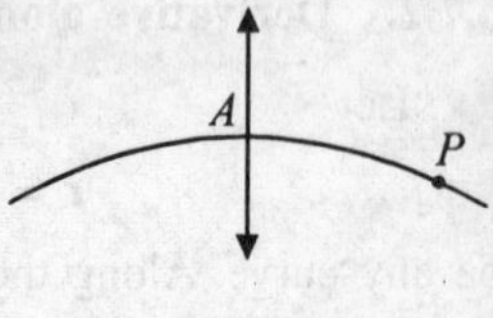

Fig. 6.7.

The direction cosines of the two sides of the normal at A (x, y, z) to the level surface

$$\phi\,(X, Y, Z) = \phi\,(x, y, z)$$

are

$$\frac{\phi_x}{\sqrt{\Sigma\phi_x^2}}, \frac{\phi_y}{\sqrt{\Sigma\phi_x^2}}, \frac{\phi_z}{\sqrt{\Sigma\phi_x^2}}; \quad \text{...(1)}$$

$$-\frac{\phi_x}{\sqrt{\Sigma\phi_x^2}}, -\frac{\phi_y}{\sqrt{\Sigma\phi_x^2}}, -\frac{\phi_z}{\sqrt{\Sigma\phi_x^2}} \quad \text{...(2)}$$

so that the direction derivatives of the function ϕ along these two sides are

$$\frac{\phi_x}{\sqrt{\Sigma\phi_x^2}}\phi_x + \frac{\phi_y}{\sqrt{\Sigma\phi_x^2}}\phi_y + \frac{\phi_z}{\sqrt{\Sigma\phi_x^2}}\phi_z = \sqrt{\Sigma\phi_x^2} = |\,\text{grad } \phi\,| \quad \text{...(3)}$$

$$-\frac{\phi_x}{\sqrt{\Sigma\phi_x^2}}\phi_x - \frac{\phi_y}{\sqrt{\Sigma\phi_x^2}}\phi_y - \frac{\phi_z}{\sqrt{\Sigma\phi_x^2}}\phi_z = -\sqrt{\Sigma\phi_z^2} = -\,|\,\text{grad } \phi\,| \quad \text{...(4)}$$

The directional derivative | grad ϕ | along the side of the normal given by (1) being positive, we see that this side of the normal is one along which ϕ is increasing.

Also we see that the direction cosines of the vector grade ϕ are the same as those given in (1).

Hence, we have the following :

6.8.1. *The vector, grade ϕ, at any point is parallel (including sense) to the side along which ϕ increases to the normal to the level surface through the point and the magnitude of grade ϕ is the directional derivative of ϕ along the same side.*

6.8.2. Again, if **a** be any unit vector so that

$$\mathbf{a} \text{ grad } \phi$$

is the directional derivative of ϕ along the same, we have

$$\mathbf{a} \text{ grad } \phi \leq | \text{ grad } \phi |,$$

for all unit vectors **a.** Thus, we see that the *direction of the vector, grad ϕ, at any point coincides with the direction along which the directional derivative of ϕ is greatest.*

This fact may also be seen otherwise as follows. Consider two level surfaces through two points A, C and let B be the point on the level surface through C where the normal at A meets the same. We have

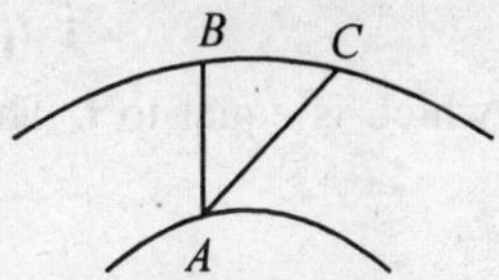

Fig. 6.8.

$$\phi(B) - \phi(A) = \phi(C) - \phi(A)$$

for $\phi(B) = \phi(C)$,

and $AB < AC$.

$$\therefore \quad \frac{\phi(B)-\phi(A)}{AB} > \frac{\phi(C)-\phi(A)}{AC}.$$

The result now follows on taking limits.

6.9. CHARACTER OF GRADIENT AS A POINT FUNCTION

Theorem. *The gradient of a scalar point function is a vector point function.*

The grad ϕ at any point (x, y, z) is the vector

$$\mathbf{i}\frac{\partial \phi}{\partial x} + \mathbf{j}\frac{\partial \phi}{\partial y} + \mathbf{k}\frac{\partial \phi}{\partial z},$$

where

$$\mathbf{i}, \mathbf{j}, \mathbf{k}$$

are three mutually perpendicular vectors and

$$\partial\phi/\partial x, \quad \partial\phi/\partial y, \quad \partial\phi/\partial z$$

are the directional derivatives of ϕ at (x, y, z) along the lines parallel to **i, j, k.** It may thus appear that the vector grad ϕ at any point depends upon the choice of the mutually perpendicular directions through the point but, as will now be seen, this is certainly not the case, *i.e.*, grad ϕ at any point is independent of the choice of perpendicular directions and is as such a vector point function.

Let **i′**, **j′**, **k′** be any other set of mutually orthogonal unit vectors. The directional derivatives of ϕ along these vectors being

$$\mathbf{i}'.\ \text{grad}\ \phi, \quad \mathbf{j}'.\ \text{grad}\ \phi, \quad \mathbf{k}'.\ \text{grad}\ \phi,$$

we see that the expression for the gradient in terms of **i′**, **j′**, **k′** is

$$\mathbf{i}'\ (\mathbf{i}'.\ \text{grad}\ \phi) + \mathbf{j}'\ (\mathbf{j}'.\ \text{grad}\ \phi) + \mathbf{k}'\ (\mathbf{k}'.\ \text{grad}\ \phi). \qquad ...(1)$$

Replacing grade ϕ by

$$\mathbf{i}\frac{\partial\phi}{\partial x} + \mathbf{j}\frac{\partial\phi}{\partial y} + \mathbf{k}\frac{\partial\phi}{\partial z},$$

We see that the coefficient of $\partial\phi/\partial x$ in (1) is

$$\mathbf{i}'\ (\mathbf{i}'\ .\ \mathbf{i}') + \mathbf{j}'\ (\mathbf{j}'\ .\ \mathbf{i}') + \mathbf{k}'\ (\mathbf{k}'\ .\ \mathbf{i}')$$

which is equal to **i**. Thus, the expression (1) is the same as

$$\mathbf{i}\frac{\partial\phi}{\partial x} + \mathbf{j}\frac{\partial\phi}{\partial y} + \mathbf{k}\frac{\partial\phi}{\partial z}.$$

Hence, the theorem.

The theorem also follows from the fact that, as shown in § 6.8.1, *the vector, grad ϕ, at any point is completely describable in terms of the level surface through the point* and this description is independent of the choice of co-ordinate axes.

6.10. THE OPERATOR ∇

It is found very convenient to introduce a symbol called *Del* or *Nabla* and written as ∇ in connection with our dealings with differential operators. We write

$$\nabla = \mathbf{i}\frac{\partial}{\partial x} + \mathbf{j}\frac{\partial}{\partial y} + \mathbf{k}\frac{\partial}{\partial z}, \qquad ...(1)$$

and state that it operates distributively. In terms of ∇, we write

$$\text{grad}\ \phi = \mathbf{i}\frac{\partial\phi}{\partial x} + \mathbf{j}\frac{\partial\phi}{\partial y} + \mathbf{k}\frac{\partial\phi}{\partial z}$$

$$= \left(\mathbf{i}\frac{\partial}{\partial x} + \mathbf{j}\frac{\partial}{\partial y} + \mathbf{k}\frac{\partial}{\partial z}\right)\phi = \nabla\phi.$$

The interest and usefulness of the introduction of this symbol lies specially in the fact that it can be *formally assumed* to have the character of a vector and as such it facilitates the manipulations with expressions involving differential operators. Thus, *formally* $\nabla\phi$ being the product of a vector ∇ by a scalar ϕ is a vector. (Refer § 6.12, Page 291).

The operator ∇ is also known as *Differential Operator*. It should be clearly understood that the expression (1) for ∇ holds only when we are dealing with cartesian rectangular co-ordinates.

6.11. OPERATOR

$$\mathbf{a} \cdot \nabla,$$

a, *being any vector.*

The operator $\mathbf{a} \cdot \nabla$ is defined by the equality

$$\mathbf{a}\cdot\nabla = \mathbf{a}\cdot\mathbf{i}\frac{\partial}{\partial x} + \mathbf{a}\cdot\mathbf{j}\frac{\partial}{\partial y} + \mathbf{a}\cdot\mathbf{k}\frac{\partial}{\partial z},$$

so that we have

$$(\mathbf{a}\cdot\nabla)\,\phi = \mathbf{a}\cdot\mathbf{i}\frac{\partial\phi}{\partial x} + \mathbf{a}\cdot\mathbf{j}\frac{\partial\phi}{\partial y} + \mathbf{a}\cdot\mathbf{k}\frac{\partial\phi}{\partial z},$$

$$(\mathbf{a}\cdot\nabla)\,\mathbf{f} = \mathbf{a}\cdot\mathbf{i}\frac{\partial\mathbf{f}}{\partial x} + \mathbf{a}\cdot\mathbf{j}\frac{\partial\mathbf{f}}{\partial y} + \mathbf{a}\cdot\mathbf{k}\frac{\partial\mathbf{f}}{\partial z}.$$

For a unit vector **a**,

$$\mathbf{a}\cdot\mathbf{i},\ \ \mathbf{a}\cdot\mathbf{j},\ \ \mathbf{a}\cdot\mathbf{k}$$

are the direction cosines of **a** and hence by § 6.5 and § 6.6.

$$(\mathbf{a}\cdot\nabla)\,\phi,\ \ (\mathbf{a}\cdot\nabla)\,\mathbf{f}$$

stand for the directional derivatives of the respective functions along the direction of the unit vector,

We have, as may be easily seen,

$$(\mathbf{a}\cdot\nabla)\,\phi = \mathbf{a}\cdot\nabla\,\phi = \mathbf{a}\cdot\text{grad}\,\phi.$$

As, however, we have not defined $\nabla\,\mathbf{f}$, where **f** is a vector, we cannot write

$$\mathbf{a}\cdot\nabla\,\mathbf{f} \text{ in place of } (\mathbf{a}\cdot\nabla)\,\mathbf{f},$$

unless, of course, we say that, by def.,

$$\mathbf{a}\cdot\nabla\,\mathbf{f} \text{ means } (\mathbf{a}\cdot\nabla)\,\mathbf{f}.$$

Corresponding to the relation $d\phi = d\mathbf{r}\cdot\nabla\phi$ obtain § 6.7.3. for scalar functions ϕ, we have the relation

$$d\mathbf{f} = (d\mathbf{r}\cdot\nabla)\,\mathbf{f} = d\mathbf{r}\cdot\nabla\,\mathbf{f}$$

for vector functions **f**.

We have,

$$d\mathbf{f} = \frac{\partial \mathbf{f}}{\partial x}dx + \frac{\partial \mathbf{f}}{\partial y}dy + \frac{\partial \mathbf{f}}{\partial z}dz$$

$$= d\mathbf{r}\cdot\mathbf{i}\frac{\partial \mathbf{f}}{\partial x} + d\mathbf{r}\cdot\mathbf{j}+\frac{\partial \mathbf{f}}{\partial y} + d\mathbf{r}\cdot\mathbf{k}\frac{\partial \mathbf{f}}{\partial z}$$

$$= d\mathbf{r}\,.\,\nabla\,\mathbf{f}.$$

EXAMPLES

Example 1. *Find grad log | **r** |.* (*Rohilkhand, 2000*)

Solution. We have $r = \sqrt{(x^2+y^2+z^2)}$

$$\log|\mathbf{r}| = \frac{1}{2}\log(x^2+y^2+z^2)$$

Now, $\dfrac{\partial}{\partial x}\log|\mathbf{r}| = \dfrac{1}{2}\cdot\dfrac{2x}{(x^2+y^2+z^2)} = \dfrac{x}{r^2}$

Similarly, $\dfrac{\partial}{\partial y}\log|\mathbf{r}| = \dfrac{y}{r^2},\ \dfrac{\partial}{\partial z}\log|\mathbf{r}| = \dfrac{z}{r^2}$

$\therefore$ $\quad$ grad $\log|\mathbf{r}| = \dfrac{1}{r^2}(\mathbf{i}x+\mathbf{j}y+\mathbf{k}z) = \dfrac{\mathbf{r}}{r^2}$

Example 2. *Prove that grad f (r) × **r** = 0.*

Solution. Let $\quad \phi = f(r) = f(x^2+y^2+z^2)^{1/2}$

$$\frac{\partial\phi}{\partial x} = f'(r)\frac{\partial r}{\partial x} = f'(r)\frac{x}{r}$$

Similarly, $\quad \dfrac{\partial\phi}{\partial y} = f'(r)\dfrac{y}{r},\ \dfrac{\partial\phi}{\partial z} = f'(r)\dfrac{z}{r}$

$\therefore$ $\quad$ grad $f(r) = \Sigma\,\mathbf{i}\dfrac{\partial\phi}{\partial x}$

$$= \frac{1}{r}f'(r)\cdot(\mathbf{i}x+\mathbf{j}y+\mathbf{k}z) = \frac{f'(r)}{r}\mathbf{r}$$

$\therefore$ $\quad$ grad $f(r)\times\mathbf{r} = \dfrac{f'(r)}{r}\mathbf{r}\times\mathbf{r} = 0.$

Example 3. *Find grad r^m where r is the distance of any point from the origin.* (*Rohilkhand 99, Kumaon 2000*)

Solution. We have

$$\phi(x, y, z) = r^m = (x^2+y^2+z^2)^{\frac{m}{2}}$$

$$\frac{\partial\phi}{\partial x} = mx\,(x^2+y^2+z^2)^{\frac{m}{2}-1} = mxr^{m-2}$$

$$\frac{\partial\phi}{\partial y} = myr^{m-2},$$

$$\frac{\partial\phi}{\partial z} = mzr^{m-2}$$

Thus,

$$\text{grad } r^m = \Sigma\,\mathbf{i}\,\frac{\partial\phi}{\partial x}$$

$$= mr^{m-2}\ \Sigma\,\mathbf{i}\,x$$

$$= mr^{m-2}\mathbf{r},$$

where, $\mathbf{r}$, is the position vector of any point. We can also write

$$\text{grad } r^m = mr^{m-1}\left(\frac{\mathbf{r}}{r}\right),$$

where $\mathbf{r}/r$ is the unit vector joining the origin to any point.

Another method. The level surfaces of the given scalar point function are spheres with centre at the origin so that if P be any point, the normal to the level surface at P is the line OP. Also as r stands for the distance measured along the normal OP, the directional derivative of r^m along OP is mr^{m-1}. Hence, by § 6.1,

$$\text{grad } r^m = mr^{m-1}\,\frac{\mathbf{r}}{r}.$$

Example 4. *If* $\phi\,(x, y) = \log\sqrt{(x^2+y^2)}$, *show that*

$$grad\ \phi = \frac{\mathbf{r}-(\mathbf{k}\cdot\mathbf{r})\,\mathbf{k}}{\{\mathbf{r}-(\mathbf{k}\cdot\mathbf{r})\,\mathbf{k}\}\cdot\{\mathbf{r}-(\mathbf{k}\cdot\mathbf{r})\,\mathbf{k}\}}$$

Solution. We have

$$\mathbf{r} = x\mathbf{i} + y\mathbf{j} + z\mathbf{k} \quad \therefore \quad \mathbf{r}\,.\,\mathbf{k} = z = \mathbf{k}\,.\,\mathbf{r} \qquad ...(1)$$

$$\phi = \frac{1}{2}\log\,(x^2+y^2) \quad \therefore \quad \frac{\partial\phi}{\partial x} = \frac{1}{2\,(x^2+y^2)}\cdot 2x = \frac{x}{x^2+y^2}$$

$$\therefore \quad \text{grad } \phi = \Sigma\,\mathbf{i}\,\frac{\partial\phi}{\partial x} = \frac{x\mathbf{i}+y\mathbf{j}+z\mathbf{k}}{x^2+y^2}$$

$$= \frac{\mathbf{r}-z\mathbf{k}}{(x\mathbf{i}+y\mathbf{j})\cdot(x\mathbf{i}+y\mathbf{j})} \qquad \text{[from (1)]}$$

$$= \frac{\mathbf{r}-z\mathbf{k}}{(r-z\mathbf{k})\cdot(r-z\mathbf{k})}$$

Now replacing z by $\mathbf{r}\,.\,\mathbf{k} = \mathbf{k}\,.\,\mathbf{r}$ in the above we get the required result.

Example 5. *If* ϕ *and* ψ *are two scalar point functions and* ϕ *is a function of* ψ, *then prove that* $\nabla\phi = \nabla\psi \dfrac{d\phi}{d\psi}$.

Solution. $\nabla\phi = \Sigma \mathbf{i} \dfrac{\partial \phi}{\partial x} = \Sigma \mathbf{i} \dfrac{d\phi}{d\psi} \cdot \dfrac{\partial \psi}{\partial x}$,

$\because$ ϕ is a function of ψ

$$= \frac{d\phi}{d\psi} \Sigma \mathbf{i} \frac{\partial \psi}{\partial x} = \frac{d\phi}{d\psi} \nabla\psi$$

$$= \nabla\psi \frac{d\phi}{d\psi}.$$

Example 6. *Prove that* (*i*) *grad* (**r** . **a**) = **a** (*ii*) *grad* [**r a b**] = **a** × **b**, *where* **a** *and* **b** *are constant vectors.*

Solution. (*i*) Let $\mathbf{r} = x\mathbf{i} + y\mathbf{j} + z\mathbf{k}$ and $\mathbf{a} = a_1\mathbf{i} + a_2\mathbf{j} + a_3\mathbf{k}$, where, a_1, a_2, a_3 are constants.

$\therefore$ $\mathbf{r} \cdot \mathbf{a} = xa_1 + ya_2 + za_3$

$\therefore$ grad $(\mathbf{r} \cdot \mathbf{a})$ = grad $(xa_1 + ya_2 + za_3)$

$$= \Sigma \mathbf{i} \frac{\partial}{\partial x} (\mathbf{i}a_1 + \mathbf{j}a_2 + \mathbf{k}a_3)$$

$$\Sigma a_1 \mathbf{i} = a_1\mathbf{i} + a_2\mathbf{j} + a_3\mathbf{k} = \mathbf{a}$$

(*ii*) $[\mathbf{r}\ \mathbf{a}\ \mathbf{b}] = \mathbf{r} \cdot (\mathbf{a} \times \mathbf{b})$

Proceed as in part (*i*) taking **a** × **b** in placed of **a**.

$\therefore$ grad $[\mathbf{r}\ \mathbf{a}\ \mathbf{b}] = \mathbf{r} \cdot (\mathbf{a} \times \mathbf{b})$

Example 7. *Find the directional derivative of* $\phi = x^2yz + 4xz^2$ *at* (*1*, – *2*, – *1*) *in the direction* $2\mathbf{i} - \mathbf{j} - 2\mathbf{k}$. *In which direction the directional derivative will be maximum and what is its magnitude. Also find a unit normal to the surface* $x^2yz + 4xz^2 = 6$ *at the point* (*1*, – *2*, – *1*), *find the equation of tangent plane and normal at the point* (*1*, – *2*, – *1*).

Solution. We have

$$\frac{\partial \phi}{\partial x} = 2xyz + 4z^2, \quad \frac{\partial \phi}{\partial y} = x^2 z, \quad \frac{\partial \phi}{\partial z} = x^2 y + 8xz.$$

$$\text{grad}\ \phi = \Sigma\, \mathbf{i} \frac{\partial \phi}{\partial x} = (2xyz + 4z^2)\, \mathbf{i} + x^2 z\, \mathbf{j} + (x^2 y + 8xz)\, \mathbf{k}.$$

or, grad $\phi = 8\mathbf{i} - \mathbf{j} - 10\mathbf{k}$ at the point (1, – 2, – 1).

Now, if $\hat{\mathbf{a}}$ be a unit vector then directional derivative of ϕ along the direction of $\hat{\mathbf{a}}$ is $\hat{\mathbf{a}} \cdot$ grad ϕ.

A unit vector in the direction (2**i** – **j** – 2**k**) is

$$\frac{2\mathbf{i}-\mathbf{j}-2\mathbf{k}}{\sqrt{(4+1+4)}}=\frac{1}{3}(2\mathbf{i}-\mathbf{j}-2\mathbf{k})$$

∴ Directional derivative

$$=\frac{1}{3}(2\mathbf{i}-\mathbf{j}-2\mathbf{k})\cdot(8\mathbf{i}-\mathbf{j}-10\mathbf{k})$$

$$=\frac{1}{3}(16+1+20)=\frac{37}{3}.$$

We know that the directional derivative is maximum in the direction of normal which is the direction of grad ϕ. Hence, directional derivative is maximum along grad $\phi = 8\mathbf{i} - \mathbf{j} - 10\mathbf{k}$. Maximum value of this directional derivative is $|\text{grad } \phi|$

$$=\sqrt{(64+1+100)}=\sqrt{(165)}.$$

A unit normal to the surface $=\dfrac{\text{grad } \phi}{|\text{grad } \phi|}$

$$=\frac{8\mathbf{i}-\mathbf{j}-10\mathbf{k}}{\sqrt{(165)}}.$$

Tangent plane to the surface at (1, – 2, – 1).

Let P be any point on the tangent plane at A (1, – 2, – 1) and the position vectors of P and A be **r** and $\mathbf{r}_0$.

∴ $\mathbf{r} = x\mathbf{i} + y\mathbf{j} + z\mathbf{k}$ and $\mathbf{r}_0 = \mathbf{i} - 2\mathbf{j} - \mathbf{k}$

$$\vec{AP}=\mathbf{r}-\mathbf{r}_0=(x-1)\,\mathbf{i}+(y+2)\,\mathbf{j}+(z+1)\,\mathbf{k}.$$

grad ϕ is normal to the surface and as such it is perpendicular to AP.

$$\therefore \quad \text{grad } \phi \cdot \vec{AP} = 0.$$

$$\Rightarrow \quad (8\mathbf{i} - \mathbf{j} - 10\mathbf{k}) \cdot [(x-1)\,\mathbf{i} + (y+2)\,\mathbf{j} + (z+1)\,\mathbf{k}] = 0$$

$$\Rightarrow \quad 8(x-1) - (y+2) - 10(z+1) = 0$$

$$\Rightarrow \quad 8x - y - 10z = 20.$$

Normal to the surface.

If ρ be any point on the normal line at A, then $(\mathbf{r} - \mathbf{r}_0)$ lies along the normal, *i.e.*, along grad ϕ, hence

$$(\mathbf{r} - \mathbf{r}_0) \times \text{grad } \phi = 0$$

$$\Rightarrow \quad \begin{vmatrix} \mathbf{i} & \mathbf{j} & \mathbf{k} \\ x-1 & y+2 & z+1 \\ 8 & -1 & -10 \end{vmatrix} = 0$$

$$\Rightarrow \mathbf{i}\{(z+1) - 10\,(y+2)\} + \mathbf{j}\{10\,(x-1) + 8\,(z+1)\}$$
$$-\,\mathrm{k}\,\{(x-1) + 8\,(y+2)\} = 0$$

Equating to zero the coefficients of **i, j** and **k**, we get

$$\frac{z+1}{10} = \frac{y+2}{1};\ \frac{z+1}{10} = \frac{x-1}{-8};\ \frac{x-1}{-8} = \frac{y+2}{1}$$

$$\therefore \qquad \frac{x-1}{-8} = \frac{y+2}{1} = \frac{z+1}{10}.$$

Note. It could be written directly as the equation of a line through (1, – 2, – 1) and with direction ratios as 8, – 1, – 10.

Example 8. *Show that the directional derivative of a scalar function of any point along any tangent line to the level surface at the point is zero.*

Solution. We know that directional derivative along any direction $\hat{\mathbf{a}}$ is $\hat{\mathbf{a}}\,.\,\text{grad}\ \phi$.

Unit vector $\hat{\mathbf{a}}$ is along the direction of tangent line as given. Also grad ϕ is along the normal, hence

$$\text{directional derivative} = \hat{\mathbf{a}}\,.\,\text{grad}\ \phi = 0.$$

Example 9. *Find the values of constants a, b, c so that the directional derivative of $\phi = axy^2 + byz + cz^2x^3$ at (1, 2, – 1) has a maximum magnitude 64 in a direction parallel to z-axis.*

Solution. The directional derivative is maximum along the normal, *i.e.* along grad ϕ.

$$\text{grad}\ \phi = (ay^2 + 3cz^2x^2)\,\mathbf{i} + (2axy + bz)\,\mathbf{j} + (by + 2czx^3)\,\mathbf{k}$$
$$= (4a + 3c)\,\mathbf{i} + (4a - b)\,\mathbf{j} + (2b - 2c)\,\mathbf{k} \text{ at } (1, 2, -1)$$

But directional derivative is maximum along z-axis. Hence, the coefficient of **i** and **j** should be zero.

$$\therefore \qquad 4a + 3c = 0 \text{ and } 4a - b = 0$$
$$\therefore \qquad \text{grad}\ \phi = (2b - 2c)\,\mathbf{k}$$

Also, maximum value of directional derivative

$$= |\,\text{grad}\ \phi\,|$$

$$\therefore \quad 2\,(b - c) = 64 \quad \Rightarrow \quad b - c = 32.$$

Solving the three equations we get

$$a = 6, \quad b = 24, \quad c = -8.$$

Example 6. *If*

$$\mathbf{F}\left(y\frac{\partial f}{\partial z} - z\frac{\partial f}{\partial y}\right)\mathbf{i} + \left(z\frac{\partial f}{\partial x} - x\frac{\partial f}{\partial z}\right)\mathbf{j} + \left(x\frac{\partial f}{\partial y} - y\frac{\partial f}{\partial x}\right)\mathbf{k}$$

prove that (*i*) $\mathbf{F} = \mathbf{r} \times \nabla f$, (*ii*) $\mathbf{F}\,.\,\mathbf{r} = 0$, (*iii*) $\mathbf{F}\,.\,\nabla f = 0$.

Solution. Given

$$\mathbf{F} = \left(y\frac{\partial f}{\partial z} - z\frac{\partial f}{\partial y} \right)\mathbf{i} + \left(z\frac{\partial f}{\partial x} - x\frac{\partial f}{\partial z} \right)\mathbf{j} + \left(x\frac{\partial f}{\partial y} - y\frac{\partial f}{\partial x} \right)\mathbf{k} \quad ...(1)$$

Also, $\nabla \mathbf{f} = \dfrac{\partial f}{\partial x}\mathbf{i} + \dfrac{\partial f}{\partial y}\mathbf{j} + \dfrac{\partial f}{\partial z}\mathbf{k}$

and $\mathbf{r} = x\mathbf{i} + y\mathbf{j} + z\mathbf{k}$.

(*i*)
$$\mathbf{r} \times \nabla \mathbf{f} = \begin{vmatrix} \mathbf{i} & \mathbf{j} & \mathbf{k} \\ x & y & z \\ \frac{\partial f}{\partial x} & \frac{\partial f}{\partial y} & \frac{\partial f}{\partial z} \end{vmatrix}$$

$$= \left(y\frac{\partial f}{\partial z} - z\frac{\partial f}{\partial y} \right)\mathbf{i} + \left(z\frac{\partial f}{\partial x} - x\frac{\partial f}{\partial z} \right)\mathbf{j} + \left(x\frac{\partial f}{\partial y} - y\frac{\partial f}{\partial x} \right)\mathbf{k}$$

$= \mathbf{F}$.

(*ii*) $\mathbf{F} \cdot \mathbf{r} = (\mathbf{r} \times \nabla \mathbf{f}) \cdot \mathbf{r}$ [by part (*i*)]

$= 0$ [scalar triple product with two equal elements **r**]

(*iii*) $\mathbf{F} \cdot \nabla \mathbf{f} = (\mathbf{r} \times \nabla \mathbf{f}) \cdot \nabla \mathbf{f}$

$= 0$, by the same cause.

EXERCISES

1. Find grad ϕ if (*i*) $\phi = \dfrac{1}{r}$; (*ii*) $\mathbf{r} \cdot \mathbf{r}$; (*iii*) $\phi = 3x^2y - y^3z^2$ at point $(1, -2, -1)$; (*iv*) $\phi = 2x^2y^3 - 3y^2z^3$ at the point $(1, -1, 1)$.

(*Rohilkhand, 98*)

2. Prove that $\nabla \int \phi(u)\, du = \phi(u) \nabla u$.

(*i*) Find the directional derivative of $\phi = x^3 + y^3 + z^3$ at the point $(1, -1, 2)$ in the direction of the vector $\mathbf{i} + 2\mathbf{j} + \mathbf{k}$.

(*ii*) Find the directional derivative of $\phi = xy + yz + zx$ in the direction of the vector $\mathbf{i} + 2\mathbf{j} + 2\mathbf{k}$ at $(1, 2, 0)$.

(*iii*) Find the directional derivative of $\phi = x^2 - 2y^2 + 4z^2$ at $(1, 1, -1)$ in the direction $2\mathbf{i} + \mathbf{j} - \mathbf{k}$. In what direction is the directional derivative from the point $(1, 1, -1)$, is maximum and what is its value.

3. (*i*) Find the unit vector normal to $\phi = x^2 - y^2 + z$ at the point $(1, -1, 2)$.

(*ii*) Find the unit vector normal to the surface $z = x^2 + y^2$ at the point $(-1, -2, 5)$.

4. Find the angle between the surfaces $x^2 + y^2 + z^2 = 9$ and $z = x^2 + y^2 - 3$ at the point (2, – 1, 2).

5. Find the values of the constants *a, b, c* so that the directional derivative of $\phi = ax^2 - by^2 + cz^2$ at (1, 1, 2) has a maximum magnitude 4 in the direction parallel to *y*-axis.

6. Find the direction along which the directional derivative of the function $xy + 2yz + 3xz$ is greatest at the point (1, 1, 1). Also find the greatest directional derivatives.

   Do the same for the function $x(y + z) - y(x + y)$ and the point (0, – 1, 2).

7. Find grad [sin (*xy*) + sin (*yz*) + sin (zx)] and $\nabla e^{(x^2+y^2+z^2)}$.

8. Find the directional derivative of the function $xy^2 + yz^2 + zx^2$ along the tangent to the curve $x = t, y = t^2, z = t^2$ at (1, 1, 1) and at (– 1, 1, – 1).

6.12. DIVERGENCE AND CURL

We have seen how, starting with a scalar point function ϕ, we can construct a vector point function grad ϕ. We shall now start with a vector point function and construct therewith two more point functions, one scalar and the other a vector, respectively called **Divergence** and **Curl** of the given function.

Def. *If* **f** *be any given continuously differentiable vector point function, then the functions*

$$\mathbf{i}\cdot\frac{\partial \mathbf{f}}{\partial x}+\mathbf{j}\cdot\frac{\partial \mathbf{f}}{\partial y}+\mathbf{k}\cdot\frac{\partial \mathbf{f}}{\partial z},$$

$$\mathbf{i}\times\frac{\partial \mathbf{f}}{\partial x}+\mathbf{j}\times\frac{\partial \mathbf{f}}{\partial y}+\mathbf{k}\times\frac{\partial \mathbf{f}}{\partial z},$$

are respectively called Divergence and Curl of **f**, *and are denoted by*

div **f**, *curl* **f**.

Clearly *div* **f** *is scalar and curl* **f** *a vector.*

In terms of the operator ∇, we write

$$\text{div}\,\mathbf{f} = \mathbf{i}\cdot\frac{\partial \mathbf{f}}{\partial x}+\mathbf{j}\cdot\frac{\partial \mathbf{f}}{\partial y}+\mathbf{k}\cdot\frac{\partial \mathbf{f}}{\partial z}$$

$$= \left(\mathbf{i}\frac{\partial}{\partial x}+\mathbf{j}\frac{\partial}{\partial y}+\mathbf{k}\frac{\partial}{\partial z}\right)\cdot\mathbf{f}$$

$$= \nabla\,.\,\mathbf{f}$$

$$\text{Curl}\,\mathbf{f} = \mathbf{i}\times\frac{\partial \mathbf{f}}{\partial x}+\mathbf{j}\times\frac{\partial \mathbf{f}}{\partial y}+\mathbf{k}\times\frac{\partial \mathbf{f}}{\partial z}$$

$$= \left(\mathbf{i}\frac{\partial}{\partial x} + \mathbf{j}\frac{\partial}{\partial y} + \mathbf{k}\frac{\partial}{\partial z}\right) \times \mathbf{f}$$

$$= \nabla \times \mathbf{f}$$

If terms of the *formal* character of ∇, as a vector, we see that the scalar $\nabla \cdot \mathbf{f}$ is the scalar product of the vector ∇ and $\mathbf{f}$ and the vector $\nabla \times \mathbf{f}$ is the vector product of the vectors ∇ and $\mathbf{f}$.

6.12.1. Expressions for div f and curl f in terms of the components of f

Let

$$\mathbf{f} = \mathbf{i}f_1 + \mathbf{i}f_2 + \mathbf{i}f_3$$

$$\therefore \quad \text{div } \mathbf{f} = \Sigma \mathbf{i} \cdot \frac{\partial \mathbf{f}}{\partial x}$$

$$= \Sigma \mathbf{i} \cdot \left[\mathbf{i}\frac{\partial f_1}{\partial x} + \mathbf{j}\frac{\partial f_2}{\partial x} + \mathbf{k}\frac{\partial f_3}{\partial x}\right]$$

$$= \frac{\partial f_1}{\partial x} + \frac{\partial f_2}{\partial y} + \frac{\partial f_3}{\partial z}.$$

and

$$\text{curl } \mathbf{f} = \Sigma \mathbf{i} \times \frac{\partial \mathbf{f}}{\partial x}$$

$$= \Sigma \mathbf{i} \times \left[\mathbf{i}\frac{\partial f_1}{\partial x} + \mathbf{j}\frac{\partial f_2}{\partial x} + \mathbf{k}\frac{\partial f_3}{\partial x}\right]$$

$$= \mathbf{i}\left(\frac{\partial f_3}{\partial y} - \frac{\partial f_2}{\partial z}\right) + \mathbf{j}\left(\frac{\partial f_1}{\partial z} - \frac{\partial f_3}{\partial x}\right) + \mathbf{k}\left(\frac{\partial f_2}{\partial x} - \frac{\partial f_1}{\partial y}\right)$$

$$= \begin{vmatrix} \mathbf{i} & \mathbf{j} & \mathbf{k} \\ \dfrac{\partial}{\partial x} & \dfrac{\partial}{\partial y} & \dfrac{\partial}{\partial z} \\ f_1 & f_2 & f_3 \end{vmatrix}$$

6.12.2. Characters of div f and curl f as Point Functions

Theorem. *The functions div* **f** *and curl* **f** *are point functions.*

We have to show that div **f** and curl **f** are independent of the choice of the orthogonal unit vectors in terms of which we may formulate the same.

Let $\mathbf{i}'$, $\mathbf{j}'$, $\mathbf{k}'$ be any other set of mutually orthogonal unit vectors. By § 6.11, Page 284, the directional derivative of **f** along the directions of the vectors $\mathbf{i}'$, $\mathbf{j}'$, $\mathbf{k}'$ are

$$(\mathbf{i}' \cdot \nabla)\, \mathbf{f}, \quad (\mathbf{j}' \cdot \nabla)\, \mathbf{f}, \quad (\mathbf{k}' \cdot \nabla)\, \mathbf{f}$$

i.e.,

$$\mathbf{i}' \cdot \mathbf{i}\frac{\partial \mathbf{f}}{\partial x} + \mathbf{i}' \cdot \mathbf{j}\frac{\partial \mathbf{f}}{\partial y} + \mathbf{i}' \cdot \mathbf{k}\frac{\partial \mathbf{f}}{\partial z}, \text{ etc.}$$

Thus, the expression for div **f** in terms of **i′**, **j′**, **k′** is

$$= \Sigma\, \mathbf{i}' \,.\left[\mathbf{i}' \cdot \mathbf{i}\frac{\partial \mathbf{f}}{\partial x} + \mathbf{i}' \cdot \mathbf{j}\frac{\partial \mathbf{f}}{\partial y} + \mathbf{i}' \cdot \mathbf{k}\frac{\partial \mathbf{f}}{\partial z}\right]$$

$$= \Sigma\, \mathbf{i}' \,.\, [(\mathbf{i}' \cdot \mathbf{i})\, \mathbf{i}' + (\mathbf{j}' \cdot \mathbf{i})\, \mathbf{j}' + (\mathbf{k}' \cdot \mathbf{i})\, \mathbf{k}'] \cdot \frac{\partial \mathbf{f}}{\partial x}$$

$$= \Sigma\, \mathbf{i} \cdot \frac{\partial \mathbf{f}}{\partial x}.$$

Thus, we see that div **f** is essentially a point function. The case of curl **f** can be treated in a similar manner.

6.13. ILLUSTRATIONS OF Curl f AND Div f.

6.13.1. Let ω be the angular velocity of a rigid body moving about a fixed point. Then the velocity **v** of any particle of the body with position vector **r** relative to the fixed point is given by

$$\mathbf{v} = \omega \times \mathbf{r}.$$

Here, **v** is a vector point function and the vector ω is the same for all the points.

We have

$$\begin{aligned} \text{curl } \mathbf{v} &= \text{curl } (\omega \times \mathbf{r}) \\ &= \Sigma\, \mathbf{i} \times \frac{\partial \mathbf{f}}{\partial x}(\omega \times \mathbf{r}) \\ &= \Sigma\, \mathbf{i} \times \left(\omega \times \frac{\partial \mathbf{r}}{\partial x}\right) = \Sigma\, \mathbf{i} \times (\omega \times \mathbf{i}) \\ &= \Sigma\, (\mathbf{i} \,.\, \mathbf{i})\, \omega - \Sigma\, (\mathbf{i} \,.\, \omega)\, \mathbf{i} = 3\omega - \omega = 2\omega. \end{aligned}$$

Thus, the *curl of the velocity of any particle of a rigid body equals twice the angular velocity of the body.*

This result justifies the term Rotation which is sometimes used for curl.

6.13.2. Let a vector point function **q** denote the velocity of any homogeneous and incompressible fluid at any time *t*.

Consider any point *P* (*x*, *y*, *z*) and enclose the same in a rectangular parallelopiped with edges parallel to the co-ordinate axes and of lengths δ*x*, δ*y*, δ*z*.

Let

$$\mathbf{q} = q_1\mathbf{i} + q_2\mathbf{j} + q_3\mathbf{k}.$$

The velocity component parallel to x-axis at any point of the face $ABCD$

$$= q_1\left(x+\frac{1}{2}\delta x, y, z\right)$$

$$= q_1(x, y, z)+\frac{1}{2}\delta x\frac{\partial q_1}{\partial x}$$

omitting powers of δx higher than one.

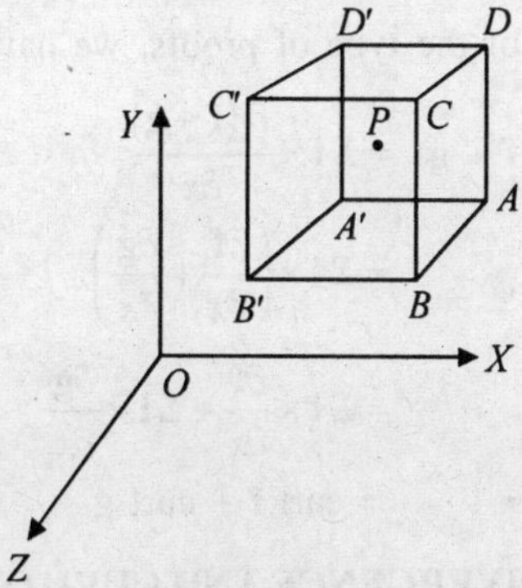

Fig. 6.9.

Thus, the mass of fluid that moves out of the face $ABCD$ in time δt

$$= \rho\left[q_1+\frac{1}{2}\delta x\frac{\partial q_1}{\partial x}\right]\delta y\,\delta z\,\delta t.$$

Similarly the mass of fluid that enters through the face $A'B'C'D'$ in time δt

$$= \rho\left[q_1-\frac{1}{2}\delta x\frac{\partial q_1}{\partial x}\right]\delta y\,\delta z\,\delta t.$$

Thus, the total mass of fluid that moves out through the face $ABCD$, $A'B'C'D'$ perpendicular to x-axis.

$$= \rho\frac{\partial q_1}{\partial x}\delta x\,\delta y\,\delta z\,\delta t.$$

Taking into consideration the other faces also, we see that the total mass of fluid that moves out of the parallelopiped in time δt

$$= \rho\left(\frac{\partial q_1}{\partial x}+\frac{\partial q_2}{\partial y}+\frac{\partial q_3}{\partial z}\right)\delta x\,\delta y\,\delta z\,\delta t$$

$$= \rho\,\delta x\,\delta y\,\delta z\,\delta t \text{ div } \mathbf{q}.$$

The volume of the parallelopiped being δx, δy, δz, we see, on taking limit, when δx, δy, δz, δt all tend to zero, that the amount of fluid per unit volume per unit time that passes through a point (x, y, z)

$$= \rho \text{ div } \mathbf{q}.$$

6.14. GRADIENT, DIVERGENCE AND CURL OF SUMS

The following results are immediately deducible from the definitions :

$$\text{grad}\,(\phi \pm \psi) = \text{grad}\,\phi \pm \text{grad}\,\psi,$$
$$\text{div}\,(\mathbf{f} \pm \mathbf{g}) = \text{div}\,\mathbf{f} \pm \text{div}\,\mathbf{g},$$
$$\text{curl}\,(\mathbf{f} \pm \mathbf{g}) = \text{curl}\,\mathbf{f} \pm \text{curl}\,\mathbf{g}.$$

Thus, each of three differential operations is distributive with respect to the sum and difference of functions.

As an illustration of the type of proofs, we have

$$\begin{aligned}\text{curl}\,(\mathbf{f} + \mathbf{g}) &= \Sigma\,\mathbf{i} \times \frac{\partial(\mathbf{f}+\mathbf{g})}{\partial x}\\ &= \Sigma\,\mathbf{i} \times \left(\frac{\partial \mathbf{f}}{\partial x} + \frac{\partial \mathbf{g}}{\partial x}\right)\\ &= \Sigma\,\mathbf{i} \times \frac{\partial \mathbf{f}}{\partial x} + \Sigma\,\mathbf{i} \times \frac{\partial \mathbf{g}}{\partial x}\\ &= \text{curl}\,\mathbf{f} + \text{curl}\,\mathbf{g}.\end{aligned}$$

6.15. GRADIENT, DIVERGENCE AND CURL OF PRODUCTS

We shall now see how the gradient, divergence and curl of products of point functions can be expressed in terms of those of the factors. Now remembering that we have defined gradient of a scalar point function and divergence and curl of a vector point function, we see that we have to consider the following cases only.

$$\text{grad}\,(\phi\psi),\ \text{grad}\,(\mathbf{f}\,.\,\mathbf{g});$$
$$\text{div}\,(\phi\mathbf{f}),\ \text{div}\,(\mathbf{f} \times \mathbf{g});$$
$$\text{curl}\,(\phi\mathbf{f}),\ \text{curl}\,(\mathbf{f} \times \mathbf{g}).$$

The following results will be obtained :

I. $\text{grad}\,(\phi\psi) = \phi\,\text{grad}\,\psi + \psi\,\text{grad}\,\phi.$

II. $\text{grad}\,(\mathbf{f}\,.\,\mathbf{g}) = \mathbf{f} \times \text{curl}\,\mathbf{g} + \mathbf{g} \times \text{curl}\,\mathbf{f} + (\mathbf{f}\,.\,\nabla)\,\mathbf{g} + (\mathbf{g}\,.\,\nabla)\,\mathbf{f}.$

(Kumaon 2000)

III. $\text{div}\,(\phi\mathbf{f}) = \phi\,\text{div}\,\mathbf{f} + \mathbf{f}\,.\,\text{grad}\,\phi.$

(Gorakhpur 98, Lucknow 98)

IV. $\text{div.}\,(\mathbf{f} \times \mathbf{g}) = \mathbf{g}\,.\,\text{curl}\,\mathbf{f} - \mathbf{f}\,.\,\text{curl}\,\mathbf{g}.$

(Ajmer 97, 99, Gorakhpur 99, Rohilkhand 2000, Kumaon 99, Lucknow 97)

V. $\text{curl}\,(\phi\mathbf{f}) = \text{grad}\,\phi \times \mathbf{f} + \phi\,\text{curl}\,\mathbf{f}.$ *(Gorakhpur 2000)*

VI. $\text{curl}\,(\mathbf{f} \times \mathbf{g}) = \mathbf{f}\,\text{div}\,\mathbf{g} - \mathbf{g}\,\text{div}\,\mathbf{f} + (\mathbf{g}\,.\,\nabla)\,\mathbf{f} - (\mathbf{f}\,.\,\nabla)\,\mathbf{g}.$

(Kolkata 96, Lucknow 97)

We shall now prove these one by one.

I. $$\text{grad}\,(\phi\psi) = \Sigma\,\mathbf{i}\,\frac{\partial(\phi\psi)}{\partial x}$$

$$= \Sigma\,\mathbf{i}\left(\phi\frac{\partial\psi}{\partial x} + \phi\frac{\partial\phi}{\partial x}\right)$$

$$= \Sigma\,\mathbf{i}\,\phi\frac{\partial\psi}{\partial x} + \Sigma\,\mathbf{i}\psi\frac{\partial\phi}{\partial x}$$

$$= \phi\,\Sigma\,\mathbf{i}\frac{\partial\psi}{\partial x} + \psi\,\Sigma\,\mathbf{i}\frac{\partial\phi}{\partial x}$$

$$= \phi\,\text{grad}\,\psi + \psi\,\text{grad}\,\phi.$$

II. $$\text{grad}\,(\mathbf{f}\,.\,\mathbf{g}) = \Sigma\,\mathbf{i}\,\frac{\partial(\mathbf{f}\,.\,\mathbf{g})}{\partial x}$$

$$= \Sigma\,\mathbf{i}\left(\mathbf{f}\cdot\frac{\partial\mathbf{g}}{\partial x} + \frac{\partial\mathbf{f}}{\partial x}\cdot\mathbf{g}\right)$$

$$= \Sigma\,\mathbf{i}\left(\mathbf{f}\cdot\frac{\partial\mathbf{g}}{\partial x}\right) + \Sigma\,\mathbf{i}\left(\frac{\partial\mathbf{f}}{\partial x}\cdot\mathbf{g}\right). \quad ...(1)$$

Now

$$\mathbf{f}\times\left(\mathbf{i}\times\frac{\partial\mathbf{g}}{\partial x}\right) = \left(\mathbf{f}\cdot\frac{\partial\mathbf{g}}{\partial x}\right)\mathbf{i} - (\mathbf{f}\,.\,\mathbf{i})\frac{\partial\mathbf{g}}{\partial x},$$

or, $$\left(\mathbf{f}\cdot\frac{\partial\mathbf{g}}{\partial x}\right)\mathbf{i} = \mathbf{f}\left(\mathbf{i}\times\frac{\partial\mathbf{g}}{\partial x}\right) + (\mathbf{f}\,.\,\mathbf{i})\frac{\partial\mathbf{g}}{\partial x}.$$

$\therefore$ $$\Sigma\left(\mathbf{f}\cdot\frac{\partial\mathbf{g}}{\partial x}\right)\mathbf{i} = \mathbf{f}\times\left(\Sigma\,\mathbf{i}\times\frac{\partial\mathbf{g}}{\partial x}\right) + \Sigma\,(\mathbf{f}\,.\,\mathbf{i})\frac{\partial\mathbf{g}}{\partial x}$$

$$= \mathbf{f}\times\text{curl}\,\mathbf{g} + (\mathbf{f}\,.\,\nabla)\,\mathbf{g}. \quad ...(2)$$

Interchanging **f** and **g**, we have

$$\Sigma\left(\mathbf{g}\cdot\frac{\partial\mathbf{f}}{\partial x}\right)\mathbf{i} = \mathbf{g}\times\text{curl}\,\mathbf{f} + (\mathbf{g}\,.\,\nabla)\,\mathbf{f}. \quad ...(3)$$

From (1), (2) and (3), we have

$$\text{grad}\,(\mathbf{f}\,.\,\mathbf{g}) = \mathbf{f}\times\text{curl}\,\mathbf{g} + \mathbf{g}\times\text{curl}\,\mathbf{f} + (\mathbf{f}\,.\,\nabla)\,\mathbf{g} + (\mathbf{g}\,.\,\nabla)\,\mathbf{f}.$$

III. $$\text{div}\,(\phi\mathbf{f}) = \Sigma\,\mathbf{i}\cdot\frac{\partial(\phi\mathbf{f})}{\partial x}$$

$$= \Sigma\,\mathbf{i}\cdot\left(\phi\frac{\partial\mathbf{f}}{\partial x} + \frac{\partial\phi}{\partial x}\mathbf{f}\right)$$

$$= \phi \, \Sigma \, \mathbf{i} \cdot \frac{\partial \mathbf{f}}{\partial x} + \left(\Sigma \, \mathbf{i} \, \frac{\partial \phi}{\partial x} \right) \cdot \mathbf{f}$$

$$= \phi \operatorname{div} \mathbf{f} + \mathbf{f} \, . \operatorname{grad} \phi.$$

IV. $$\operatorname{div} (\mathbf{f} \times \mathbf{g}) = \Sigma \, \mathbf{i} \cdot \frac{\partial (\mathbf{f} \times \mathbf{g})}{\partial x}$$

$$= \Sigma \, \mathbf{i} \cdot \left(\frac{\partial \mathbf{f}}{\partial x} \times \mathbf{g} + \mathbf{f} \times \frac{\partial \mathbf{g}}{\partial x} \right)$$

$$= \Sigma \, \mathbf{i} \cdot \frac{\partial \mathbf{f}}{\partial x} \times \mathbf{g} + \Sigma \, \mathbf{i} \cdot \mathbf{f} \times \frac{\partial \mathbf{g}}{\partial x}$$

$$= \Sigma \, \mathbf{i} \times \frac{\partial \mathbf{f}}{\partial x} \cdot \mathbf{g} - \Sigma \, \mathbf{i} \times \frac{\partial \mathbf{g}}{\partial x} \cdot \mathbf{f}$$

$$= \left(\Sigma \, \mathbf{i} \times \frac{\partial \mathbf{f}}{\partial x} \right) \cdot \mathbf{g} - \left(\Sigma \, \mathbf{i} \times \frac{\partial \mathbf{g}}{\partial x} \right) \cdot \mathbf{f}$$

$$= \operatorname{curl} \mathbf{f} \, . \, \mathbf{g} - \operatorname{curl} \mathbf{g} \, . \, \mathbf{f}.$$

V. $$\operatorname{curl} (\phi \, \mathbf{f}) = \Sigma \, \mathbf{i} \times \frac{\partial (\phi \mathbf{f})}{\partial x}$$

$$= \Sigma \, \mathbf{i} \times \left(\frac{\partial \phi}{\partial x} \mathbf{f} + \phi \frac{\partial \mathbf{f}}{\partial x} \right)$$

$$= \Sigma \, \mathbf{i} \times \frac{\partial \phi}{\partial x} \times \mathbf{f} + \Sigma \, \mathbf{i} \times \phi \frac{\partial \mathbf{f}}{\partial x}$$

$$= \Sigma \, \mathbf{i} \times \frac{\partial \phi}{\partial x} \times \mathbf{f} + \left(\Sigma \, \mathbf{i} \times \frac{\partial \mathbf{f}}{\partial x} \right) \phi$$

$$= \operatorname{grad} \phi \times \mathbf{f} + \phi \operatorname{curl} \mathbf{f}.$$

VI. $$\operatorname{curl} (\mathbf{f} \times \mathbf{g}) = \Sigma \, \mathbf{i} \times \frac{\partial (\mathbf{f} \times \mathbf{g})}{\partial x}$$

$$= \Sigma \, \mathbf{i} \times \left(\mathbf{f} \times \frac{\partial \mathbf{g}}{\partial x} + \frac{\partial \mathbf{f}}{\partial x} \times \mathbf{g} \right)$$

$$= \Sigma \, \mathbf{i} \times \left(\mathbf{f} \times \frac{\partial \mathbf{g}}{\partial x} \right) + \Sigma \, \mathbf{i} \times \left(\frac{\partial \mathbf{f}}{\partial x} \times \mathbf{g} \right)$$

$$= \Sigma \left(\mathbf{i} \cdot \frac{\partial \mathbf{g}}{\partial x} \right) \mathbf{f} - \Sigma \, (\mathbf{i} \cdot \mathbf{f}) \frac{\partial \mathbf{g}}{\partial x} + \Sigma \, (\mathbf{i} \cdot \mathbf{g}) \frac{\partial \mathbf{f}}{\partial x} - \left(\Sigma \mathbf{i} \cdot \frac{\partial \mathbf{f}}{\partial x} \right) \mathbf{g}.$$

$$= \left(\Sigma \mathbf{i} \cdot \frac{\partial \mathbf{g}}{\partial x}\right) \mathbf{f} - \left(\Sigma \mathbf{i} \cdot \frac{\partial \mathbf{f}}{\partial x}\right) \mathbf{g} + \Sigma (\mathbf{i} \cdot \mathbf{g}) \frac{\partial \mathbf{f}}{\partial x} - \Sigma (\mathbf{i} \cdot \mathbf{f}) \frac{\partial \mathbf{g}}{\partial x}$$

$$= \mathbf{f} \text{ div } \mathbf{g} - \mathbf{g} \text{ div } \mathbf{f} + (\mathbf{g} \cdot \nabla) \mathbf{f} - (\mathbf{f} \cdot \nabla) \mathbf{g}.$$

6.16. SECOND ORDER DIFFERENTIAL OPERATORS

The gradient of a scalar point function being a vector point function, we may again form its divergence and curl,

i.e.,
$$\text{div grad } \phi = \nabla \cdot \nabla \phi,$$
$$\text{curl grad } \phi = \nabla \times \nabla \phi.$$

Also curl of a vector point function being a vector point function, we may again form its divergence and curl, *i.e.*,

$$\text{div curl } \mathbf{f} = \nabla \cdot \nabla \times \mathbf{f},$$
$$\text{curl curl } \mathbf{f} = \nabla \times \nabla \times \mathbf{f}.$$

Finally divergence of a vector point function being a scalar, we may have

$$\text{grad div } \mathbf{f} = \nabla \nabla \cdot \mathbf{f}.$$

We shall now prove the following :

I. $\text{div grad } \phi = \nabla \cdot \nabla \phi = \dfrac{\partial^2 \phi}{\partial x^2} + \dfrac{\partial^2 \phi}{\partial y^2} + \dfrac{\partial^2 \phi}{\partial z^2} \cdot$

II. $\text{curl grad } \phi = \nabla \times \nabla \times \phi = 0.$ (*Gorakhpur, 99*)

III. $\text{div curl } \mathbf{f} = \nabla \cdot \nabla \times \mathbf{f} = 0.$

IV, V. $\text{grad div } \mathbf{f} = \text{curl curl } \mathbf{f} + \Sigma \dfrac{\partial^2 \mathbf{f}}{\partial x^2},$

i.e.,
$$\nabla \nabla \cdot \mathbf{f} = \nabla \times \nabla \times \mathbf{f} + \Sigma \frac{\partial^2 \mathbf{f}}{\partial x^2} \cdot$$

These will now be proved one by one.

I.
$$\nabla \phi = \Sigma \mathbf{i} \frac{\partial \phi}{\partial x} = \mathbf{i} \frac{\partial \phi}{\partial x} + \mathbf{j} \frac{\partial \phi}{\partial y} + \mathbf{k} \frac{\partial \phi}{\partial z} \cdot$$

$\therefore$
$$\nabla \cdot \nabla \phi = \Sigma \mathbf{i} \cdot \frac{\partial}{\partial x} \left(\mathbf{i} \frac{\partial \phi}{\partial x} + \mathbf{j} \frac{\partial \phi}{\partial y} + \mathbf{k} \frac{\partial \phi}{\partial z} \right)$$
$$= \frac{\partial^2 \phi}{\partial x^2} + \frac{\partial^2 \phi}{\partial y^2} + \frac{\partial^2 \phi}{\partial z^2} \cdot$$

II.
$$\text{curl grad } \phi = \Sigma \mathbf{i} \times \frac{\partial}{\partial x} \left(\mathbf{i} \frac{\partial \phi}{\partial x} + \mathbf{j} \frac{\partial \phi}{\partial y} + \mathbf{k} \frac{\partial \phi}{\partial z} \right)$$

$$= \Sigma \mathbf{i} \times \left(\mathbf{i} \frac{\partial^2 \phi}{\partial x^2} + \mathbf{j} \frac{\partial^2 \phi}{\partial x \partial y} + \mathbf{k} \frac{\partial^2 \phi}{\partial x \partial z} \right)$$

$$= \Sigma \left(\mathbf{k} \frac{\partial^2 \phi}{\partial x \partial y} - \mathbf{j} \frac{\partial^2 \phi}{\partial x \partial z} \right)$$

$$= \mathbf{0},$$

terms cancelling in pairs.

III. $$\text{div . curl } \mathbf{f} = \Sigma \mathbf{i} \cdot \frac{\partial}{\partial x} \left(\mathbf{i} \times \frac{\partial \mathbf{f}}{\partial x} + \mathbf{j} \times \frac{\partial \mathbf{f}}{\partial y} + \mathbf{k} \times \frac{\partial \mathbf{f}}{\partial z} \right)$$

$$= \Sigma \mathbf{i} \cdot \left(\mathbf{i} \times \frac{\partial^2 \mathbf{f}}{\partial x^2} + \mathbf{j} \times \frac{\partial^2 \mathbf{f}}{\partial x \partial y} + \mathbf{k} \times \frac{\partial^2 \mathbf{f}}{\partial x \partial z} \right)$$

$$= \Sigma \left(\mathbf{i} \times \mathbf{i} \cdot \frac{\partial^2 \mathbf{f}}{\partial x^2} + \mathbf{i} \times \mathbf{j} \cdot \frac{\partial^2 \mathbf{f}}{\partial x \partial y} + \mathbf{i} \times \mathbf{k} \cdot \frac{\partial^2 \mathbf{f}}{\partial x \partial z} \right)$$

$$= \Sigma \left(\mathbf{k} \cdot \frac{\partial^2 \mathbf{f}}{\partial x \partial y} - \mathbf{j} \cdot \frac{\partial^2 \mathbf{f}}{\partial x \partial z} \right)$$

$$= 0.$$

IV, V. $$\nabla \times (\nabla \times \mathbf{f}) = \Sigma \mathbf{i} \times \frac{\partial}{\partial x} \left(\mathbf{i} \times \frac{\partial \mathbf{f}}{\partial x} + \mathbf{j} \times \frac{\partial \mathbf{f}}{\partial y} + \mathbf{k} \times \frac{\partial \mathbf{f}}{\partial z} \right)$$

$$= \Sigma \mathbf{i} \times \left(\mathbf{i} \times \frac{\partial^2 \mathbf{f}}{\partial x^2} + \mathbf{j} \times \frac{\partial^2 \mathbf{f}}{\partial x \partial y} + \mathbf{k} \times \frac{\partial^2 \mathbf{f}}{\partial x \partial z} \right)$$

$$= \Sigma \left[\left(\mathbf{i} \cdot \frac{\partial^2 \mathbf{f}}{\partial x^2} \mathbf{i} - \mathbf{i} \cdot \mathbf{i} \frac{\partial^2 \mathbf{f}}{\partial x^2} \right) + \left(\mathbf{i} \cdot \frac{\partial^2 \mathbf{f}}{\partial x \partial y} \mathbf{j} - \mathbf{i} \cdot \mathbf{j} \frac{\partial^2 \mathbf{f}}{\partial x \partial y} \right) + \left(\mathbf{i} \cdot \frac{\partial^2 \mathbf{f}}{\partial x \partial z} \mathbf{k} - \mathbf{i} \cdot \mathbf{k} \frac{\partial^2 \mathbf{f}}{\partial x \partial z} \right) \right]$$

$$= \Sigma \left(\mathbf{i} \cdot \frac{\partial^2 \mathbf{f}}{\partial x^2} \mathbf{i} + \mathbf{i} \cdot \frac{\partial^2 \mathbf{f}}{\partial x \partial y} \mathbf{j} + \mathbf{i} \cdot \frac{\partial^2 \mathbf{f}}{\partial x \partial z} \mathbf{k} \right) - \Sigma \frac{\partial^2 \mathbf{f}}{\partial x^2} \cdot$$

Also $$\nabla \nabla \cdot \mathbf{f} = \Sigma \mathbf{i} \frac{\partial}{\partial x} \left(\mathbf{i} \cdot \frac{\partial \mathbf{f}}{\partial x} + \mathbf{j} \cdot \frac{\partial \mathbf{f}}{\partial y} + \mathbf{k} \cdot \frac{\partial \mathbf{f}}{\partial z} \right)$$

$$= \Sigma \mathbf{i} \left(\mathbf{i} \cdot \frac{\partial^2 \mathbf{f}}{\partial x^2} + \mathbf{j} \cdot \frac{\partial^2 \mathbf{f}}{\partial x \partial y} + \mathbf{k} \cdot \frac{\partial^2 \mathbf{f}}{\partial x \partial z} \right)$$

$$= \Sigma \left(\mathbf{i} \cdot \frac{\partial^2 \mathbf{f}}{\partial x^2} \mathbf{i} + \mathbf{i} \cdot \frac{\partial^2 \mathbf{f}}{\partial x \partial y} \mathbf{j} + \mathbf{i} \cdot \frac{\partial^2 \mathbf{f}}{\partial x \partial z} \mathbf{k} \right)$$

$$= \nabla \times (\nabla \times \mathbf{f}) + \Sigma \frac{\partial^2 \mathbf{f}}{\partial x^2}.$$

$$\therefore \quad \text{grad div } \mathbf{f} = \text{curl curl } \mathbf{f} + \Sigma \frac{\partial^2 \mathbf{f}}{\partial x^2}$$

6.17. THE LAPLACIAN OPERATOR, ∇^2

Looking upon ∇ formally as a vector, we have

$$\nabla^2 = \nabla \cdot \nabla = \left(\Sigma \mathbf{i} \frac{\partial}{\partial x} \right) \cdot \left(\Sigma \mathbf{i} \frac{\partial}{\partial x} \right)$$

$$= \frac{\partial^2}{\partial x^2} + \frac{\partial^2}{\partial y^2} + \frac{\partial^2}{\partial z^2}.$$

Now, by def., the operator ∇^2, called Laplacian, is given by

$$\nabla^2 \phi = \frac{\partial^2 \phi}{\partial x^2} + \frac{\partial^2 \phi}{\partial y^2} + \frac{\partial^2 \phi}{\partial z^2}.$$

$$\nabla^2 \mathbf{f} = \frac{\partial^2 \mathbf{f}}{\partial x^2} + \frac{\partial^2 \mathbf{f}}{\partial y^2} + \frac{\partial^2 \mathbf{f}}{\partial z^2}.$$

Note on the Role of ∇ as a symbolic vector. The proper role of the symbol ∇ is that of a differential operator and the symbol has actually been introduced as such. However we should notice that it helps greatly in reproducing the various formulae concerning differential operations, if ∇ is *thought* of as a vector and this role of ∇ is constantly borne in mind.

Thus, in the first instance we see that the vectorial character of ∇ is consistent with the following statements :

(*i*) $\nabla \phi$ is a vector.

(*ii*) $\nabla \cdot \mathbf{f}$ is a scalar.

(*iii*) $\nabla \times \mathbf{f}$ is a vector.

(*iv*) $\nabla \mathbf{f}$ is meaningless.

(*v*) curl grad $\phi = \nabla \times \nabla\phi = 0$.

(*vi*) div curl $\mathbf{f} = \nabla \cdot \nabla \times \mathbf{f} = 0$.

(*vii*) div grad $\phi = \nabla \cdot \nabla\phi = \nabla^2\phi$.

(*viii*) curl curl $\mathbf{f} = \nabla \times (\nabla \times \mathbf{f})$

$= \nabla \nabla \cdot \nabla \mathbf{f} - \nabla \cdot \nabla \mathbf{f}$

$=$ grad div . $\mathbf{f} - \nabla^2\mathbf{f}$.

The following suggested procedure involving the role of ∇ as a vector proves very helpful in reproducing the various formulae concerning the operation of ∇ on a product of two functions (§ 6.15).

I. Regarding the two factors separately as constants, express the result of the opeartion of ∇ as a sum of two terms.

II. Transform each of the two terms, taking care that the operator always appears before (and not after) a function and bearing in mind whether the result of the operation is a vector or a scalar. The properties of triple products may also be sometimes needed to carry out the transformation.

III. Restore the change so far as regarding the functions as constants is concerned.

We shall now see how the three-step procedure suggested above can actually be employed to reproduce the various formulae of § 6.15.

I. $\text{grad}\,(\phi\psi) = \nabla\,(\phi\psi)$

$$= \nabla\,(\phi_c\psi + \phi\psi_c)$$

$$= \phi_c\,\nabla\psi + \psi_c\,\nabla\phi$$

$$= \phi\nabla\psi + \psi\nabla\phi.$$

II. $\text{grad}\,(\mathbf{f} \cdot \mathbf{g}) = \nabla\,(\mathbf{f} \cdot \mathbf{g})$

$$= \nabla\,(\mathbf{f}_c \cdot \mathbf{g}) + \nabla\,(\mathbf{f} \cdot \mathbf{g}_c)$$

$$= \mathbf{f}_c \times (\nabla \times \mathbf{g}) + (\mathbf{f}_c \cdot \nabla)\,\mathbf{g} + \mathbf{g}_c \times (\nabla \times \mathbf{f}) + (\mathbf{g}_c \cdot \nabla)\,\mathbf{f}\ ?$$

$$= \mathbf{f} \times (\nabla \times \mathbf{g}) + \mathbf{g} \times (\nabla \times \mathbf{f}) + (\mathbf{f} \cdot \nabla)\,\mathbf{g} + (\mathbf{g} \cdot \nabla)\,\mathbf{f}\ ?$$

III. $\text{div}\,(\phi\mathbf{f}) = \nabla \cdot (\phi\mathbf{f})$ (*Ajmer 98*)

$$= \nabla \cdot (\phi_c\mathbf{f}) + \nabla \cdot (\phi\mathbf{f}_c)$$

$$= \phi_c\nabla \cdot \mathbf{f} + \nabla\phi \cdot \mathbf{f}_c$$

$$= \phi\nabla \cdot \mathbf{f} + \nabla\phi \cdot \mathbf{f}.$$

IV. $\text{div}\,(\mathbf{f} \times \mathbf{g}) = \nabla \cdot (\mathbf{f} \times \mathbf{g})$

$$= \nabla \cdot (\mathbf{f}_c \times \mathbf{g}) + \nabla \cdot (\mathbf{f} \times \mathbf{g}_c)$$

$$= -\nabla \times \mathbf{g} \cdot \mathbf{f}_c + \nabla \times \mathbf{f} \cdot \mathbf{g}_c$$

$$= -\nabla \times \mathbf{g} \cdot \mathbf{f} + \nabla \times \mathbf{f} \cdot \mathbf{g}.$$

V. $\text{curl}\,(\phi\mathbf{f}) = \nabla \times (\phi\mathbf{f})$

$$= \nabla \times (\phi_c\mathbf{f}) + \nabla \times (\phi\mathbf{f}_c)$$

$$= \phi_c\nabla \times \mathbf{f} + \nabla\phi \times \mathbf{f}_c$$

$$= \phi\nabla \times \mathbf{f} + \nabla\phi \times \mathbf{f}.$$

VI. $\text{curl}\,(\mathbf{f} \times \mathbf{g}) = \nabla \times (\mathbf{f} \times \mathbf{g})$

$$= \nabla \times (\mathbf{f}_c \times \mathbf{g}) + \nabla \times (\mathbf{f} \times \mathbf{g}_c)$$

$$= \nabla \cdot \mathbf{g}\,\mathbf{f}_c - (\mathbf{f}_c \cdot \nabla)\,\mathbf{g} + (\mathbf{g}_c \cdot \nabla)\,\mathbf{f} - \nabla\mathbf{f}\,\mathbf{g}_c$$

$$= \mathbf{f}\,\nabla \cdot \mathbf{g} - \mathbf{g}\nabla \cdot \mathbf{f} + (\mathbf{g} \cdot \nabla)\,\mathbf{f} - (\mathbf{f} \cdot \nabla)\,\mathbf{g}.$$

EXAMPLES

Example 1. *If* $\mathbf{f} = x^2z\mathbf{i} - 2y^3z^2\mathbf{j} + xy^2z\mathbf{k}$. *Find div* $\mathbf{f}$, *curl* $\mathbf{f}$, *at* $(1, -1, 1)$.

Solution. $\text{div}\,\mathbf{f} = \Sigma \mathbf{i} \cdot \dfrac{\partial \mathbf{f}}{\partial x} = \nabla \cdot \mathbf{f}$

$$= \left(\mathbf{i}\frac{\partial}{\partial x} + \mathbf{j}\frac{\partial}{\partial y} + \mathbf{k}\frac{\partial}{\partial z}\right) \cdot \left(x^2z\mathbf{i} - 2y^3z^2\mathbf{j} + xy^2z\mathbf{k}\right)$$

$$= \frac{\partial}{\partial x}\left(x^2z\right) - \frac{\partial}{\partial y}\left(2y^3z^2\right) + \frac{\partial}{\partial x}\left(xy^2z\right)$$

$$= 2xz - 6y^2z^2 + xy^2$$

$$= -3 \text{ at } (1, -1, 1)$$

$$\text{curl}\,\mathbf{f} = \nabla \times \mathbf{f} = \begin{vmatrix} \mathbf{i} & \mathbf{j} & \mathbf{k} \\ \dfrac{\partial}{\partial x} & \dfrac{\partial}{\partial y} & \dfrac{\partial}{\partial z} \\ x^2z & -2y^3z^2 & xy^2z \end{vmatrix}$$

$$= \mathbf{i}\,(2xyz + 4y^3z) + \mathbf{j}\,(x^2 - y^2z) + \mathbf{k}\,(0 - 0)$$

$$= -6\mathbf{i} \text{ at } (1, -1, 1).$$

Example 2. *Prove that* $r^n\mathbf{r}$ *is irrotational. Find n when it is solenoidal.*

(Ajmer 97, Gorakhpur 98, Rohilkhand 98)

Solution. *A vector is said to be irrotational if curl* $\mathbf{f} = 0$ *and solenoidal if div* $\mathbf{f} = 0$.

$$r^n\mathbf{r} = r^nx\mathbf{i} + r^ny\mathbf{j} + r^nz\mathbf{k} = f_1\mathbf{i} + f_2\mathbf{j} + f_3\mathbf{k}$$

$$\therefore \qquad \text{div}\, r^n\mathbf{r} = \frac{\partial f_1}{\partial x} + \frac{\partial f_2}{\partial y} + \frac{\partial f_3}{\partial z}$$

Now, $\quad f_1 = r^n x \Rightarrow \dfrac{\partial f_1}{\partial x} = r^n + xnr^{n-1}\dfrac{\partial r}{\partial x}$

$$= r^n + nxr^{n-1}\frac{x}{r} = r^n + nr^{n-2}x^2 \text{ etc.}$$

$$\therefore \quad \text{div}\, r^n\mathbf{r} = \Sigma\frac{\partial f_1}{\partial x} = 3r^n + nr^{n-2}\,\Sigma x^2$$

$$= 3r^n + nr^{n-2}r^2 = (n+3)\,r^n$$

For $r^n\mathbf{r}$ to be solenoidal, $(n + 3)\, r^n = 0$.

$\Rightarrow \quad n + 3 = 0 \Rightarrow n = -3.$

$$\text{curl } \mathbf{f} = \begin{vmatrix} \mathbf{i} & \mathbf{j} & \mathbf{k} \\ \dfrac{\partial}{\partial x} & \dfrac{\partial}{\partial y} & \dfrac{\partial}{\partial z} \\ r^n x & r^n y & r^n z \end{vmatrix}$$

coeff. of **i** is $\dfrac{\partial}{\partial y}\left(r^n z\right) - \dfrac{\partial}{\partial z}\left(r^n y\right)$

$$= z.nr^{n-1}\frac{\partial r}{\partial y} - y.nr^{n-1}\frac{\partial r}{\partial z}$$

$$= nr^{n-1}\left[z\cdot\frac{y}{r} - y\cdot\frac{z}{r}\right] = 0$$

Similarly, coeff. of **j** = coeff. of **k** = 0

$\therefore$ curl $r^n\mathbf{r} = 0$.

$\Rightarrow$ $r^n\mathbf{r}$ is a irrotational vector function.

Example 3. *Prove that* $\nabla^2 f(r) = f''(r) + \dfrac{2}{r} f'(r)$. *(Ajmer 97)*

Solution. Let $\phi = f(r)$; $\therefore$ $\nabla^2 f(r) = \nabla^2\phi = \Sigma\dfrac{\partial^2\phi}{\partial x^2}\cdot$

$$\frac{\partial\phi}{\partial x} = f'(r)\frac{\partial r}{\partial x} = \frac{x}{r} f'(r)$$

$$\therefore \quad \frac{\partial^2\phi}{\partial x^2} = \frac{r\left\{1\cdot f'(r) + x f''(r)\cdot\dfrac{x}{r}\right\} - x f'(r)\cdot\dfrac{x}{r}}{r^2}$$

$$= \frac{1}{r} f'(r) + \frac{x^2}{r^2} f''(r) - \frac{x^2}{r^3} f'(r)$$

$$\therefore \quad \Sigma\frac{\partial^2\phi}{\partial x^2} = \frac{3}{r} f'(r) + \frac{\Sigma x^2}{r^2} f''(r) - \frac{\Sigma x^2}{r^3} f'(r)$$

$$= \frac{3}{r} f'(r) + f''(r) - \frac{1}{r} f'(r)$$

$$= f''(r) + \frac{2}{r} f'(r).$$

Example 4. *Prove that* $\nabla^2\left(\dfrac{x}{r^2}\right) = \dfrac{-2x}{x^4}\cdot$

Solution. $$\frac{\partial^2}{\partial x^2}\left(\frac{x}{r^2}\right)=\frac{\partial}{\partial x}\left[\frac{\partial}{\partial x}\left(\frac{x}{r^2}\right)\right]$$

$$=\frac{\partial}{\partial x}\left[1\cdot\frac{1}{r^2}-\frac{2x}{r^3}\cdot\frac{x}{r}\right] \qquad \left[\because \frac{\partial r}{\partial x}=\frac{x}{r}\right]$$

$$=\frac{\partial}{\partial x}\left[\frac{1}{r^2}-\frac{2x^2}{r^4}\right]$$

$$=\left[-\frac{2}{r^3}\cdot\frac{x}{r}-\frac{4x}{r^4}+\frac{8x^2}{r^5}\cdot\frac{x}{r}\right]$$

$$=\left[-\frac{2}{r^4}x-\frac{4x}{r^4}+\frac{8x^3}{r^6}\right]$$

$$=-\frac{6x}{r^4}+\frac{8x^3}{r^6}=-2x\left[\frac{3}{r^4}-\frac{4x^2}{r^6}\right] \qquad ...(1)$$

$$\frac{\partial^2}{\partial y^2}\left(\frac{x}{r^2}\right)=\frac{\partial}{\partial y}\left[\frac{\partial}{\partial y}\left(\frac{x}{r^2}\right)\right]$$

$$=\frac{\partial}{\partial y}\left[-\frac{2x}{r^3}\cdot\frac{\partial r}{\partial y}\right]$$

$$=\frac{\partial}{\partial y}\left[-\frac{2x}{r^3}\cdot\frac{y}{r}\right]=\frac{\partial}{\partial y}\left[-\frac{2xy}{r^4}\right]$$

$$=-2x\left[\frac{1}{r^4}-\frac{4y}{r^5}\cdot\frac{y}{r}\right]=-2x\left[\frac{1}{r^4}-\frac{4y^2}{r^6}\right] \qquad ...(2)$$

Similarly, $$\frac{\partial^2}{\partial z^2}\left(\frac{x}{r^2}\right)=-2x\left[\frac{1}{r^4}-\frac{4z^2}{r^6}\right] \qquad ...(3)$$

$$\therefore \quad \nabla^2\left(\frac{x}{r^2}\right)=\Sigma\frac{\partial^2}{\partial x^2}\left(\frac{x}{r^2}\right)$$

$$=-2x\left[\frac{3}{r^4}+\frac{1}{r^4}+\frac{1}{r^4}-\frac{4}{r^6}\left(x^2+y^2+z^2\right)\right]$$

by (1), (2) and (3)

$$=-2x\left[\frac{5}{r^4}-\frac{4}{r^4}\right]=-\frac{2x}{r^4}.$$

Example 5. *If* **f** *and* **g** *are irrotational then show that* **f** × **g** *is a solenoidal vector.*

Solution. Given that curl **f** = 0, curl **g** = 0.

$$\text{div}\,(\mathbf{f}\times\mathbf{g}) = (\text{curl}\,\mathbf{f})\cdot\mathbf{g} + (\text{curl}\,\mathbf{g})\cdot\mathbf{f}$$
$$= 0 + 0 = 0.$$

Hence, **f** × **g** is solenoidal.

Example 6. *Prove that div* [(**r** × **a**) × **b**) = − 2**b** . **a**.

(*Gorakhpur 99, Lucknow 97*)

Solution.
$$\mathbf{r}\times\mathbf{a} = \begin{vmatrix} \mathbf{i} & \mathbf{j} & \mathbf{k} \\ x & y & z \\ a_1 & a_2 & a_3 \end{vmatrix}$$
$$= \mathbf{i}\,(ya_3 - za_2) + \mathbf{j}\,(za_1 - xa_3) + \mathbf{k}\,(xa_2 - ya_1)$$

$$\therefore \quad (\mathbf{r}\times\mathbf{a})\times\mathbf{b} = \begin{vmatrix} \mathbf{i} & \mathbf{j} & \mathbf{k} \\ ya_3 - za_2 & za_1 - xa_3 & xa_2 - ya_1 \\ b_1 & b_2 & b_3 \end{vmatrix}$$
$$= \Sigma\,\mathbf{i}\left[b_3(za_1 - xa_3) - b_2(xa_3 - ya_1)\right]$$
$$= f_1\mathbf{i} + f_2\mathbf{j} + f_3\mathbf{k}$$

$$\therefore \quad \frac{\partial f_1}{\partial x} = (-b_3a_3 - b_2a_2) \text{ etc.}$$

Now,
$$\text{div}\,.\,\mathbf{f} = \frac{\partial f_1}{\partial x} + \frac{\partial f_2}{\partial y} + \frac{\partial f_3}{\partial z}$$
$$= (-b_3a_3 - b_2a_2) + (-b_2a_2 - b_1a_1) + (-b_1a_1 - b_3a_3)$$
$$= -2\,(a_1b_1 + a_2b_2 + a_3b_3)$$
$$= -2\,(\mathbf{a}\cdot\mathbf{b}) = -2\,(\mathbf{b}\cdot\mathbf{a}).$$

Example 7. *Prove that div* $\left(\frac{f(r)}{r}\mathbf{r}\right) = \frac{1}{r^2}\frac{d}{dr}\left[r^2 f(r)\right]$.

Solution.
$$\frac{f(r)}{r}\mathbf{r} = \frac{f(r)}{r}(\mathbf{i}x + \mathbf{j}y + \mathbf{k}z)$$
$$= \frac{f(r)}{r}x\mathbf{i} + \frac{f(r)}{r}y\mathbf{j} + \frac{f(r)}{r}z\mathbf{k}$$
$$= f_1\mathbf{i} + f_2\mathbf{j} + f_3\mathbf{k}$$

$$\therefore \quad \text{div}\ \frac{f(r)}{r}\mathbf{r} = \frac{\partial f_1}{\partial x} + \frac{\partial f_2}{\partial y} + \frac{\partial f_3}{\partial z} = \Sigma \frac{\partial f_1}{\partial x}$$

$$f_1 = \frac{f(r)}{r}x, \ \therefore \ \frac{\partial f_1}{\partial x} = \frac{f(r)}{r}\cdot 1 + x\cdot\frac{rf'(r) - f(r)}{r^2}\cdot\frac{\partial r}{\partial x}$$

$$= \frac{f(r)}{r} + x\cdot\frac{rf'(r) - f(r)}{r^2}\cdot\frac{x}{r}$$

$$= \frac{f(r)}{r} + \frac{x^2}{r^2}f'(r) - \frac{x^2}{r^3}f(r)$$

$$\therefore \qquad \Sigma\frac{\partial f_1}{\partial x} = \frac{3f(r)}{r} + \frac{\Sigma x^2}{r^2}f'(r) - \frac{\Sigma x^2}{r^3}f(r)$$

$$= \frac{3f(r)}{r} + f'(r) - \frac{1}{r}f(r)$$

$$= \frac{2f(r)}{r} + f'(r) = \frac{2rf(r) + r^2 f'(r)}{r^2}$$

$$= \frac{1}{r^2}\frac{d}{dr}\left[r^2 f(r)\right].$$

Example 8. *Find div grad r^m and verify that*

$$\nabla \times \nabla\, r^m = 0.$$

Solution. We have

$$\text{grad}\ r^m = mr^{m-2}\,\mathbf{r}$$

$$= mr^{m-2}(\mathbf{i}x + \mathbf{j}y + \mathbf{k}z)$$

$$\therefore \qquad \text{div grad}\ r^m = \frac{\partial}{\partial x}\left(mr^{m-2}x\right) + \frac{\partial}{\partial y}\left(mr^{m-2}y\right) + \frac{\partial}{\partial z}\left(mr^{m-2}z\right).$$

Also

$$\frac{\partial}{\partial x}\left(mr^{m-2}x\right) = m(m-2)\,r^{m-3} + \frac{\partial r}{\partial x}x + mr^{m-2}$$

$$= m(m-2)\,r^{m-4}x^2 + mr^{m-2}, \text{ for } \partial r/\partial x = x/r.$$

Similarly,

$$\frac{\partial}{\partial y}\left(mr^{m-2}y\right) = m(m-2)\,r^{m-1}y^2 + mr^{m-2},$$

$$\frac{\partial}{\partial z}\left(mr^{m-2}z\right) = m(m-2)\,r^{m-4}z^2 + mr^{m-2}.$$

$$\therefore \quad \text{div grad } r^m = m(m-2)\, r^{m-4} r^2 + 3mr^{m-2}.$$

$$= m(m+1)\, r^{m-2}.$$

$$\text{Again curl grad } r^m = \begin{vmatrix} \mathbf{i} & \mathbf{j} & \mathbf{k} \\ \dfrac{\partial}{\partial x} & \dfrac{\partial}{\partial y} & \dfrac{\partial}{\partial z} \\ mr^{m-2}x & mr^{m-2}y & mr^{m-2}z \end{vmatrix}$$

Thus, the coefficient of **i**, in curl grad r^m

$$= \frac{\partial}{\partial y}\left(mr^{m-2}z\right) - \frac{\partial}{\partial z}\left(mr^{m-2}y\right)$$

$$= m(m-2)\, r^{m-3} \frac{\partial r}{\partial y} z - m(m-2)\, r^{m-3} \frac{\partial r}{\partial z} y$$

$$= m(m-2)\, r^{m-3} \frac{yz}{r} - m(m-2)\, r^{m-3} \frac{zy}{r} = 0.$$

Hence, curl grad $r^m = \nabla \times \nabla r^m = 0.$

Example 9. *Show that*

$$curl\ \frac{\mathbf{a} \times \mathbf{r}}{r^3} = -\frac{\mathbf{a}}{r^3} + \frac{3\mathbf{r}}{r^3}(\mathbf{a} \,.\, \mathbf{r}),$$

where **a** *is a constant vector.*

(*Kolkata 99, Gorakhpur 98, 2000, Lucknow 97*)

Solution. Now,

$$\text{curl}\, \frac{\mathbf{a} \times \mathbf{r}}{r^3} = \Sigma \mathbf{i} \times \frac{\partial}{\partial x}\left(\frac{\mathbf{a} \times \mathbf{r}}{r^3}\right)$$

Also

$$\frac{\partial}{\partial x}\left(\frac{\mathbf{a} \times \mathbf{r}}{r^3}\right) = -\frac{3}{r^4}\frac{\partial r}{\partial x}(\mathbf{a} \times \mathbf{r}) + \frac{1}{r^3}\left(\mathbf{a} \times \frac{\partial \mathbf{r}}{\partial x}\right)$$

$$= -\frac{3x}{r^5}(\mathbf{a} \times \mathbf{r}) + \frac{1}{r^3}\mathbf{a} \times \mathbf{i}.$$

$$\therefore \quad \mathbf{i} \times \frac{\partial}{\partial x}\left(\frac{\mathbf{a} \times \mathbf{r}}{r^3}\right) = -\frac{3x}{r^5}\left[(\mathbf{i} \,.\, \mathbf{r})\, \mathbf{a} - (\mathbf{i} \,.\, \mathbf{a})\, \mathbf{r}\right] + \frac{1}{r^3}\left[(\mathbf{i} \,.\, \mathbf{i})\mathbf{a} - (\mathbf{i} \,.\, \mathbf{a})\, \mathbf{i}\right]$$

$$= -\frac{3x^2}{r^5}\mathbf{a} + \frac{3\mathbf{r}}{r^5}(\mathbf{i}x \cdot \mathbf{a}) + \frac{1}{r^3}\mathbf{a} - \frac{1}{r^3}(\mathbf{i} \cdot \mathbf{a})\, \mathbf{i}.$$

$$\therefore \quad \Sigma \mathbf{i} \times \frac{\partial}{\partial x}\left(\frac{\mathbf{a} \times \mathbf{r}}{r^3}\right) = -\frac{3\mathbf{a}}{r^5} r^2 + \frac{3\mathbf{r}}{r^5} \mathbf{r} \cdot \mathbf{a} + \frac{3}{r^3} \mathbf{a} - \frac{1}{r^3} \mathbf{a}$$

$$= -\frac{\mathbf{a}}{r^3} + \frac{3\mathbf{r}}{r^5} \mathbf{a} \,.\, \mathbf{r}.$$

The question could also be done by employing the results of § 6.15 V, VI.

Example 6. *If* **F** *and* ϕ *are point functions, show that the components of the former normal and tangential to the surface* $\phi = 0$ *at any point thereof are*

$$\frac{(\mathbf{F} \,.\, \nabla\phi)\, \nabla\phi}{(\nabla\phi)^2} \quad and \quad \frac{\nabla\phi \times (\mathbf{F} \times \nabla\phi)}{(\nabla\phi)^2}.$$

Solution. The unit vector normal to the surface $\phi = 0$ being $\nabla\phi \,/\, |\nabla\phi|$, the magnitude of the component of **F** along this normal

$$= \mathbf{F} \cdot \frac{\nabla\phi}{|\nabla\phi|} = \frac{(\mathbf{F} \cdot \nabla\phi)}{|\nabla\phi|}$$

Thus, the vector **F** is equal to the sum of vector

$$\frac{(\mathbf{F} \cdot \nabla\phi)}{|\nabla\phi|} \frac{\nabla\phi}{|\nabla\phi|} = \frac{(\mathbf{F} \cdot \nabla\phi)}{(\nabla\phi)^2} \nabla\phi \qquad \text{...(1)}$$

along the normal and the vector

$$\mathbf{F} - \frac{\mathbf{F} \cdot \nabla\phi}{(\nabla\phi)^2} \nabla\phi \qquad \text{...(2)}$$

along the tangent. This latter

$$= \frac{\mathbf{F}\,(\nabla\phi \cdot \nabla\phi) - (\mathbf{F} \cdot \nabla\phi)\, \nabla\phi}{(\nabla\phi)^2}$$

$$= \frac{\nabla\phi \times (\mathbf{F} \times \nabla\phi)}{(\nabla\phi)^2}$$

Hence, the result.

EXERCISES

(In the following, **r** denotes the position vector of any point $r = |\mathbf{r}|$ and **a, b,** denote constant vectors. Also x, y, z denote rectangular cartesian co-ordinates).

1. If

$$\mathbf{f} = \mathbf{i}\,(x + y + 1) + \mathbf{j} + \mathbf{k}\,(-x - y),$$

prove that

$$\mathbf{f} \,.\, \text{curl}\, \mathbf{f} = 0.$$

2. Find div **F** and curl **F** where

$$\mathbf{F} = \text{grad } (x^3 + y^3 + z^3 - 3xyz).$$

3. If $\phi = x/r^3$, show that $\nabla^2\phi = 0$.

4. Show that

(*i*) div $\mathbf{r} = 3$. (*Ajmer 98, Kumaon 2000*)

(*ii*) curl $\mathbf{r} = 0$. (*Ajmer 98, Kumaon 2000*)

(*iii*) $\nabla^2 r^{-1} = 0$. (*Gorakhpur 99*)

(*iv*) div $(r^n\mathbf{r}) = (n + 3)\, r^n$.

5. Show that

(*i*) div $(\mathbf{r} \times \mathbf{a}) = 0$, (*ii*) curl $(\mathbf{r} \times \mathbf{a}) = -2\mathbf{a}$

(*iii*) grad $(\mathbf{r} \cdot \mathbf{a}) = \mathbf{a}$.

6. Show that

(*i*) grad $[\mathbf{r}\ \mathbf{a}\ \mathbf{b}] = \mathbf{a} \times \mathbf{b}$ (*ii*) div $[(\mathbf{r} \times \mathbf{a}) \times \mathbf{b}] = 0$. (*Ajmer 98*)

(*iii*) curl $[(\mathbf{r} \times \mathbf{a}) \times \mathbf{b}] = 2\mathbf{b} \times \mathbf{a}$.

(*iv*) grad $[(\mathbf{r} \times \mathbf{a}) \cdot (\mathbf{r} \times \mathbf{b})] = (\mathbf{b} \times \mathbf{r}) \times \mathbf{a} + (\mathbf{a} \times \mathbf{r}) \times \mathbf{b}$.

7. Evaluate

$$\nabla \cdot (\mathbf{a}\ \mathbf{b} \cdot \mathbf{r}) \text{ and } \nabla \times [\mathbf{a} \times (\mathbf{b} \times \mathbf{r})].$$

8. Expand

$$\nabla\left(\mathbf{a} \cdot \frac{\mathbf{r}}{r^n}\right),\ \nabla \cdot \left(\mathbf{a} \times \frac{\mathbf{r}}{r^n}\right),\ \nabla \times \left(\mathbf{a} \times \frac{\mathbf{r}}{r^n}\right).$$

9. Show that

$$\text{curl } \mathbf{r}\phi\ (r) = 0 \text{ and } (\mathbf{a} \cdot \mathbf{r})\ \mathbf{a} = 0.$$

Also find

$$\text{curl } \mathbf{a}\phi\ (r) \text{ and } (\nabla\phi \cdot \nabla)\phi.$$

6. If

$$\mathbf{f} = (\mathbf{a} \times \mathbf{r})\ r^n,$$

show that

$$\text{div } \mathbf{f} = 0,\ \text{curl } \mathbf{f} = (n + 2)\ r^n\ \mathbf{a} - nr^{n-2}\ (\mathbf{a} \cdot \mathbf{r})\ \mathbf{r}.$$

11. A vector function **f** is the product of a scalar function and the gradient of a scalar function, show that

$$\mathbf{f} \cdot \text{curl } \mathbf{f} = 0.$$

12. If $\mathbf{f} = \nabla\ (\mathbf{a} \cdot \nabla r^{-1})$, show that div $\mathbf{f} = 0$, and $\mathbf{f} = \text{curl } \mathbf{g}$, where

$$\mathbf{g} = -\ \mathbf{a} \times \nabla\ (r^{-1}).$$

13. Show that

$$\text{curl } (\psi\nabla\phi) = \nabla\phi \times \nabla\psi = -\text{ curl } (\phi\nabla\psi)$$

where ϕ and ψ are two scalar functions.

14. ϕ_1 and ϕ_2 are two scalar functions such that

$$\left(\nabla^2 + k^2\right)\phi_1 = 0 \text{ and } \left(\nabla^2 + k^2\right)\phi_2 = 0$$

and $\mathbf{f} = \nabla \times [\mathbf{r}\phi_1 + \mathbf{r} \times \nabla\phi_2]$, show that div $\mathbf{f} = 0$, $(\nabla^2 + k^2)\,\mathbf{f} = 0$.

15. ρ and p are two scalar point functions such that ρ is a function of p; show that

$$\nabla\rho = \frac{d\rho}{dp}\nabla p.$$

16. Show that

$$\mathbf{q} \cdot \nabla\mathbf{q} = \frac{1}{2}\nabla\mathbf{q}^2 - \mathbf{q} \times \text{curl } \mathbf{q}.$$

17. Show that

$$\text{div. curl curl } (\mathbf{a}\phi) + \nabla^2 \text{ div } (\mathbf{a}\phi) = \mathbf{a} \,.\, \text{grad } \nabla^2\phi.$$

where ϕ is a scalar point function.

6.18. DIFFERENTIAL OPERATIONS IN TERMS OF CURVILINEAR CO-ORDINATES

Curvilinear co-ordinates. Consider three continuously differentiable scalar functions

$$u = \phi_1(x, y, z), \quad v = \phi_2(x, y, z), \quad w = \phi_3(x, y, z)$$

defined in any region R of space; x, y, z denoting the rectangular cartesian co-ordinates of any point P of R. These three functions associate a triad of numbers u, v, w to each point of the region and we suppose that these functions are such that to two different points of the region correspond different triads. Thus, to the aggregate of the points P of the region R, corresponds an aggregate of triads u, v, w such that to each point P of the region corresponds one and only one triad (u, v, w). We say that (u, v, w) *are the curvilinear co-ordinates of P* in as much as the point uniquely determines and is uniquely determined by the triad.

6.18.1. Co-ordinate Surfaces and Curves

The loci of the equations

$$u = u_0, \quad v = v_0, \quad w = w_0$$

[Curvilinear co-ordinates]

which are the same as those of the equations

$$\phi_1(x, y, z) = u_0, \quad \phi_2(x, y, z) = v_0, \quad \phi_3(x, y, z) = w_0,$$

[Cartesian co-ordinates]

u_0, v_0, w_0, being constants, are called co-ordinate surfaces. Through each point there passes a system of three co-ordinate surfaces.

The three curves of intersection of three co-ordinate surfaces through any point, taken in pairs, are called *co-ordinate curves* through the point. The three co-ordinate curves through any point (u_0, v_0, w_0) are

$$v = v_0,\ w = w_0,\ w = w_0,\ u = u_0;\ u = u_0,\ v = v_0.$$

6.18.2. Orthogonal Curvilinear Co-ordinates

A system of co-ordinates is said to be orthogonal, if the three co-ordinate surfaces through every point are mutually orthogonal, *i.e.*, if the three tangent planes, and accordingly the three normals to the three surfaces, through every point are orthogonal in pairs. The line of intersection of the tangent planes to any two co-ordinate surfaces is the normal to the third. The tangent to any co-ordinate curve through a point, being the line of intersection of the tangent planes to the two co-ordinate surfaces through the curve, lies along the normal to the third surface. We shall concerned with orthogonal co-ordinates only.

6.18.3. Condition for Orthogonality

Suppose that on solving

$$u = \phi_1(x, y, z),\quad v = \phi_2(x, y, z),\quad w = \phi_3(x, y, z)$$

for x, y, z we obtain

$$x = \psi_1(u, v, w),\quad y = \psi_2(u, v, w),\quad z = \psi_3(u, v, w).$$

We write

$$\begin{aligned}\mathbf{r} &= \mathbf{i}x + \mathbf{j}y + \mathbf{k}z\\ &= \mathbf{i}\psi_1(u, v, w) + \mathbf{j}\psi_2(u, v, w) + \mathbf{k}\psi_3(u, v, w)\\ &= \mathbf{f}(u, v, w), \text{ say.}\end{aligned}$$

The co-ordinate curve $v = v_0$, $w = w_0$ through (u_0, v_0, w_0) is the same as

$$\mathbf{r} = \mathbf{f}(u, v_0, w_0);$$

u being the parameter.

The tangent to this co-ordinate curve is parallel to the vector

$$\frac{\partial \mathbf{r}}{\partial u}.$$

The tangents to the other two co-ordinate curves are parallel to the vectors

$$\frac{\partial \mathbf{r}}{\partial v},\ \frac{\partial \mathbf{r}}{\partial w}.$$

Thus, for the orthogonality of the curvilinear co-ordinate system, we must have

$$\frac{\partial \mathbf{r}}{\partial u}\cdot\frac{\partial \mathbf{r}}{\partial v} = 0,\quad \frac{\partial \mathbf{r}}{\partial v}\cdot\frac{\partial \mathbf{r}}{\partial w} = 0,\quad \frac{\partial \mathbf{r}}{\partial w}\cdot\frac{\partial \mathbf{r}}{\partial u} = 0,$$

which are the required conditions.

6.19. DIFFERENTIAL OF LENGTH

We have

$$d\mathbf{r} = \frac{\partial \mathbf{r}}{\partial u} du + \frac{\partial \mathbf{r}}{\partial v} dv + \frac{\partial \mathbf{r}}{\partial w} dw. \qquad \textit{(Lucknow 98)}$$

$$\therefore \quad d\mathbf{r} \cdot d\mathbf{r} = \left(\frac{\partial \mathbf{r}}{\partial u}\right)^2 (du)^2 + \left(\frac{\partial \mathbf{r}}{\partial v}\right)^2 (dv)^2 + \left(\frac{\partial \mathbf{r}}{\partial w}\right)^2 (dw)^2 +$$

$$2\frac{\partial \mathbf{r}}{\partial u} \cdot \frac{\partial \mathbf{r}}{\partial u} du\, dv + 2\frac{\partial \mathbf{r}}{\partial v} \cdot \frac{\partial \mathbf{r}}{\partial w} dv\, dw + 2\frac{\partial \mathbf{r}}{\partial w} \cdot \frac{\partial \mathbf{r}}{\partial u} dw\, du.$$

Hence,

$$(ds)^2 = \left(\frac{\partial \mathbf{r}}{\partial u}\right)^2 (du)^2 + \left(\frac{\partial \mathbf{r}}{\partial v}\right)^2 (dv)^2 + \left(\frac{\partial \mathbf{r}}{\partial w}\right)^2 (dw)^2 \qquad ...(1)$$

by virtue of the conditions for the orthogonality of the curvilinear co-ordinates.

If ds_1 denote the differential of length along the co-ordinate curve

$$v = v_0,\ w = w_0$$

so that

$$dv = 0, \quad dw = 0,$$

we have

$$(ds_1)^2 = \left(\frac{\partial \mathbf{r}}{\partial u}\right)^2 (du)^2. \qquad ...(2)$$

Similarly, for the other two coordinate curves.

6.20. FUNDAMENTAL TRIAD OF MUTUALLY ORTHOGONAL UNIT VECTORS THROUGH ANY POINT

By § 6.1.1, the vectors

$$\nabla u, \nabla v, \nabla w$$

lie along the normals to the co-ordinate surfaces which are the level surfaces of the functions u, v, w.

Because of the assumed orthogonality of the curvilinear co-ordinate system, we have

$$\nabla u \,.\, \nabla v = 0,\ \nabla v \,.\, \nabla w = 0,\ \nabla w \,.\, \nabla u = 0.$$

We suppose that the mutually orthogonal unit vectors

$$\frac{\nabla u}{|\nabla u|}, \frac{\nabla v}{|\nabla v|}, \frac{\nabla w}{|\nabla w|},$$

in this order, form a right handed system, for otherwise we could change the order of consideration of the three surfaces.

Now, by § 6.1.1, $|\nabla u|$ is the directional derivatives of u along the direction of the normal to the surface $u = u_0$, *i.e.*, along the tangent to the curve $v = v_0$, $w = w_0$. A ds_1 denotes the differential to length along this curve, we see that

$$|\nabla u| = \frac{du}{ds_1}$$

$$= \frac{1}{\sqrt{\left(\frac{\partial \mathbf{r}}{\partial u}\right)^2}}. \qquad \text{[By (2) above]}$$

We write

$$h_1 = \frac{1}{|\nabla u|} = \sqrt{\left(\frac{\partial \mathbf{r}}{\partial u}\right)^2},$$

$$h_2 = \frac{1}{|\nabla v|} = \sqrt{\left(\frac{\partial \mathbf{r}}{\partial v}\right)^2},$$

$$h_3 = \frac{1}{|\nabla w|} = \sqrt{\left(\frac{\partial \mathbf{r}}{\partial w}\right)^2},$$

and see that

$$\mathbf{a} = h_1 \nabla u, \quad \mathbf{b} = h_2 \nabla v, \quad \mathbf{c} = h_3 \nabla w$$

is a right handed system of mutually orthogonal unit vectors. These systems of vectors are different for different points.

6.21. DIFFERENTIAL OPERATORS IN TERMS OF ORTHOGONAL CURVILINEAR CO-ORDINATES

6.21.1. Gradient

Consider any scalar point function

$$\phi\,(u, v, w)$$

given in terms of orthogonal curvilinear co-ordinates u, v, w.

We think of u, v, w as functions of rectangular cartesian co-ordinates x, y, z. We have

$$\frac{\partial \phi}{\partial x} = \frac{\partial \phi}{\partial u}\frac{\partial u}{\partial x} + \frac{\partial \phi}{\partial v}\frac{\partial v}{\partial x} + \frac{\partial \phi}{\partial w}\frac{\partial w}{\partial x},$$

$$\frac{\partial \phi}{\partial y} = \frac{\partial \phi}{\partial u}\frac{\partial u}{\partial y} + \frac{\partial \phi}{\partial v}\frac{\partial v}{\partial y} + \frac{\partial \phi}{\partial w}\frac{\partial w}{\partial y},$$

$$\frac{\partial \phi}{\partial z} = \frac{\partial \phi}{\partial u}\frac{\partial u}{\partial z} + \frac{\partial \phi}{\partial v}\frac{\partial v}{\partial z} + \frac{\partial \phi}{\partial w}\frac{\partial w}{\partial z}.$$

Multiplying with the vectors **i, j, k** and adding, we get

$$\nabla\phi = \frac{\partial\phi}{\partial u}\nabla u + \frac{\partial\phi}{\partial v}\nabla v + \frac{\partial\phi}{\partial w}\nabla w$$

$$= \frac{1}{h_1}\frac{\partial\phi}{\partial u}\mathbf{a} + \frac{1}{h_2}\frac{\partial\phi}{\partial v}\mathbf{b} + \frac{1}{h_3}\frac{\partial\phi}{\partial w}\mathbf{c},$$

which is the required expression. Thus, the components of grad ϕ along the vectors **a, b, c** are

$$\frac{1}{h_1}\frac{\partial\phi}{\partial u}, \frac{1}{h_2}\frac{\partial\phi}{\partial v}, \frac{1}{h_3}\frac{\partial\phi}{\partial w}$$

respectively.

6.21.2. Divergence

Consider any vector point function

$$\mathbf{f}(u, v, w).$$

Let

$$\mathbf{f} = f_1\mathbf{a} + f_2\mathbf{b} + f_3\mathbf{c},$$

so that f_1, f_2, f_3 are the components of **f** along **a, b, c**.

We have

$$\mathbf{f} = f_1\mathbf{b}\times\mathbf{c} + f_2\mathbf{c}\times\mathbf{a} + f_3\mathbf{a}\times\mathbf{b}$$

$$= f_1h_2h_3\,\nabla v\times\nabla w + f_2h_3h_1\,\nabla w\times\nabla u + f_3h_1h_2\,\nabla u\times\nabla v.$$

$$\therefore\quad \nabla\cdot\mathbf{f} = \nabla\cdot(f_1h_2h_3\,\nabla v\times\nabla w) + \nabla\cdot(f_2h_3h_1\,\nabla w\times\nabla u) + \nabla\cdot(f_3h_1h_2\,\nabla u\times\nabla v) \quad (\S\ 6.14) \quad ...(1)$$

By § 6.15, III,

$$\nabla\cdot(f_1h_2h_3\,\nabla v\times\nabla w) = f_1h_2h_3\,.\,\nabla(\nabla v\times\nabla w) + \nabla v\times\nabla w\,.\,\nabla(f_1h_2h_3) \quad ...(2)$$

By § 6.15, IV,

$$\nabla\cdot(\nabla v\times\nabla w) = \nabla w\,.\,\text{curl}\,\nabla v - \nabla v\,.\,\text{curl}\,\nabla w = 0, \quad ...(3)$$

for by § 6.16,

$$\text{curl}\,\nabla v = \text{curl grad}\,v = \mathbf{0},\ \text{curl}\,\nabla w = \mathbf{0}.$$

Also as in § 6.21,

$$\nabla(f_1h_2h_3) = \frac{\partial(f_1h_2h_3)}{\partial u}\nabla u + \frac{\partial(f_1h_2h_3)}{\partial v}\nabla v + \frac{\partial(f_1h_2h_3)}{\partial w}\nabla w. \quad ...(4)$$

From (2), (3) and (4)

$\nabla \cdot (f_1 h_2 h_3 \nabla v \times \nabla w)$

$$= \nabla v \times \nabla w \cdot \nabla u + \frac{\partial (f_1 h_2 h_3)}{\partial u}$$

$$= \frac{\mathbf{b} \times \mathbf{c} \cdot \mathbf{a}}{h_1 h_2 h_3} \frac{\partial (f_1 h_2 h_3)}{\partial u} = \frac{1}{h_1 h_2 h_3} \frac{\partial (f_1 h_2 h_3)}{\partial u}. \quad ...(5)$$

Finally, from (1) and (5),

$$\nabla . \mathbf{f} = \frac{1}{h_1 h_2 h_3} \left[\frac{\partial (f_1 h_2 h_3)}{\partial u} + \frac{\partial (f_2 h_3 h_1)}{\partial v} + \frac{\partial (f_3 h_1 h_2)}{\partial w} \right],$$

which is the required expression for div **f**. *(Lucknow 97)*

6.21.3. Curl

Consider a vector point function **f** (u, v, w).

Let

$$\mathbf{f} = f_1 \mathbf{a} + f_2 \mathbf{b} + f_3 \mathbf{c} = h_1 f_1 \nabla u + h_2 f_2 \nabla v + h_3 f_3 \nabla w.$$

$$\therefore \quad \nabla \times \mathbf{f} = \nabla \times (h_1 f_1 \nabla u) + \nabla \times (h_2 f_2 \nabla v) + \nabla \times (h_3 f_3 \nabla w)$$

(§ 6.14) ...(1)

By § 6.16, V,

$$\nabla \times (h_1 f_1 \nabla u) = \nabla (h_1 f_1) \times \nabla v + h_1 f_1 \nabla \times \nabla u$$

$$= \nabla (h_1 f_1) \times \nabla u, \text{ for } \nabla \times \nabla u = 0, \quad (\S\ 6.16, \text{ II})$$

$$= \left[\frac{\partial (h_1 f_1)}{\partial u} \nabla u + \frac{\partial (h_1 f_1)}{\partial v} \nabla v + \frac{\partial (h_1 f_1)}{\partial w} \nabla w \right] \times \nabla u$$

$$= \frac{\partial (h_1 f_1)}{\partial v} \nabla v \times \nabla u + \frac{\partial (h_1 f_1)}{\partial w} \nabla w \times \nabla u$$

$$= \frac{1}{h_1 h_2} \frac{\partial (h_1 f_1)}{\partial v} \mathbf{b} \times \mathbf{a} + \frac{1}{h_3 h_1} \frac{\partial (h_1 f_1)}{\partial w} \mathbf{c} \times \mathbf{a}$$

$$= -\frac{1}{h_1 h_2} \frac{\partial (h_1 f_1)}{\partial v} \mathbf{c} + \frac{1}{h_3 h_1} \frac{\partial (h_1 f_1)}{\partial w} \mathbf{b}. \quad ...(2)$$

From (1) and (2),

$$\nabla \times \mathbf{f} = \frac{1}{h_2 h_3} \left[\frac{\partial (h_3 f_3)}{\partial v} - \frac{\partial (h_2 f_2)}{\partial w} \right] \mathbf{a} + \frac{1}{h_3 h_1} \left[\frac{\partial (h_1 f_1)}{\partial w} - \frac{\partial (h_3 f_3)}{\partial u} \right] \mathbf{b} + \frac{1}{h_1 h_2} \left[\frac{\partial (h_2 f_2)}{\partial u} - \frac{\partial (h_1 f_1)}{\partial v} \right] \mathbf{c}.$$

$$= \frac{1}{h_1 h_2 h_3} \begin{vmatrix} h_1 \mathbf{a} & h_2 \mathbf{b} & h_3 \mathbf{c} \\ \dfrac{\partial}{\partial u} & \dfrac{\partial}{\partial v} & \dfrac{\partial}{\partial w} \\ h_1 f_1 & h_2 f_2 & h_3 f_3 \end{vmatrix}$$

6.22.4. Laplacian, ∇^2

We have

$$\nabla^2\phi = \nabla \,.\, \nabla\phi = \nabla \cdot \left[\frac{1}{h_1}\frac{\partial \phi}{\partial u}\mathbf{a} + \frac{1}{h_2}\frac{\partial \phi}{\partial v}\mathbf{b} + \frac{1}{h_3}\frac{\partial \phi}{\partial w}\mathbf{c}\right] \qquad (\S\ 6.22)$$

$$= \frac{1}{h_1h_2h_3}\left[\frac{\partial}{\partial u}\left(\frac{h_2h_3}{h_1}\frac{\partial \phi}{\partial u}\right) + \frac{\partial}{\partial v}\left(\frac{h_3h_1}{h_2}\frac{\partial \phi}{\partial v}\right) + \frac{\partial}{\partial w}\left(\frac{h_1h_2}{h_3}\frac{\partial \phi}{\partial w}\right)\right].$$

6.22.5. The Operator u.

∇; **u** being any vector.

We have seen that

$$\text{grad}\ \phi = \nabla\phi = \frac{1}{h_1}\frac{\partial \phi}{\partial u}\mathbf{a} + \frac{1}{h_2}\frac{\partial \phi}{\partial v}\mathbf{b} + \frac{1}{h_3}\frac{\partial \phi}{\partial w}\mathbf{c},$$

so that the directional derivatives of ϕ along the directions of **a, b, c** are

$$\frac{1}{h_1}\frac{\partial \phi}{\partial u},\ \frac{1}{h_2}\frac{\partial \phi}{\partial v},\ \frac{1}{h_3}\frac{\partial \phi}{\partial w}$$

respectively. Thus, if

$$\mathbf{u} = u_1\mathbf{a} + u_2\mathbf{b} + u_3\mathbf{c},$$

be any unit vector, the directional derivative $\mathbf{u} \,.\, \nabla\phi$ of ϕ along **u** is

$$= \frac{u_1}{h_1}\frac{\partial \phi}{\partial u} + \frac{u_2}{h_2}\frac{\partial \phi}{\partial v} + \frac{u_3}{h_3}\frac{\partial \phi}{\partial w}$$

$$= \left(\mathbf{u}\cdot\frac{\mathbf{a}}{h_1}\frac{\partial}{\partial u} + \mathbf{u}\cdot\frac{\mathbf{b}}{h_2}\frac{\partial}{\partial v} + \mathbf{u}\cdot\frac{\mathbf{c}}{h_3}\frac{\partial}{\partial w}\right)\phi.$$

From this it follows that if **u** be any vector whatsoever, we have

$$\mathbf{u}\cdot\nabla = \frac{\mathbf{u.a}}{h_1}\frac{\partial}{\partial u} + \frac{\mathbf{u.b}}{h_2}\frac{\partial}{\partial v} + \frac{\mathbf{u.c}}{h_3}\frac{\partial}{\partial w}.$$

6.23. SPECIAL CURVILINEAR SYSTEMS. SPHERICAL POLAR AND CYLINDRICAL POLAR SYSTEMS

Spherical Polar. The equations of transformation between the rectangular cartesian co-ordinates x, y, z and the spherical polar co-ordinates r, θ, ϕ are

$$x = r\sin\theta\cos\phi,\ y = r\sin\theta\cos\phi,\ z = r\cos\theta.$$

The co-ordinate surface $r = r_0$ is the sphere with its centre at the origin and radius r_0; the co-ordinate surface $\theta = \theta_0$ is the right circular semi-cone with its vertex at the origin and semi-vertical angle θ_0; finally the co-ordinate surface $\phi = \phi_0$ is the semi-plane through z-axis inclined at angle ϕ_0 to the XOZ plane.

We have

$$dx = \frac{\partial x}{\partial r} dr + \frac{\partial x}{\partial \theta} d\theta + \frac{\partial x}{\partial \phi} d\phi$$

$$= \sin\theta \cos\phi \, dr + r\cos\theta\cos\phi \, d\theta - r\sin\theta\sin\phi \, d\phi,$$

$$dy = \sin\theta\sin\phi \, dr + r\cos\theta\sin\phi \, d\theta + r\sin\theta\cos\phi \, d\phi,$$

$$dz = \cos\theta \, dr - r\sin\theta \, d\theta + 0 \, d\phi.$$

$$\therefore \quad (ds)^2 = (dx)^2 + (dy)^2 + (dz)^2$$

$$= (dr)^2 + r^2 (d\theta)^2 + r^2 \sin^2\theta \, (d\phi)^2.$$

Thus, for this system

$$h_1 = 1, \quad h_2 = r, \quad h_3 = r\sin\theta.$$

If ψ be any scalar point function, we obtain

$$\nabla^2 \psi = \frac{1}{r^2 \sin\theta}\left[\frac{\partial}{\partial r}\left(r^2 \sin\theta \frac{\partial \psi}{\partial r}\right) + \frac{\partial}{\partial \theta}\left(\sin\theta \frac{\partial \psi}{\partial \theta}\right) + \frac{\partial}{\partial \phi}\left(\frac{1}{\sin\theta}\frac{\partial \psi}{\partial \phi}\right)\right]$$

[§ 6.22.4]

$$= \frac{\partial^2 \psi}{\partial r^2} + \frac{1}{r^2}\frac{\partial^2 \psi}{\partial \theta^2} + \frac{1}{\sin\theta}\frac{\partial^2 \psi}{\partial \phi^2} + \frac{2}{r}\frac{\partial \psi}{\partial r} + \frac{\cot\theta}{r^2}\frac{\partial \psi}{\partial \theta}.$$

Cylindrical Polar. The equations of transformation between rectangular cartesian co-ordinates x, y, z and cylindrical polar co-ordinates r, θ, z are

$$x = r\cos\theta, \quad y = r\sin\theta, \quad z = z.$$

We may show that

$$(ds)^2 = (dr)^2 + r^2 (d\theta)^2 + (dz)^2,$$

so that

$$h_1 = 1, \quad h_2 = r, \quad h_3 = 1.$$

Thus, in this case if ψ be any scalar point function,

$$\nabla^2 \psi = \frac{1}{r}\left[\frac{\partial}{\partial r}\left(r\frac{\partial \psi}{\partial r}\right) + \frac{\partial}{\partial \theta}\left(\frac{1}{r}\frac{\partial \psi}{\partial \theta}\right) + \frac{\partial}{\partial z}\left(r\frac{\partial \psi}{\partial z}\right)\right]$$

$$= \frac{\partial^2 \psi}{\partial r^2} + \frac{1}{r^2}\frac{\partial^2 \psi}{\partial \theta^2} + \frac{\partial^2 \psi}{\partial z^2} + \frac{1}{r}\frac{\partial \psi}{\partial r}.$$

Ex. Describe the co-ordinate surfaces in the case of cylindrical polar co-ordinates.

EXAMPLES

Example 1. *If u_1, u_2, u_3 are orthogonal curvilinear co-ordinates, show that $\partial\mathbf{r}/\partial u_1$, $\partial\mathbf{r}/\partial u_2$, $\partial\mathbf{r}/\partial u_3$ and ∇u_1, ∇u_2, ∇u_3 are reciprocal systems of vectors.*

Solution. We know that $\nabla u_1 = e_1/h_1$, $\nabla u_2 = e_2/h_2$, $\nabla u_3 = e_3/h_3$.

$\therefore \quad [\nabla u_1 + \nabla u_2 + \nabla u_3] = (\nabla u_1 \times \nabla u_2) . \nabla u_3$

$$= \left(\frac{1}{h_1}\mathbf{e}_1 \times \frac{1}{h_2}\mathbf{e}_2\right) \times \frac{1}{h_3}\mathbf{e}_3$$

$$= \frac{1}{h_1 h_2 h_3}\mathbf{e}_3 \cdot \mathbf{e}_3 = \frac{1}{h_1 h_2 h_3} \qquad \text{...(1)}$$

Also, $$\frac{\partial \mathbf{r}}{\partial u_1} = h_1\mathbf{e}_1, \quad \frac{\partial \mathbf{r}}{\partial u_2} = h_2\mathbf{e}_2, \quad \frac{\partial \mathbf{r}}{\partial u_3} = h_3\mathbf{e}_3 \qquad \text{...(2)}$$

Now, $$\frac{\nabla u_2 \times \nabla u_3}{[\nabla u_1 \ \nabla u_2 \ \nabla u_3]} = \frac{(1/h_2)\,\mathbf{e}_2 \times (1/h_3)\,\mathbf{e}_3}{1/h_1 h_2 h_3}$$

$$= h_1\mathbf{e}_1 = \frac{\partial \mathbf{r}}{\partial u_1} \qquad \text{by (2)}$$

Proceeding similarly, we get

$$\frac{\partial \mathbf{r}}{\partial u_2} = \frac{\nabla u_3 \times \nabla u_1}{[\nabla u_1 \ \nabla u_2 \ \nabla u_3]}, \quad \frac{\partial \mathbf{r}}{\partial u_3} = \frac{\nabla u_1 \times \nabla u_2}{[\nabla u_1 \ \nabla u_2 \ \nabla u_3]}$$

Hence, $\partial\mathbf{r}/\partial u_1$, $\partial\mathbf{r}/\partial u_2$, $\partial\mathbf{r}/\partial u_3$ *and* ∇u_1, ∇u_2, ∇u_3 form reciprocal system of vectors.

Example 2. *Express the vector $\mathbf{f} = z\mathbf{i} - 2x\mathbf{j} + y\mathbf{k}$ in (i) cylindrical polar, (ii) spherical polar co-ordinates.* *(Lucknow, 97)*

Solution. (*i*) Cartesian and cylindrical polar co-ordinates (x, y, z) and (ρ, ϕ, z) are related as follows :

$x = \rho \cos \phi, \ \ y = \rho \sin \phi, \ \ z = z$

$\mathbf{i} = \cos \phi \ \mathbf{e}_\rho - \sin \phi \ \mathbf{e}_\phi, \ \ \mathbf{j} = \sin \phi \ \mathbf{e}_\rho + \cos \phi \ \mathbf{e}_\phi, \ \ k = k.$

$\therefore \quad z\mathbf{i} - 2x\mathbf{j} + y\mathbf{k}$

$$= z\,(\cos \phi \ \mathbf{e}_\rho - \sin \phi \ \mathbf{e}_\phi) - 2\rho \cos \phi \,(\sin \phi \ \mathbf{e}_\rho + \cos \phi \ \mathbf{e}_\phi) + \rho \sin \phi \ \mathbf{k}$$

$$= z\,(\cos \phi - 2\rho \sin \phi \cos \phi)\ \mathbf{e}_\rho - (z \sin \phi + 2\rho \cos^2 \phi)\ \mathbf{e}_\rho + \rho \sin \phi \ \mathbf{k}.$$

(*ii*) Cartesian and spherical polar co-ordinates (x, y, z) and (r, θ, ϕ) are related as follows :

$$x = r \sin\theta \cos\phi, \quad y = r \sin\theta \sin\phi, \quad z = r\cos\phi$$

and $\mathbf{i} = \sin\theta\cos\phi\,\mathbf{e}_r + \cos\theta\cos\phi\,\mathbf{e}_\theta - \sin\phi\,\mathbf{e}_\phi$,

$\mathbf{k} = \cos\theta\,\mathbf{e}_r - \sin\theta\,\mathbf{e}_\theta$

$\therefore \quad z\mathbf{i} - 2x\mathbf{j} + y\mathbf{k}$

$$= r\cos\theta\,(\sin\theta\cos\phi\,\mathbf{e}_r + \cos\theta\cos\phi\,\mathbf{e}_\theta - \sin\phi\,\mathbf{e}_\phi)$$
$$- 2r\sin\theta\cos\theta\,(\sin\theta\cos\phi\,\mathbf{e}_r + \cos\theta\sin\phi\,\mathbf{e}_\theta$$
$$+ \cos\phi\,\mathbf{e}_\phi) + r\,\sin\theta\cos\phi\,(\cos\theta\,\mathbf{e}_r - \sin\theta\sin\,\mathbf{e}_\theta)$$
$$= (r\sin\theta\cos\phi - 2r\sin^2\theta\sin\phi\cos\theta + r\sin\theta\sin\phi\cos\theta)$$
$$\mathbf{e}_r + (r\cos^2\theta\cos\phi - 2r\cos\theta\cos\phi\sin\theta\sin\phi$$
$$- r\sin^2\theta\sin\phi)\,\mathbf{e}_\phi$$
$$- (r\cos\theta\cos\phi - 2r\sin\theta\cos^2\phi)\,\mathbf{e}_\phi.$$

Example 3. *Transform the function* $\mathbf{f} = \rho\,\mathbf{e}_\rho + \rho\,\mathbf{e}_\phi$ *from cylindrical to cartesian system.*

Solution. Given $\quad \mathbf{f} = \rho\,\mathbf{e}_\rho + \rho\,\mathbf{e}_\phi$

Cylindrical and cartesian co-ordinates are related by

$$x = \rho\cos\phi, \quad y = \rho\sin\phi, \quad z = z$$

and $\quad \mathbf{e}_\rho = \cos\phi\,\mathbf{i} + \sin\phi\,\mathbf{j}, \ \mathbf{e}_\phi = -\sin\phi\,\mathbf{i} + \cos\phi\,\mathbf{j}$

$\therefore \quad \mathbf{f} = \rho\,(\cos\phi\,\mathbf{i} + \sin\phi\,\mathbf{j}) + \rho\,(-\sin\phi\,\mathbf{i} + \cos\phi\,\mathbf{j})$

$$= x\mathbf{i} + y\mathbf{j} - y\mathbf{i} + x\mathbf{j}$$
$$= (x - y)\,\mathbf{i} + (x + y)\,\mathbf{j}.$$

EXERCISES

1. Show that the cylindrical and spherical co-ordinate systems (bases) are self-reciprocal.
2. Express wave equation $\dfrac{\partial^2 u}{\partial t^2} = c^2\nabla^2 u$ in spherical co-ordinates if u is independent of ϕ.
3. Prove that in any orthogonal curvilinear co-ordinate system div curl $\mathbf{f} = 0$ and curl grad $\mathbf{f} = 0$.
4. If $u_1 = 2x + 3$, $u_2 = y - 4$, $u_3 = z + 2$, show that u_1, u_2, u_3 are orthogonal.
5. If $x = u_1u_2\cos u_3$, $y = u_1u_2\sin u_3$, $z = \dfrac{1}{2}\left(u_1^2 - u_2^2\right)$, obtain h_1, h_2, h_3 and the unit vectors $\mathbf{e}_1, \mathbf{e}_2, \mathbf{e}_3$. Determine grad ψ, div $\mathbf{F}$, curl $\mathbf{F}$ and $\nabla^2\psi$ for the above system of co-ordinates.
6. If (r, θ, ϕ) are spherical co-ordinates, show that
   (*i*) $\nabla(1/r) = \nabla \times (\cos\theta\,\nabla\phi)$ (*ii*) $\nabla\phi = \nabla \times (r\,\mathrm{cosec}\,\theta\,\nabla\theta)$

7. Express the following vectors in terms of (*i*) cylindrical polar (*ii*) spherical polar co-ordinates.
 (*a*) $\mathbf{f} = 3y\mathbf{i} + x^2\mathbf{j} - z^2\mathbf{k}$
 (*b*) $\mathbf{f} = xyz^2\mathbf{i} + yz\mathbf{j} + xy\mathbf{k}$
 (*c*) $\mathbf{f} = xy\mathbf{i} + 2yz\,\mathbf{j} + (z^2 - yz)\,\mathbf{k}$.

8. Transform the following function from spherical to cartesian system :
 $\mathbf{f} = 3ar^2 \sin\theta \cos\phi\, \mathbf{e}_r + 2a^2 r \cos\theta \sin\phi\, \mathbf{e}_\theta + r^3\, \mathbf{e}_\phi$.

OBJECTIVE QUESTIONS

For each of the following questions, four alternatives are given for the answer. Only one of them is correct. Choose the correct alternative.

1. Value of grad r is
 (*a*) $\mathbf{r}$ (*b*) $\hat{\mathbf{r}}$ (*c*) $-\dfrac{\mathbf{r}}{r^2}$ (*d*) $\dfrac{\mathbf{r}}{r^2}$

2. Value of $\nabla\,(1/r)$ is
 (*a*) $\mathbf{r}$ (*b*) $\hat{\mathbf{r}}$ (*c*) $-\dfrac{\mathbf{r}}{r^2}$ (*d*) $\dfrac{-\mathbf{r}}{r^3}$

3. Value of $\mathbf{a}\cdot\nabla\left(\dfrac{1}{r}\right)$ is
 (*a*) $-\dfrac{\mathbf{a}\,.\,\mathbf{r}}{r^2}$ (*b*) $\dfrac{\mathbf{a}\,.\,\mathbf{r}}{r^2}$ (*c*) $-\dfrac{\mathbf{a}\,.\,\mathbf{r}}{r^3}$ (*d*) $\dfrac{\mathbf{a}\,.\,\mathbf{r}}{r^3}$

4. Value of $\mathbf{f}\,(u)\,\nabla u$ is
 (*a*) $u\,\nabla\mathbf{f}\,(u)$ (*b*) $\nabla u\,\mathbf{f}\,(u)$
 (*c*) $\nabla\int\mathbf{f}\,(u)\,du$ (*d*) None of these

5. Value of grad $(\mathbf{r}\,.\,\mathbf{a})$ is
 (*a*) $\mathbf{r}$ (*b*) $\mathbf{a}$ (*c*) 0 (*d*) $\hat{\mathbf{a}}\,.\,\hat{\mathbf{r}}$

6. A vector whose divergence is zero, is called
 (*a*) Solenoidal vector (*b*) Irrotational vector
 (*c*) Polar vector (*d*) Axial vector

7. If curl $\mathbf{a} = 0$ then $\mathbf{a}$ is
 (*a*) Solenoidal (*b*) Irrotational
 (*c*) Polar (*d*) Axial

8. Value of div $\mathbf{r}$ is
 (*a*) 0 (*b*) 1 (*c*) 3 (*d*) 2

9. Value of curl $\mathbf{r}$ is
 (*a*) 0 (*b*) $\hat{\mathbf{r}}$ (*c*) $\mathbf{r}$ (*d*) None of these

6. Value of curl grad **f** is

(*a*) $\nabla^2\mathbf{f}$ (*b*) 0 (*c*) grad div **f** (*d*) None of these

11. Value of div $\hat{\mathbf{r}}$ is

(*a*) $1/r$ (*b*) $2/r$ (*c*) $-1/r$ (*d*) $1/r^2$

12. Value of div ($\mathbf{r} \times \mathbf{a}$), where **a** is a constant vector is

(*a*) 0 (*b*) $-2\mathbf{a}$ (*c*) $2\mathbf{a}$ (*d*) **a**

13. If **a** and **b** are irrotational, then $\mathbf{a} \times \mathbf{b}$ is

(*a*) irrotational

(*b*) solenoidal

(*c*) irrotational and solenoidal both

(*d*) neither irrotational and solenoidal

14. If **v** is a vector having fixed direction, then value of curl **v** is

(*a*) 0 (*b*) along **v**

(*c*) perpendicular to **v** (*d*) None of these

15. For the circular motion of a particle, we have $\mathbf{v} = \omega \times \mathbf{r}$, assuming that ω is constant, we have div **v** =

(*a*) 2ω (*b*) ω

(*c*) -2ω (*d*) 0

16. For the surface $\phi(x, y, z) = c$, c being a constant, grad ϕ represents

(*a*) position vector (*b*) vector tangent

(*c*) vector normal (*d*) None of these

17. Two surfaces $\phi(x, y, z) = c_1$ and $\phi(x, y, z) = c_2$ will cut orthogonally, if

(*a*) $\nabla\phi \cdot \nabla\psi = 0$ (*b*) $\nabla^2\phi = 0, \nabla^2\psi = 0$

(*c*) $\nabla^2\phi = 0, \nabla^2\psi \neq 0$ (*d*) $\nabla^2\phi \neq 0, \nabla^2\psi = 0$

18. Curl of the velocity of any particle of a rigid body is equal to

(*a*) the linear velocity vector of the body

(*b*) the twice the linear velocity vector of the body

(*c*) the angular velocity vector of the body

(*d*) the twice the angular velocity vector of the body

19. Function $\mathbf{f} = \rho\mathbf{e}_\rho + \rho\mathbf{e}_\phi$, when transformed from cylindrical to cartesian system, yields

(*a*) $(x + y)\,\mathbf{i} + (x - y)\,\mathbf{j}$ (*b*) $(x - y)\,\mathbf{i} + (x + y)\,\mathbf{j}$

(*c*) $(x + y)\,\mathbf{i} - (x - y)\,\mathbf{j}$ (*d*) $(x - y)\,\mathbf{i} - (x + y)\,\mathbf{j}$

$$\xi_1,\ \xi_2 \ldots,\ \xi_{p-1}, \xi_p, \ldots, \xi_n.$$

Consider now the sum S of the scalar products given by

$$S = \sum_{p=1}^{n} \mathbf{F}(\xi_p) \cdot (\mathbf{r}_p - \mathbf{r}_{p-1}) = \sum_{p=1}^{n} \mathbf{F}(\xi_p) \cdot \delta \mathbf{r}_p,$$

where the summation extends to all the sub-arcs and

$$\delta \mathbf{r}_p = \mathbf{r}_p - \mathbf{r}_{p-1}.$$

Then the limit of this sum, if it exists and is independent of the arbitrary points ξ_p when n tends to infinity and each of

$$|\mathbf{r}_p - \mathbf{r}_{p-1}|$$

tends to zero, is called the *tangential line integral* of $\mathbf{F}(\mathbf{r})$ along C and is denoted by the symbol

$$\int \mathbf{F}(\mathbf{r}) \cdot d\mathbf{r} \text{ or simply } \int_C \mathbf{F} \cdot d\mathbf{r}.$$

Existence and Scalar Cartesian Expression. Introduce any system of rectangular axes.

Let

$$\mathbf{F}(\mathbf{r}) = \mathbf{F}(x, y, z) = \mathbf{i}\, F_1(x, y, z) + \mathbf{j}\, F_2(x, y, z) + \mathbf{k}\, F_3(x, y, z)$$

and

$$\mathbf{r}_p = \mathbf{i}x_p + \mathbf{j}y_p + \mathbf{k}z_p,$$

$$\xi_p = \mathbf{i}\alpha_p + \mathbf{j}\beta_p + \mathbf{k}\gamma_p.$$

$$\therefore \quad S = \Sigma \mathbf{F}(\xi_p) \cdot (\mathbf{r}_p - \mathbf{r}_{p-1})$$

$$= \Sigma[\mathbf{i}F_1(\alpha_p, \beta_p, \gamma_p) + \mathbf{j}F_2(\alpha_p, \beta_p, \gamma_p) + \mathbf{k}F_3(\alpha_p, \beta_p, \gamma_p)].$$
$$[\mathbf{i}(x_p - x_{p-1}) + \mathbf{j}(y_p - y_{p-1}) + \mathbf{k}(z_p - z_{p-1})]$$

$$= \Sigma F_1(\alpha_p, \beta_p, \gamma_p)(x_p - x_{p-1}) + \Sigma F_2(\alpha_p, \beta_p, \gamma_p)(y_p - y_{p-1}) + $$
$$+ \Sigma F_3(\alpha_p, \beta_p, \gamma_p)(z_p - z_{p-1})$$

$$= S_1 + S_2 + S_3, \text{ say.}$$

As $\mathbf{F}(x, y, z)$ is continuous, therefore, $F_1(x, y, z)$, $F_2(x, y, z)$, $F_3(x, y, z)$ are also continuous.

From Scalar Calculus, we know that under the given conditions, the sums S_1, S_2, S_3 tend to limits denoted by the symbols

$$\int_C F_1(x, y, z)\, dx,\ \int_C F_2(x, y, z)\, dy,\ \int_C F_3(x, y, z)\, dz.$$

Hence, lim S, also exists. Also we see that

$$\int_C \mathbf{F} \cdot d\mathbf{r} = \int_C (F_1 dx + F_2 dy + F_3 dz). \qquad \ldots(1)$$

If s denote the actual length of any point (x, y, z) on the curve from a fixed point, we have

$$\int_C (F_1 dx + F_2 dy + F_3 dz) = \int_{s_1}^{s_2} \left(F_1 \frac{dx}{ds} + F_2 \frac{dy}{ds} + F_3 \frac{dz}{ds} \right) ds$$

$$= \int_{s_1}^{s_2} \mathbf{F} \cdot \mathbf{t} \, ds. \qquad \text{...(2)}$$

Here s_1, s_2 are the values of s, for the extremities A, B of the arc C and

$$\mathbf{t} = \mathbf{i}\frac{dx}{ds} + \mathbf{j}\frac{dy}{ds} + \mathbf{k}\frac{dz}{ds},$$

is the unit tangent vector to the curve.

We also have

$$\int_C \mathbf{F} \cdot d\mathbf{r} = \int_C (F_1 dx + F_2 dy + F_3 dz)$$

$$= \int_a^b \left(F_1 \frac{dx}{dt} + F_2 \frac{dy}{dt} + F_3 \frac{dz}{dt} \right) dt$$

$$= \int_a^b \mathbf{F} \cdot \frac{d\mathbf{r}}{dt} dt. \qquad \text{...(3)}$$

The equations (2) and (3) express the line integral in terms of ordinary definite integrals of single variable.

7.1.2. Other Types of Line Integrals.

$$\int_C \mathbf{F} \times d\mathbf{r}, \int_C \phi d\mathbf{r},$$

where $\mathbf{F}$ is a continuous vector and ϕ a continuous scalar point function.

In terms of the notation of the preceding § 7.1.1, we consider the vector sum

$$S = \sum_{p=1}^{n} \mathbf{F}(\xi_p) \times (\mathbf{r}_p - \mathbf{r}_{p-1})$$

$$= \sum_{p=1}^{n} \mathbf{F}(\xi_p) \times \delta \mathbf{r}_p.$$

The limit of $\mathbf{S}$, when n tends to infinity and each of

$$| \mathbf{r}_p - \mathbf{r}_{p-1} |$$

tends to zero, if it exists and does not depend on the choice of ξ_p, is denoted by the symbol

$$\int_C \mathbf{F}(\mathbf{r}) \times d\mathbf{r} \text{ or simply } \int_C \mathbf{F} \times d\mathbf{r}.$$

Existence and expression in terms of components.

We have

$$\Sigma \mathbf{F}(\xi_p) \times (\mathbf{r}_p - \mathbf{r}_{p-1})$$

$$= \Sigma[\mathbf{i}F_1(\alpha_p, \beta_p, \gamma_p) + \mathbf{j}F_2(\alpha_p, \beta_p, \gamma_p) + \mathbf{k}F_3(\alpha_p, \beta_p, \gamma_p)] \times [\mathbf{i}(x_p - x_{p-1}) + \mathbf{j}(y_p - y_{p-1}) + \mathbf{k}(z_p - z_{p-1})]$$

$$= \Sigma\{[F_2(z_p - z_{p-1}) - F_3(y_p - y_{p-1})]\,\mathbf{i} + [F_3(x_p - x_{p-1}) - F_1(z_p - z_{p-1})]\,\mathbf{j} + [F_1(y_p - y_{p-1}) - F_2(x_p - x_{p-1})]\,\mathbf{k}\},$$

where, for the sake of brevity, we have written F_1, F_2, F_3 instead of $F_1(\alpha_p, \beta_p, \gamma_p)$, $F_2(\alpha_p, \beta_p, \gamma_p)$ and $F_3(\alpha_p, \beta_p, \gamma_p)$ respectively.

From scalar calculus, we know that under the given conditions

$$\Sigma\,[F_2(\alpha_p, \beta_p, \gamma_p)(z_p - z_{p-1}) - F_3(\alpha_p, \beta_p, \gamma_p)(y_p - y_{p-1})] \to \int_C (F_2dz - F_3dy),$$

$$\Sigma\,[F_3(\alpha_p, \beta_p, \gamma_p)(x_p - x_{p-1}) - F_1(\alpha_p, \beta_p, \gamma_p)(z_p - z_{p-1})] \to \int_C (F_3dx - F_1dz),$$

$$\Sigma\,[F_1(\alpha_p, \beta_p, \gamma_p)(y_p - y_{p-1}) - F_2(\alpha_p, \beta_p, \gamma_p)(x_p - x_{p-1})] \to \int_C (F_1dy - F_2dx).$$

Hence, lim **S** exists and we have

$$\int_C \mathbf{F} \times d\mathbf{r} = \mathbf{i}\int_C (F_2dz - F_3dy) + \mathbf{j}\int_C (F_3dx - F_1dz) + \mathbf{k}\int_C (F_1dy - F_2dx).$$

As before, it can also be seen that

$$\int_C \mathbf{F} \times d\mathbf{r} = \int_{s_1}^{s_2} \mathbf{F} \times \mathbf{t}\, ds.$$

Finally, in terms of the notation adopted above,

$$\int_C \phi d\mathbf{r} = \lim \Sigma\phi(\xi_p)(\mathbf{r}_p - \mathbf{r}_{p-1})$$

and we may show that

$$\int_C \phi d\mathbf{r} = \mathbf{i}\int_C \phi dx + \mathbf{j}\int_C \phi dy + \mathbf{k}\int \phi dz.$$

7.2. CIRCULATION, IRROTATIONAL VECTOR POINT FUNCTIONS

Def. 1. *The line integral*

$$\int_C \mathbf{F} \cdot d\mathbf{r}$$

of a continuous vector point function **F** *along a closed curve C is called the circulation of* **F** *along C.*

Def. 2. *A vector point function is said to be irrotational in a region, if its circulation along every closed curve in the region is zero.*

We shall now prove a theorem which gives a necessary and sufficient condition for a vector function to be irrotational.

7.2.1. Theorem.

The necessary and sufficient condition for a continuous vector point function to be irrotational in a simply connected region R is that it is the gradient of a scalar point function.

The condition is necessary. Let **F** be irrotational in a simply connected region R. Take any two points A and P in the region and join them by any two curves ALP and AMP in the region. Let these be denoted by C_1 and C_2 and let C denote the closed curve $ALPMA$.

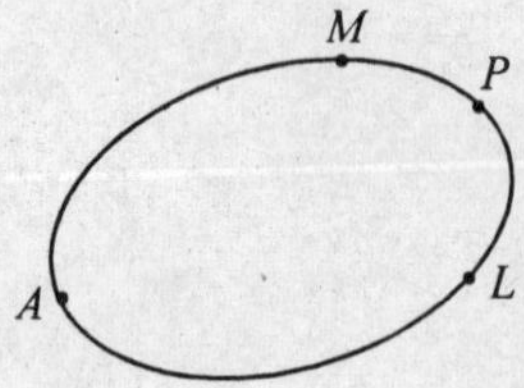

Fig. 7.2.

We have

$$0 = \int_C \mathbf{F} \cdot d\mathbf{r} = \int_{C_1} \mathbf{F} \cdot d\mathbf{r} + \int_{-C_2} \mathbf{F} \cdot d\mathbf{r}$$

$$= \int_{C_1} \mathbf{F} \cdot d\mathbf{r} - \int_{C_2} \mathbf{F} \cdot d\mathbf{r}.$$

$$\therefore \quad \int_{C_1} \mathbf{F} \cdot d\mathbf{r} = \int_{C_2} \mathbf{F} \cdot d\mathbf{r}.$$

Thus, the line integral

$$\int \mathbf{F} \cdot d\mathbf{r},$$

along any curve joining A to P depends only on the point B if we keep A fixed. We write

$$\phi(P) = \int_A^P \mathbf{F} \cdot d\mathbf{r}.$$

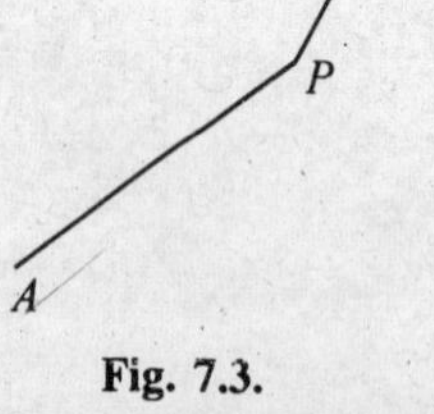

Fig. 7.3.

Take a point Q near P and join PQ by a straight line. Let **a** be the unit vector along AQ. We have

$$\phi(Q) = \int_A^Q \mathbf{F} \cdot d\mathbf{r} = \int_A^P \mathbf{F} \cdot d\mathbf{r} + \int_P^Q \mathbf{F} \cdot d\mathbf{r}.$$

$$\therefore \qquad \phi(Q) - \phi(P) = \int_P^Q \mathbf{F} \cdot d\mathbf{r} = \int_0^{PQ} \mathbf{F} \cdot \frac{d\mathbf{r}}{ds} ds,$$

where s denotes the distance of any point on PQ from P.

$$\therefore \qquad \phi(Q) - \phi(P) = \int_0^{PQ} \mathbf{F} \cdot \mathbf{a}\, ds$$

$$= [\mathbf{F}(R) \,.\, \mathbf{a}]\, PQ, \text{ by Mean Value Theorem}$$

R being a point of the line PQ.

$$\therefore \qquad \frac{\phi(Q) - \phi(P)}{PQ} = \mathbf{F}(R) \cdot \mathbf{a}$$

Let $Q \to P$ so that $R \to P$.

$$\therefore \qquad \lim \frac{\phi(Q) - \phi(P)}{PQ} = \mathbf{F}(P) \cdot \mathbf{a}.$$

But

$$\lim \frac{\phi(Q) - \phi(P)}{PQ} = \mathbf{a} \nabla \phi,$$

being the directional derivative of ϕ along the direction of $\mathbf{a}$. Thus, for any point P,

$$\mathbf{a} \,.\, \nabla \phi = \mathbf{a} \,.\, \mathbf{F}$$

or

$$\mathbf{a} \,.\, (\nabla \phi - \mathbf{F}) = 0$$

i.e.,

$$\nabla \phi = \mathbf{F};$$

since $\mathbf{a}$ is a any arbitrary vector.

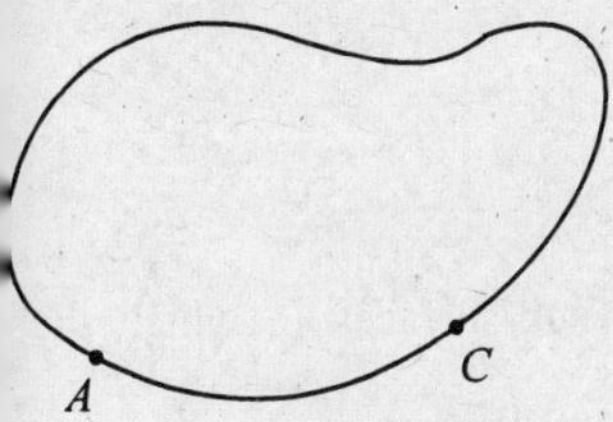

Fig. 7.4.

The condition is sufficient.

Suppose now that

$$\mathbf{F} = \nabla \phi.$$

If C be any closed curve,

$$\int_C \mathbf{F} \cdot d\mathbf{r} = \int_C \left(\frac{\partial \phi}{\partial x} \frac{dx}{ds} + \frac{\partial \phi}{\partial y} \frac{dy}{ds} + \frac{\partial \phi}{\partial z} \frac{dz}{ds} \right) ds$$

$$= \int_C \frac{d\phi}{ds} ds = |\, \phi \,|_A^A = \phi(A) - \phi(A) = 0.$$

A being any point on the curve C.

Note 1. The student should notice that the condition that line integral along every closed curve is zero is equivalent to the condition that the line integral along a curve joining any two given points is independent of the particular curve and depends on the position vectors of end points only.

Note. 2. Later on we shall obtain another form of necessary and sufficient condition for a vector point function to be irrotational.

7.3. SURFACE INTEGRALS

A surface $\mathbf{r} = \mathbf{f}(u, v)$ is called smooth if $\mathbf{f}(u, v)$ possesses continuous first order partial derivatives. In the following, it will always be understood that any surface which we consider is smooth or it consists of a finite number of connected smooth pieces.

7.3.1. Normal Surface Integral

Let $\mathbf{F}(\mathbf{r})$ be a continuous vector point function and $\mathbf{r} = \mathbf{f}(u, v)$ a surface.

Consider any portion S of the surface, closed or not. We shall distinguish between the two sides of the surface and suppose that the particular side considered is a part of the name S.

Divide S into any finite number of sub-surfaces and consider any sub-surface. Take a point ξ_p, in the sub-surface and let $\mathbf{n}_p$ denote the unit vector along the normal to the sub-surface at ξ_p drawn on the side of S.

Let δS_p denote the magnitude of the area of the sub-surface. Then

$$\delta \mathbf{a}_p = \mathbf{n}_p \, \delta S_p,$$

denotes the vector area of the sub-surface.

Consider now the sum

$$\Sigma \, \mathbf{F}(\xi_p) \cdot \delta \mathbf{a}_p = \Sigma \, \mathbf{F}(\xi_p) \cdot \mathbf{n}_p \, \delta S_p,$$

where the summation extends to the various sub-surfaces in which the surface S has been divided.

The limit of this sum when the number of sub-surfaces tends to infinity and the area of each sub-surface tends to zero is called the *normal surface integral* of $\mathbf{F}(\mathbf{r})$ over S and is denoted by

$$\int_S \mathbf{F}(\mathbf{r}) \cdot d\mathbf{a} = \int_S \mathbf{F}(\mathbf{r}) \cdot \mathbf{n} \, dS$$

or simply

$$\int_S \mathbf{F} \cdot d\mathbf{a} = \int_S \mathbf{F} \cdot \mathbf{n} \, dS.$$

The value of the integral will change in sign but not in magnitude when we carry the integral on one side than on the other, for in that case, in each term the sign of $\mathbf{n}_p$ would change.

Existence and Scalar Cartesian Expression

Let

$$\mathbf{F}(\mathbf{r}) = \mathbf{F}(x, y, z) = \mathbf{i}\, F_1(x, y, z) + \mathbf{j}\, F_2(x, y, z) + \mathbf{k}\, F_3(x, y, z),$$
$$\xi_p = \mathbf{i}\alpha_p + \mathbf{j}\beta_p + \mathbf{k}\gamma_p,$$
$$\delta\mathbf{a}_p = \mathbf{n}_p\delta S_p = \mathbf{i}\delta A_p + \mathbf{j}\delta B_p + \mathbf{k}\delta C_p.$$

Here δA_p, δB_p, δC_p, denote the projections of the vector area $\mathbf{n}_p\delta S_p$ on the three co-ordinate planes.

$$\therefore \quad \Sigma\mathbf{F}(\xi_p) \,.\, \delta A_p = \Sigma\, [F_1(\alpha_p, \beta_p, \gamma_p)\, \delta A_p + F_2(\alpha_p, \beta_p, \gamma_p)\, \delta B_p + F_3(\alpha_p, \beta_p, \gamma_p)\, \delta C_p] \quad ...(1)$$

Now from Scalar Calculus, the sum on the right of (1) tends, under the conditions stated, to the limit.

$$\int_S \int \left[F_1(x, y, z)\, dydz + F_2(x, y, z)\, dzdx + F_3(x, y, z)\, dxdy\right] \quad ...(2)$$

Thus,

$$\int_S \mathbf{F} \cdot d\mathbf{a} = \int_S \mathbf{F} \cdot \mathbf{n}\, dS.$$

exists and equals (2).

By the rule of change of variables for integrals, the integral (2) can be re-written as an ordinary double integral

$$\iint \left\{ F_1 \frac{\partial (y, z)}{\partial (u, v)} + F_2 \frac{\partial (z, x)}{\partial (u, v)} + F_3 \frac{\partial (x, y)}{\partial (u, v)} \right\} dudv$$

carried over the region in the *u-v* plane which corresponds to the surface S given by $\mathbf{r} = \mathbf{f}\,(u, v)$.

Here $\dfrac{\partial (y, z)}{\partial (u, v)} = \begin{vmatrix} \dfrac{\partial y}{\partial u} & \dfrac{\partial y}{\partial v} \\ \dfrac{\partial z}{\partial u} & \dfrac{\partial z}{\partial v} \end{vmatrix}$ and so on.

7.3.2. Other Forms of Surface Integrals

$$\int_S \mathbf{F} \times d\mathbf{a}, \int_S \phi d\mathbf{a}$$

where $\mathbf{F}$ is a continuous vector and ϕ a continuous scalar point function.

In terms of the notation of the preceding sub-section § 7.3.1, we have

$$\int_S \mathbf{F} \times d\mathbf{a} = \lim \Sigma\mathbf{F}\left(\xi_p\right) \times \delta\mathbf{a}_p$$

and

$$\int_S \phi d\mathbf{a} = \lim \Sigma\, \phi\left(\xi_p\right) \delta \mathbf{a}_p.$$

It can be shown that under the conditions stated, these integrals exist and

$$\int_S \mathbf{F} \times d\mathbf{a} = \mathbf{i} \int_S \int \left(F_2 dydz - F_3 dzdx\right) +$$

$$\mathbf{j} \int_S \int \left(F_3 dydz - F_1 dxdy\right) + \mathbf{k} \int_S \int \left(F_1 dzdx - F_2 dydx\right)$$

$$\int_S \phi d\mathbf{a} = \mathbf{i} \int_S \int \phi dydz + \mathbf{j} \int_S \int \phi dzdx + \mathbf{k} \int_S \int \phi dxdy.$$

The proofs are left as exercises for the reader.

7.4. FLUX ACROSS A SURFACE : SOLENOIDAL VECTOR POINT FUNCTIONS

Def. 1. The normal surface integral

$$\int_S \mathbf{F} \cdot d\mathbf{a},$$

of a continuous vector point function **F** *over a closed surface S is called the Flux of* **F** *across S.*

Def. 2. *A vector point function is said to be solenoidal in a region if its flux across every closed surface in the region is zero.*

Note. We shall later on obtain a necessary and sufficient condition for a vector function to be solenoidal in a region.

7.5. VOLUME INTEGRALS

$$\int_V \mathbf{F}\, dv.$$

Consider a continuous vector point function $\mathbf{F}(\mathbf{r})$ and a region V enclosed by a surface $\mathbf{r} = \mathbf{f}\,(u, v)$.

Divide the region into a finite number of sub-regions and consider any sub-region. Let δv_p denote the volume of this sub-region and ξ_p the position vector of any point of the same.

Consider the sum

$$\Sigma\, \mathbf{F}\,(\xi_p)\, \delta v_p,$$

where the summation extends to all the sub-regions. The limit of this sum if it exists when the number of sub-regions tends to infinity and the volume of each sub-region tends to zero is denoted by the symbol

$$\int_V \mathbf{F}(\mathbf{r})\,dv \text{ or briefly } \int_V \mathbf{F}\,dv.$$

Introducing rectangular cartesian co-ordinates, we may see that the limit, in question, exists and

$$\int_V \mathbf{F}\,dv = \mathbf{i}\iiint_V F_1 dxdydz + \mathbf{j}\iiint_V F_2 dxdydz + \mathbf{k}\iiint_V F_3 dxdydz.$$

EXAMPLES

Example 1. *Evaluate*

$$\int_C \mathbf{F}\cdot d\mathbf{r},$$

where

$$\mathbf{F} = c\,[-3a \sin^2\theta \cos\theta\,\mathbf{i} + a\,(2\sin\theta - 3\sin^3\theta)\,\mathbf{j} + b\sin 2\theta\,\mathbf{k}]$$

and the curve C is given by

$$\mathbf{r} = a\cos\theta\,\mathbf{i} + a\sin\theta\,\mathbf{j} + b\theta\,\mathbf{k};$$

θ varying from π/4 to π/2.

Solution. We have

$$\frac{d\mathbf{r}}{d\theta} = -a\sin\theta\mathbf{i} + a\cos\theta\mathbf{j} + b\mathbf{k}$$

$$\therefore \quad \int_C \mathbf{F}\cdot d\mathbf{r} = \int_{\pi/4}^{\pi/2} \mathbf{F}\cdot\frac{d\mathbf{r}}{d\theta}\,d\theta \qquad \text{[Refer (3), § 7.1.1]}$$

$$= c\left(a^2+b^2\right)\int_{\pi/4}^{\pi/2} \sin 2\theta\,d\theta = \frac{1}{2}c\left(a^2+b^2\right).$$

Example 2. *Evaluate*

$$\int_C \mathbf{F}\cdot d\mathbf{r}$$

where $\mathbf{F} = (x^2 + y^2)\,\mathbf{i} - 2xy\mathbf{j}$ *and the curve C is the rectangle in the xy-plane bounded by y = 0, x = a, y = b, x = 0.*

Solution. In the *xy*-plane $z = 0$.

$$\therefore \quad \mathbf{r} = x\mathbf{i} + y\mathbf{j}$$

$$d\mathbf{r} = dx\mathbf{i} + dy\mathbf{j}$$

$$\therefore \quad \int_C \mathbf{F}\cdot d\mathbf{r}$$

$$= \int_C \left\{\left(x^2+y^2\right)dx - 2xy\,dy\right\} \qquad ...(1)$$

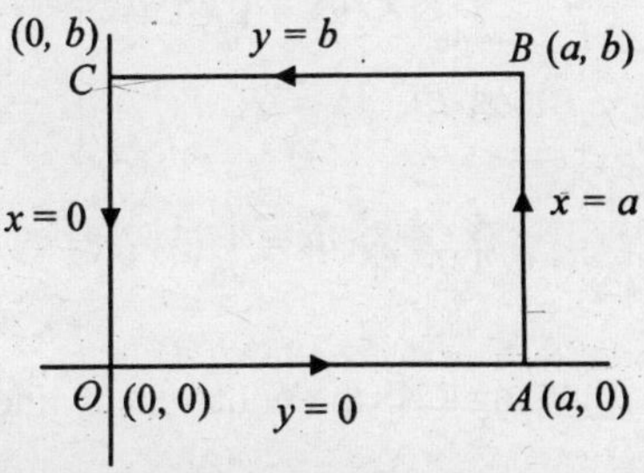

Fig. 7.5.

$$\int_C \mathbf{F}\cdot d\mathbf{r} = \int_{OA} \mathbf{F}\cdot d\mathbf{r} + \int_{AB} \mathbf{F}\cdot d\mathbf{r} + \int_{BC} \mathbf{F}\cdot d\mathbf{r} + \int_{CO} \mathbf{F}\cdot d\mathbf{r} \qquad ...(2)$$

Along OA, $y = 0$ $\therefore$ $dy = 0$ and x varies from 0 to a.

Along AB, $x = a$ $\therefore$ $dx = 0$ and y varies from 0 to b.

Along BC, $y = b$ $\therefore$ $dy = 0$ and x varies from a to 0.

Along CO, $x = 0$ $\therefore$ $dx = 0$ and y varies from b to 0.

Hence, from (1) and (2),

$$\int_C \mathbf{F}\cdot d\mathbf{r} = \int_0^a x^2 dx - \int_0^b 2ay\, dy + \int_a^0 \left(x^2+b^2\right) dx + \int_b^0 0.dy$$

$$= \frac{a^3}{3} - 2a\frac{b^2}{2} + \left[\frac{x^3}{3} + b^2 x\right]_a^0$$

$$= \frac{a^3}{3} - ab^2 - \frac{a^3}{3} - b^2 a = -2ab^2.$$

Example 3. *If* $\mathbf{F} = \left(2x^2+y^2\right)\mathbf{i} + (3y-4x)\mathbf{j}$, *evaluate* $\int_C \mathbf{F}\cdot d\mathbf{r}$ *around the triangle ABC whose vertices are A (0, 0); B = (2, 0); C = (2, 1).*

Solution. $\int_C \mathbf{F}\cdot d\mathbf{r}$

$$= \int_C \left(2x^2+y^2\right) dx + (3y-4x)\, dy \qquad ...(1)$$

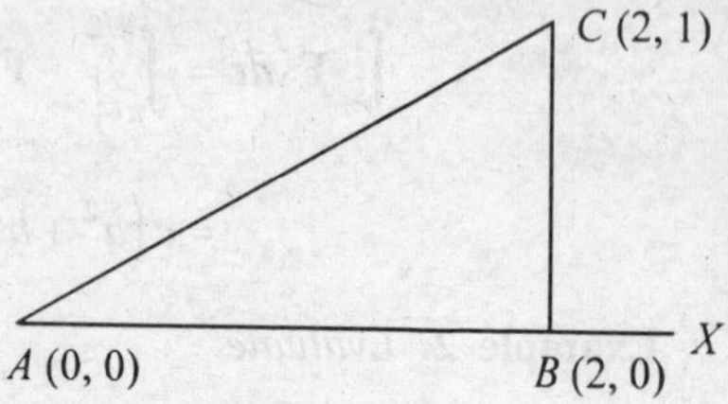

Fig. 7.6.

$r = x\mathbf{i} + y\mathbf{j}$

so, $d\mathbf{r} = dx\mathbf{i} + dy\mathbf{j}$

Along AB, $y = 0$, $\therefore$ $dy = 0$ and x varies from 0 to 2.

$$\therefore \int_{AB} \mathbf{F}\cdot d\mathbf{r} = \int_0^2 2x^2 dx = \frac{2}{3}\left[x^3\right]_0^2 = \frac{16}{3}.$$

Along BC, $x = 2$, $\therefore$ $dx = 0$ and y varies from 0 to 1.

$$\therefore \int_{BC} \mathbf{F}\cdot d\mathbf{r} = \int_0^1 (3y-8)\, dy = \left[\frac{3y^2}{2} - 8y\right]_0^1 = -\frac{13}{2}$$

Along CA, the equation of lines is $y = \frac{1}{2}x$ or $x = 2y$.

$\therefore$ $dx = 2dy$ and y varies from 1 to 0.

$$\therefore \quad \int_C \mathbf{F}\cdot d\mathbf{r} = \int_1^0 \left[2(2y)^2 + y^2\right]2dy + \left[3y - 4(2y)\right]dy$$

$$= \int_1^0 \left(18y^2 - 5y\right)dy = \left[6y^3 - \frac{5y^2}{2}\right]_1^0 = -\frac{7}{2}$$

$$\therefore \quad \int_C \mathbf{F}\cdot d\mathbf{r} = \frac{16}{3} - \frac{13}{2} - \frac{7}{2} = -\frac{14}{3}$$

Example 4. *Find the circulation of* **F** *round the curve C, where* $\mathbf{F} = \left(2x + y^2\right)\mathbf{i} + (3y - 4x)\mathbf{j}$ *and C is the curve* $y = x^2$ *from (0, 0) to (1, 1) and the curve* $y^2 = x$ *from (1, 1) to (0, 0).*

Solution.

$$\int_C \mathbf{F}\cdot d\mathbf{r}$$

$$= \int_C \left[\left(2x + y^2\right)dx + (3y - 4x)\,dy\right]$$

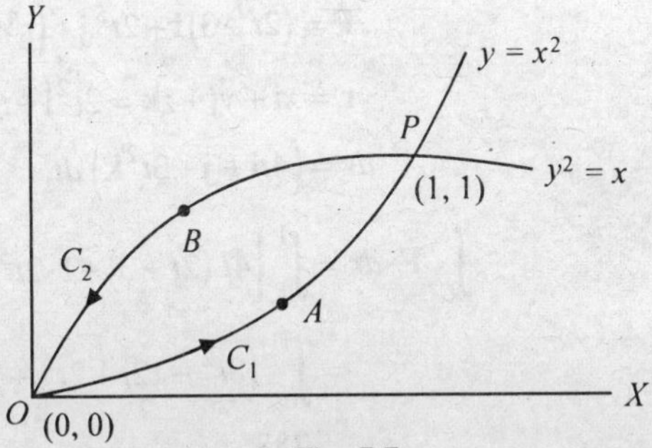

Fig. 7.7.

We have to evaluate this integral along two curves C_1 and C_2.

For C_1, $y = x^2 \quad \therefore \quad dy = 2xdx$ and x varies from 0 to 1.

For C_2, $x = y^2 \quad \therefore \quad dx = 2ydy$ and y varies from 1 to 0.

$$\int_{C_1} \mathbf{F}\cdot d\mathbf{r} = \int_0^1 \left(2x + x^4\right)dx + \left(3x^2 - 4x\right)2xdx$$

$$= \int_0^1 \left(2x - 8x^2 + 6x^3 + x^4\right)dx$$

$$= 2\cdot\frac{1}{2} - 8\cdot\frac{1}{3} + 6\cdot\frac{1}{4} + \frac{1}{5} = \frac{1}{30}$$

$$\int_{C_2} \mathbf{F}\cdot d\mathbf{r} = \int_1^0 \left(2y^2 + y^2\right)2y\,dy + \left(3y - 4y^2\right)dy$$

$$= \int_1^0 \left(3y - 4y^2 + 6y^3\right)dy$$

$$= -\left[\frac{3}{2} - \frac{4}{3} + \frac{6}{4}\right] = \frac{5}{3}$$

$$\therefore \quad \int_C \mathbf{F}\cdot d\mathbf{r} = \int_{C_1} \mathbf{F}\cdot d\mathbf{r} + \int_{C_2} \mathbf{F}\cdot d\mathbf{r}$$

$$= \frac{1}{30} - \frac{5}{3} = -\frac{49}{30}.$$

Example 5. *If* $\mathbf{F} = (2y+3)\mathbf{i} + xz\,\mathbf{j} + (yz - x)\,\mathbf{k}$, *evaluate* $\int_C \mathbf{F}\cdot d\mathbf{r}$ *along the following paths C :*

(*a*) $x = 2t^2,\ y = t,\ z = t^3$ *from* $t = 0$ *to* $t = 1$.

(*b*) *The straight lines from (0, 0, 0) to (0, 0, 1) then to (0, 1, 1) and then to (2, 1, 1).*

(*c*) *The straight line joining (0, 0, 0) to (2, 1, 1).*

Solution. (*a*) On putting the values of *x, y, z* in terms of *t*.

$$\mathbf{F} = (2t+3)\mathbf{i} + 2t^5\mathbf{j} + \left(t^4 - 2t^2\right)\mathbf{k}$$

$$\mathbf{r} = x\mathbf{i} + y\mathbf{j} + z\mathbf{k} = 2t^2\mathbf{i} + t\mathbf{j} + t^3\mathbf{k}$$

$$\therefore \quad d\mathbf{r} = \left(4t\mathbf{i} + \mathbf{j} + 3t^2\mathbf{k}\right) dt.$$

$$\int_C \mathbf{F}\cdot d\mathbf{r} = \int_0^1 \left[4t\,(2t+3) + 1\cdot 2t^5 + 3t^2\left(t^4 - 2t^2\right)\right] dt$$

$$= \int_0^1 \left(8t^2 + 12t + 2t^5 + 3t^6 - 6t^4\right) dt$$

$$= \frac{288}{35}.$$

(*b*) $$\int_C \mathbf{F}\cdot d\mathbf{r} = \int_C (F_1 dx + F_2 dy + F_3 dz)$$

$$= \int \left[(2y+3)\,dx + xz\,dy + (yz - x)\,dz\right] \qquad \text{...(1)}$$

Along C_1 the line joining (0, 0, 0) to (0, 0, 1), $x = 0,\ y = 0$.

$\therefore\ dx = 0,\ dy = 0$ and z varies from 0 to 1.

$$\therefore \quad \int_C \mathbf{F}\cdot d\mathbf{r} = \int_0^1 (yz - x)\,dz = 0, \ \because\ x = 0,\ y = 0.$$

Along C_2 the line joining (0, 0, 1) to (0, 1, 1), $x = 0,\ z = 1$.

$\therefore\ dx = 0,\ dz = 0$ and y varies from 0 to 1.

$$\therefore \quad \int_2 \mathbf{F}\cdot d\mathbf{r} = \int_0^1 xz\,dy = 0, \ \because\ x = 0.$$

Along C_3, the line joining (0, 1, 1) to (2, 1, 1), $y = 1,\ z = 1$.

$\therefore\ dy = 0,\ dz = 0$ and x varies from 0 to 2.

$$\therefore \quad \int_{C_3} \mathbf{F}\cdot d\mathbf{r} = \int_0^2 (2y+3)\,dx = \int_0^2 5dx$$

$$= 10.$$

$$\therefore \quad \int_C \mathbf{F}\cdot d\mathbf{r} = \int_{C_1} \mathbf{F}\cdot d\mathbf{r} + \int_{C_2} \mathbf{F}\cdot d\mathbf{r} + \int_{C_3} \mathbf{F}\cdot d\mathbf{r}$$

$$= 0 + 0 + 10 = 10.$$

(*c*) The parametric equations of the line joining (0, 0, 0) to (2, 1, 1) is $\frac{x}{2} = \frac{y}{1} = \frac{z}{1} = t.$

or $\qquad x = 2t, \;\; y = t, \;\; z = t$ and t varies from 0 to 1.

$$\therefore \quad \int_C \mathbf{F}\cdot d\mathbf{r} = \int_C (2y+3)\,dx + xz\,dy + (yz - x)\,dz$$

$$= \int_0^1 \left[(2t+3)\,2 + 2t^2 + \left(t^2 - 2t\right)\right] dt$$

$$= \int_0^1 \left(3t^2 + 2t + 6\right) dt$$

$$= \left[t^3 + t^2 + 6t\right]_0^1 = 8.$$

Example 6. *Evaluate*

$$\int_S \frac{\mathbf{r}}{r^3} \cdot d\mathbf{a},$$

where S denotes the sphere of radius a with centre at the origin.

Solution. At any point of S,

$$\mathbf{n} = \frac{\mathbf{r}}{r},$$

for the normal lies along the line joining the centre to the point.

$$\therefore \quad \int_S \frac{\mathbf{r}}{r^3} \cdot d\mathbf{a} = \int_S \frac{\mathbf{r}}{r^3} \cdot \frac{\mathbf{r}}{r}\, dS$$

$$= \int_S \frac{\mathbf{r}^2}{r^4}\, dS = \frac{1}{a^2} \int_S dS = \frac{1}{a} \times 4\pi a = 4\pi.$$

Example 7. *Evaluate the surface integral*

$$\int_S (yz\mathbf{i} + zx\mathbf{j} + xy\mathbf{k}) \cdot d\mathbf{a},$$

where S is the surface of the sphere $x^2 + y^2 + z^2 = 1$ *in the first octant.*

Solution. The sphere is parametrically given by

$$x = \sin\theta \cos\phi, \;\; y = \sin\theta \sin\phi, \;\; z = \cos\theta$$

and the part in the first octant corresponds to

$$0 \le \theta \le \frac{1}{2}\pi, \;\; 0 \le \phi \le \frac{1}{2}\pi.$$

Thus, the required surface integral

$$= \iint_S (yz\, dy\, dz + zx\, dz\, dx + xy\, dx\, dy)$$

$$= \int_0^{\frac{1}{2}\pi} \int_0^{\frac{1}{2}\pi} \left[yz \frac{\partial(y,z)}{\partial(\theta,\phi)} + zx \frac{\partial(z,x)}{\partial(\theta,\phi)} + xy \frac{\partial(x,y)}{\partial(\theta,\phi)} + \right] d\theta\, d\phi$$

Now

$$\frac{\partial(y,z)}{\partial(\theta,\phi)} = \begin{vmatrix} \dfrac{\partial y}{\partial\theta} & \dfrac{\partial y}{\partial\phi} \\ \dfrac{\partial z}{\partial\theta} & \dfrac{\partial z}{\partial\phi} \end{vmatrix} = \begin{vmatrix} \cos\theta\sin\phi & \sin\theta\cos\phi \\ -\sin\theta & 0 \end{vmatrix}$$

$$= \sin\theta\cos^2\phi.$$

$$\frac{\partial(z,x)}{\partial(\theta,\phi)} = \begin{vmatrix} \dfrac{\partial z}{\partial\theta} & \dfrac{\partial x}{\partial\phi} \\ \dfrac{\partial x}{\partial\theta} & \dfrac{\partial x}{\partial\phi} \end{vmatrix} = \begin{vmatrix} -\sin\theta & 0 \\ \cos\theta\cos\phi & -\sin\theta\sin\phi \end{vmatrix}$$

$$= \sin^2\theta\sin\phi.$$

$$\frac{\partial(x,y)}{\partial(\theta,\phi)} = \begin{vmatrix} \dfrac{\partial x}{\partial\theta} & \dfrac{\partial x}{\partial\phi} \\ \dfrac{\partial y}{\partial\theta} & \dfrac{\partial y}{\partial\phi} \end{vmatrix} = \begin{vmatrix} \cos\theta\cos\phi & -\sin\theta\sin\phi \\ \cos\theta\sin\phi & -\sin\theta\cos\phi \end{vmatrix}$$

$$= \sin\theta\cos\theta.$$

$\therefore$ integral

$$= \int_0^{\frac{1}{2}\pi} \int_0^{\frac{1}{2}\pi} [\sin\theta\cos\theta\sin\phi\sin^2\theta\cos\phi +$$

$$\sin\theta\cos\theta\cos\phi\sin^2\theta\sin\phi + \sin^2\theta\sin\phi\cos\phi\sin\theta\cos\theta]\, d\theta\, d\phi$$

$$= \int_0^{\frac{1}{2}\pi} \int_0^{\frac{1}{2}\pi} \left[2\sin^3\theta\cos\theta\sin\phi\cos\phi + \sin^3\theta\cos\theta\sin\phi\cos\phi\right] d\theta\, d\phi$$

$$= \frac{7}{20}.$$

Example 8. *If* $\mathbf{f} = y\mathbf{i} + (x - 2xz)\,\mathbf{j} - xy\mathbf{k}$, *evaluate* $\int_S (\nabla \times \mathbf{f}) \cdot \mathbf{n}\, dS$, *where S is the surface of the sphere* $x^2 + y^2 + z^2 = a^2$ *above the xy-plane.*

(Rohilkhand 2000)

Solution. Let $\mathbf{F} = \nabla \times \mathbf{f} = \begin{vmatrix} \mathbf{i} & \mathbf{j} & \mathbf{k} \\ \dfrac{\partial}{\partial x} & \dfrac{\partial}{\partial y} & \dfrac{\partial}{\partial z} \\ y & x - 2xz & -xy \end{vmatrix}$

$$= x\mathbf{i} + y\mathbf{j} - 2z\mathbf{k}$$

$\phi = c$ is given by $x^2 + y^2 + z^2 = a^2$

grad ϕ is along the normal and grad $\phi = \Sigma \mathbf{i} \dfrac{\partial \phi}{\partial x}$

$$= 2x\mathbf{i} + 2y\mathbf{j} + 2z\mathbf{k}$$

$\therefore \quad \mathbf{n}$ = a unit normal $= \dfrac{2x\mathbf{i} + 2y\mathbf{j} + 2z\mathbf{k}}{\sqrt{(4x^2 + 4y^2 + 4z^2)}}$

$$= \frac{x\mathbf{i} + y\mathbf{j} + z\mathbf{k}}{a}$$

$$\therefore \quad \mathbf{F} \cdot \mathbf{n} = (x\mathbf{i} + y\mathbf{j} - 2z\mathbf{k}) \cdot \frac{(x\mathbf{i} + y\mathbf{j} + z\mathbf{k})}{a}$$

$$= \frac{x^2 + y^2 + 2z^2}{a}$$

Also, we know that

$$\int_S \mathbf{F} \cdot \mathbf{n} \, dS = \iint_{S_3} \mathbf{F} \cdot \mathbf{n} \frac{dx\, dy}{\mathbf{n} \cdot \mathbf{k}}$$

where S_3 is the projection of S on xy-plane.

$$\mathbf{n} \cdot \mathbf{k} = \frac{x\mathbf{i} + y\mathbf{j} + z\mathbf{k}}{a} \cdot \mathbf{k} = \frac{z}{a}$$

$$= \frac{\sqrt{(a^2 - x^2 - y^2)}}{a}$$

Also $\quad \mathbf{F} \cdot \mathbf{n} = \dfrac{x^2 + y^2 - 2z^2}{a} = \dfrac{x^2 + y^2 - 2(a^2 - x^2 - y^2)}{a}$

$$= \frac{3(x^2 + y^2) - 2a^2}{a}$$

$$\therefore \quad \int_S \mathbf{F} \cdot \mathbf{n} \, dS = \iint_{S_3} \frac{3(x^2 + y^2) - 2a^2}{a} \cdot \frac{dx\, dy}{\sqrt{(a^2 - x^2 - y^2)}} a \qquad ...(1)$$

Now, S_3 is the projection of $x^2+y^2+z^2=a^2$ in the xy-plane and is given by

$$x^2 + y^2 = a^2.$$

In order to integrate (1) put $x = r\cos\theta$, $y = r\sin\theta$.

$\therefore$ $dxdy = rd\theta dr$ and limits of θ are from 0 to 2π and that of r from 0 to a.

$$\therefore \quad \int_S \mathbf{F}\cdot\mathbf{n}\,dS = \int_0^{2\pi}\int_0^a \frac{3r^2-2a^2}{\sqrt{(a^2-r^2)}}\,rd\theta dr$$

$$= 2\pi\int_0^a \frac{3r^2-2a^2}{\sqrt{(a^2-r^2)}}\,rdr.$$

Put $a^2 - r^2 = t$, $\therefore$ $-2rdr = 2tdt$

$$\therefore \quad \int_S \mathbf{F}\cdot\mathbf{n}\,dS = 2\pi\int_0^a \frac{3(a^2-t^2)-2a^2}{t}(-t)\,dt$$

$$= 2\pi\int_0^a (a^2-3t^2)\,dt$$

$$= 2\pi(a^3-a^3) = 0.$$

Example 9. *Evaluate* $\int_V (2x+y)\,dV$, *where V is the closed region bounded by the cylinder* $z = 4 - x^2$ *and the plane* $x = 0$, $y = 0$, $y = 2$ *and* $z = 0$.

Solution. The given cylinder $z = 4 - x^2$ meets the x-axis ($y = 0, z = 0$) at $x^2 = 4$ or $x = 2$, *i.e.*, at the point (2, 0, 0). It meets z-axis at $z = 4$, *i.e.*, at (0, 0, 4). Therefore, the limits of integration are as under.

$$z = 4 \text{ to } z = 4 - x^2,\quad y = 0 \text{ to } y = 2 \text{ and } x = 0 \text{ to } x = 2.$$

Also $dV = dx\,dy\,dz$.

$$\therefore \quad \int_V (2x+y)\,dV = \iiint (2x+y)\,dx\,dy\,dz$$

$$= \iint (2x+y)\left[z\right]_0^{4-x^2} dx\,dy$$

$$= \int_{x=0}^{2}\int_{y=0}^{2} (2x+y)(4-x^2)\,dx\,dy$$

$$= \int_0^2 \left[2x(4-x^2)y + (4-x^2)\frac{y^2}{2}\right]_{y=0}^{2} dx$$

$$= \int_0^2 \left[4x\left(4-x^2\right)+2\left(4-x^2\right)\right] dx$$

$$= 2\int_0^2 \left(4+8x-x^2-2x^3\right) dx$$

$$= \frac{80}{3}.$$

EXERCISES

1. Evaluate

$$\int_C \mathbf{F}\cdot d\mathbf{r} \text{ when } \mathbf{F} = xy\mathbf{i} + yz\,\mathbf{j} + zx\mathbf{k}$$

where C is the curve $\mathbf{r} = \mathbf{i}t + \mathbf{j}t^2 + \mathbf{k}t^3$; t varying from -1 to $+1$.

2. Calculate

$$\int_C \left[\left(x^2+y^2\right)\mathbf{i} + \left(x^2-y^2\right)\mathbf{j}\right]\cdot d\mathbf{r}$$

where C is the curve

(*i*) $y^2 = x$ joining (0, 0) to (1, 1).

(*ii*) $x^2 = y$ joining (0, 0) to (1, 1).

(*iii*) consisting of two straight lines joining (0, 0) to (1, 0) and (1, 0) to (1, 1).

(*iv*) consisting of three straight lines joining (0, 0) to (2, – 2); (2, – 2) to (0, – 1) and (0, – 1) to (1, 1).

3. Show that

$$\int_C \left(\frac{y}{x^2+y^2}\mathbf{i} + \frac{x}{x^2+y^2}\mathbf{j}\right)\cdot d\mathbf{r} = 2\pi,$$

where C is the circle $x^2 + y^2 = 1$ in the z-plane described in the counter-clockwise sense.

4. Evaluate

$$\int_C \frac{-y^3\mathbf{i} + x^3\mathbf{j}}{\left(x^2+y^2\right)^2}\cdot d\mathbf{r},$$

where C is the

(*i*) boundary of the square $x = \pm a,\ y = \pm a$;

(*ii*) the circle $x^2 + y^2 = a^2$; both in the counter-clockwise sense.

5. Show that

$$\int_C \left[\left(y^2+z^2\right)\mathbf{i} + \left(z^2+x^2\right)\mathbf{j} + \left(x^2+y^2\right)\mathbf{k}\right]\cdot d\mathbf{r} = -2\pi ab^2$$

where the curve C is that part for which $z \geq 0$ of the intersection of the surfaces $x^2 + y^2 + z^2 = 2ax$ and $x^2 + y^2 = 2bx$, $a > b > 0$; the curve begins at the origin and runs at first in the positive octant,

$$\left[x = b(1+\cos\theta),\ y = b\sin\theta,\ z = 2\sqrt{(a-b)\,b}\cos\frac{\theta}{2}\cdot(-\pi \le \theta \le \pi)\right].$$

6. Evaluate

$$\int_S \mathbf{r}\cdot d\mathbf{a}.$$

where S is the part of the surface of the ellipsoid

$$\frac{x^2}{a^2}+\frac{y^2}{b^2}+\frac{z^2}{c^2}=1.$$

lying above the z-plane, the normal at any point being directed outward.

7. Evaluate

$$\int_S \left(x^3\mathbf{i}+y^3\mathbf{j}+z^3\mathbf{k}\right)\cdot d\mathbf{a}$$

where S is the surface of the sphere

$$x^2 + y^2 + z^2 = 1.$$

8. Evaluate

$$\int_S \mathbf{F}\cdot\mathbf{n}\, dS$$

where $\mathbf{F} = 4xz\mathbf{i} - y^2\mathbf{j} + yz\mathbf{k}$ where S is the surface of the cube bounded by $x = 0,\ x = 1,\ y = 0,\ y = 1,\ z = 0,\ z = 1.$

9. If

$$\mathbf{F} = \left(2x^2 - 3z\right)\mathbf{i} - 2xy\mathbf{j} - 4x\mathbf{k}$$

then evaluate $\int_V \nabla\cdot\mathbf{F}\, dV$ and $\iiint_V \nabla\times\mathbf{F}\, dV$, where V is closed region bounded by the plane $x = 0,\ y = 0,\ z = 0$ and $2x + 2y + z = 4.$

(Kumaon, 98)

10. Evaluate

$$\int_S \phi\, \mathbf{n}\, dS,$$

where $\phi = \frac{3}{8}xyz$ and S is the surface of the cylinder $x^2 + y^2 = 16$ included in the first octant between $z = 0$ and $z = 5.$

7.6. REDUCTION OF VOLUME TO SURFACE INTEGRALS

Gauss's Divergence Theorem. *If* **F** *is a continuously differentiable vector point function and S is a closed surface enclosing a region V, then*

$$\int_S \mathbf{F}\cdot\mathbf{n}\, dS = \int_V \operatorname{div}\mathbf{F}\, dv. \qquad ...(1)$$

where **n** *is the unit outward drawn normal vector.*

(Ajmer 98, Gorakhpur 99, Lucknow 97)

Briefly, the theorem states that the normal surface integral of a function **F** over the boundary of a closed region is equal to the volume of div **F** taken throughout the region.

To start with, we shall suppose that the region V is such that it is possible to choose a rectangular cartesian co-ordinate system such that each line parallel to any co-ordinate axis which has internal points in common with the region meets the boundary S in two points.

In terms of cartesian co-ordinates, the equation (1) is stated as follows :

$$\iint_S (F_1 dy\,dz + F_2 dz\,dx + F_3 dx\,dy)$$

$$= \iiint_V \left(\frac{\partial F_1}{\partial x} + \frac{\partial F_2}{\partial y} + \frac{\partial F_3}{\partial z}\right) dx\,dy\,dz$$

Consider now the volume integral

$$\iiint_V \frac{\partial F_3}{\partial z}\,dx\,dy\,dz.$$

Let R_3 denote the projection of the region V, on the OXY plane. Every line through a point $(x, y, 0)$ of R_3 meets the boundary S in two points. Let the z, co-ordinates of these points be

$$z = \phi\,(x, y), \quad z = \psi\,(x, y)$$

where

$$\phi\,(x, y) \geq \psi\,(x, y).$$

Now

$$\iiint_V \frac{\partial F_3}{\partial z}\,dx\,dy\,dz = \iint_{R_3}\left[\int_\psi^\phi \frac{\partial F_3}{\partial z}\,dz\right] dx\,dy$$

$$= \iint_{R_3}\left[F_3(x, y, \phi) - F_3(x, y, \psi)\right] dx\,dy$$

$$= \int_{R_3}\!\!\int F_3(x, y, \phi)\,dx\,dy -$$

$$\iint_{R_3} F_3(x, y, \psi)\,dx\,dy \qquad \text{...(1)}$$

Let us now denote the parts of the surface S corresponding to $z = \phi$ (x, y) and $z = \psi$ (x, y) by S_1 and S_2 respectively.

If **n** denotes the *outward* drawn unit normal vector at any point of S, we have

$$\iint_{R_3} F_3(x, y, \phi)\,dx\,dy = \int_{S_1} F_3\,\mathbf{n}\,.\,\mathbf{k}\,dS \qquad \text{...(2)}$$

$$\iint_{R_3} F_3(x, y, \psi)\,dx\,dy = -\int_{S_2} F_3\,\mathbf{n}\,.\,\mathbf{k}\,dS \qquad \text{..(3)}$$

for the outward drawn normal at any point of S_1 makes an acute angle with positive direction of z-axis and that at any point of S_2 makes an obtuse angle with z-axis.

From (1), (2) and (3), we have

$$\iiint_V \frac{\partial F_3}{\partial z} dx\, dy\, dz = \int_{S_1} F_3 \, \mathbf{n} \,.\, \mathbf{k} \, dS + \int_{S_2} F_3 \, \mathbf{n} \,.\, \mathbf{k} \, dS$$

$$= \int_S F_3 \, \mathbf{n} \,.\, \mathbf{k} \, dS. \qquad ...(4)$$

Similarly it may be proved that

$$\iiint_V \frac{\partial F^2}{\partial y} dxdydz = \int_S F_2 \mathbf{n} \cdot \mathbf{j} ds \qquad ...(5)$$

$$\iiint_V \frac{\partial F_1}{\partial x} dx\, dy\, dz = \int_S F_1 \, \mathbf{n} \,.\, \mathbf{i} \, dS. \qquad ...(6)$$

Adding (4), (5) and (6), we get

$$\iiint_V \left(\frac{\partial F_1}{\partial x} + \frac{\partial F_2}{\partial y} + \frac{\partial F_3}{\partial z} \right) dx\, dy\, dz = \int_S (F_1 \mathbf{i} + F_2 \mathbf{j} + F_3 \mathbf{k}) \cdot \mathbf{n} \, dS$$

i.e.,

$$\int_V \operatorname{div} \mathbf{F} \, dv = \int_S \mathbf{F} \cdot \mathbf{n} \, dS = \int_S \mathbf{F} \cdot d\mathbf{a}. \qquad ...(7)$$

We now extend the result to the case of a general region.

Suppose that the region is such that it can be split up into a finite number of sub-regions each of which is met by a line parallel to any co-ordinate axis in two points only. We shall now apply the theorem to each sub-region and add the corresponding results. The volume integrals over separate sub-regions will combine to give the volume integral over the whole region V. Also each boundary surface of two sub-regions appears twice as parts of the two regions and the two corresponding surface integrals over the same will cancel, for the two normals at any point are oppositely directed. Thus, the soul of the surface integrals also combine to give a surface integral over the entire surface S. Thus, in the general case also we have

$$\int_V \operatorname{div} \mathbf{F} \, dv = \int_S \mathbf{F} \cdot \mathbf{n} \, dS = \int_S \mathbf{F} \cdot d\mathbf{a}.$$

Note. The theorem also holds good for a region V (shaded) such as that shown in the figure. It is bounded by two closed surface S_1 and S_2 one of which lies within the other. It should, however, be understood that the *outward* drawn normal at any point means the normal directed away from the region. Thus, the normals at points of S_1 and S_2 are to be directed as shown in the figure.

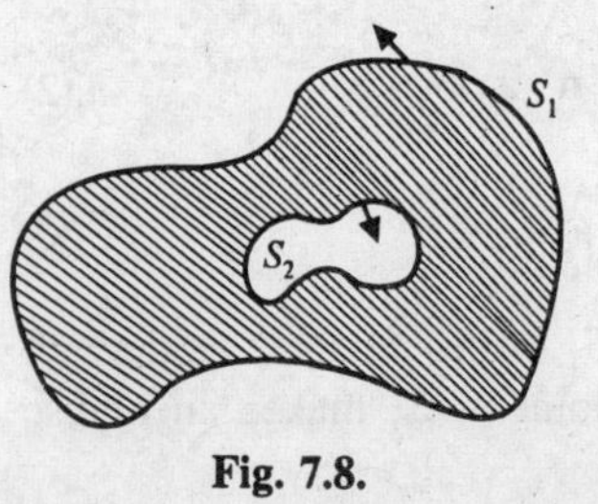

Fig. 7.8.

The theorem can be applied to regions enclosed by several surfaces.

Cor. 1. The Gauss's theorem obtained above enables us to express a normal surface integral as a volume integral. We shall now see that with the help of this theorem we are able to express the other two types of surface integrals also as volume integrals.

It will be shown that

(*i*) $\int_S \mathbf{F} \times \mathbf{n}\, dS = \int_S \mathbf{F} \times d\mathbf{a} = -\int_V \text{curl}\, \mathbf{F}\, dv$

(*ii*) $\int_S \phi\, \mathbf{n}\, dS = \int_S \phi\, d\mathbf{a} = \int_V \nabla\phi\, dr$

We write

$$\mathbf{f} = \mathbf{a} \times \mathbf{F},$$

where **a** is any constant vector and apply Gauss's theorem to the vector function **f**. Thus, we have

$$\int_S \mathbf{a} \times \mathbf{F} \cdot \mathbf{n}\, dS = \int_V \text{div}\,(\mathbf{a} \times \mathbf{F})\, dv. \qquad ...(1)$$

But

$$\mathbf{a} \times \mathbf{F} \,.\, \mathbf{n} = \mathbf{a} \,.\, \mathbf{F} \times \mathbf{n}, \qquad ...(2)$$

and

$$\text{div}\,(\mathbf{a} \times \mathbf{F}) = \nabla \,.\, (\mathbf{a} \times \mathbf{F}) = -\,\mathbf{a} \,.\, \nabla \times \mathbf{F}. \qquad ...(3)$$

From (1), (2) and (3),

$$\int_S \mathbf{a} \cdot \mathbf{F} \times \mathbf{n}\, dS = -\int_V \mathbf{a} \cdot \nabla \times \mathbf{F}\, dv.$$

or

$$\mathbf{a} \cdot \int_S \mathbf{F} \times \mathbf{n}\, dS = -\,\mathbf{a} \cdot \int_V \nabla \times \mathbf{F}\, dv$$

or

$$\mathbf{a} \cdot \left[\int_S \mathbf{F} \times \mathbf{n}\, dS + \int_V \nabla \times \mathbf{F}\, dv\right] = 0 \qquad ...(4)$$

As, however, **a** is any arbitrary vector, we have from (4)

$$\int_S \mathbf{F} \times \mathbf{n}\, dS + \int_V \nabla \times \mathbf{F}\, dv = 0,$$

i.e.,

$$\int_S \mathbf{F} \times \mathbf{n}\, dS = -\int_V \nabla \times \mathbf{F}\, dv = -\int_V \text{curl}\, \mathbf{F}\, dv.$$

Thus, we have proved (*i*).

To prove (*ii*) we write

$$\mathbf{f} = \mathbf{a}\phi$$

where **a** is a constant vector. Applying Gauss's theorem to the function **f**, we have

$$\int_S \mathbf{f} \cdot \mathbf{n}\, dS = \int_S \text{div}\, \mathbf{f}$$

i.e.,

$$\int_S \mathbf{a}\phi \cdot \mathbf{n}\, dS = \int_V \text{div}\,(\mathbf{a}\phi)\, dv = \int_V \nabla \cdot (\mathbf{a}\phi)\, dv$$

or

$$\mathbf{a} \cdot \int_S \phi\mathbf{n}\, dS = \mathbf{a} \cdot \int_V \nabla\phi\, dv$$

or

$$\mathbf{a} \cdot \left[\int_S \phi\mathbf{n}\, dS - \int_V \nabla\phi\, dv\right] = 0.$$

The vector **a** being arbitrary, we obtain

$$\int_S \phi\mathbf{n}\, dS - \int_V \nabla\phi\, dv = 0,$$

or

$$\int_S \phi\mathbf{n}\, dS - \int_V \nabla\phi\, dv,$$

which is (*ii*).

Note. The three results obtained above can be re-written as follows :

$$\int_S \mathbf{n} \cdot \mathbf{F}\, dS = \int_V \nabla \cdot \mathbf{F}\, dv,$$

$$\int_S \mathbf{n} \times \mathbf{F}\, dS = \int_V \nabla \times \mathbf{F}\, dv,$$

$$\int_S \mathbf{n}\phi\, dS = \int_V \nabla\phi\, dv.$$

It will thus, be seen that each of these results can be written by just changing **n** to ∇.

The integrand in each case has been so written that **n** appears *before* the function and not after.

Cor. 2. With the help of the results obtained above, we can state the following integral expressions for grad ϕ, div **F** and curl **F**. These expressions are also often adopted as definitions for the differential operators.

$$\text{grad}\,\phi = \lim_{v\to 0} \frac{\int_S \mathbf{n}\phi\, dS}{V},$$

$$\text{div}\,\mathbf{F} = \lim_{v\to 0} \frac{\int_S \mathbf{n} \cdot \mathbf{F}\, dS}{V},$$

$$\text{curl}\,\mathbf{F} = \lim_{v\to 0} \frac{\int_S \mathbf{n} \times \mathbf{F}\, dS}{V}.$$

Take any point P and enclose the same in a small region volume V enclosed by a surface S.

As proved in Cor. 1, above

$$\int_S \mathbf{n}\phi \, dS = \int_V \nabla\phi \, dv$$

$$= V\left[(\nabla\phi)_0 + \varepsilon\right],$$

where $(\nabla\phi)_0$ denotes the value of $\nabla\phi$ at P and $\varepsilon \to 0$ as $V \to 0$.

$$\therefore \qquad \lim_{v\to 0} \frac{\int_S \mathbf{n}\phi \, ds}{V} = (\nabla\phi)_0.$$

Similarly for div **F** and curl **F**.

7.7. PHYSICAL INTERPRETATION OF GAUSS'S THEOREM

Let the vector point function **q**, denote the velocity vector of an incompressible fluid of unit density and let S denote any closed surface drawn in the fluid. By Gauss's Theorem,

$$\int_S \mathbf{q} \cdot \mathbf{n} \, ds = \int_V \text{div } \mathbf{q} \, dv. \qquad ...(1)$$

Now **q . n** is the component of the velocity at any point of S in the direction of the outward drawn normal so that **q . n** δS denotes the amount of fluid that flows out in unit time through the element δS. Hence *the left hand side of (1) denotes the amount of fluid flowing across the surface S in unit time from the inside to the outside.* This amount may be positive, negative or zero.

Now the total amount flowing outwards must be continuously supplied so that inside the region we must have sources producing fluid. We have already seen that div **q** at any point denotes the amount of fluid per unit time per unit volume that goes through any point. Thus div **q** may be thought of as the *source-intensity* of the incompressible fluid at any point P. Hence *the right hand side of (1) denotes the amount of fluid per unit time supplied by the sources within S.* Thus, the equality (1) appears intuitively evident.

EXAMPLES

Example 1. *Show that*

$$\frac{1}{3}\int_S \mathbf{r} \cdot d\mathbf{a} = V.$$

where V is the volume enclosed by the surface S.

Solution. We have

$$\int_S \frac{1}{3}\mathbf{r}\cdot d\mathbf{a} = \int_V \operatorname{div}\frac{1}{3}\mathbf{r}\,dv$$

$$= \int_V dv = V, \text{ for div } \mathbf{r} = 3.$$

Example 2. *If* $\vec{OA} = a\mathbf{i}$, $\vec{OB} = a\mathbf{j}$, $\vec{OC} = a\mathbf{k}$, *form three coterminous edges of a cube and S denotes the surface of the cube, evaluate*

$$\int_S \left\{\left(x^3 - yz\right)\mathbf{i} - 2x^2y\mathbf{j} + 2\mathbf{k}\right\}\cdot\mathbf{n}\,dS,$$

by expressing it as a volume integral. Also verify the result by direct evaluation of the surface integral.

Solution. Let

$$\mathbf{F} = \left(x^3 - yz\right)\mathbf{i} - 2x^2y\mathbf{j} + 2\mathbf{k}$$

$$\therefore \quad \operatorname{div}\mathbf{F} = \frac{\partial\left(x^3 - yz\right)}{\partial x} + \frac{\partial\left(-2x^2y\right)}{\partial y} + \frac{\partial(2)}{\partial z}$$

$$= 3x^2 - 2x^2 = x^2.$$

$$\therefore \quad \int_S \mathbf{F}\cdot\mathbf{n}\,dS = \int_V \operatorname{div}\mathbf{F}\,dv = \iiint_V x^2 dx\,dy\,dz$$

$$= \left[\int_0^a dz\right]\left[\int_0^a dy\right]\left[\int_0^a x^2 dx\right]$$

$$= a\cdot a\cdot\frac{1}{2}a^3 = \frac{1}{2}a^5.$$

We shall now evaluate the surface integral directly. The surface *S* consists of *six* faces.

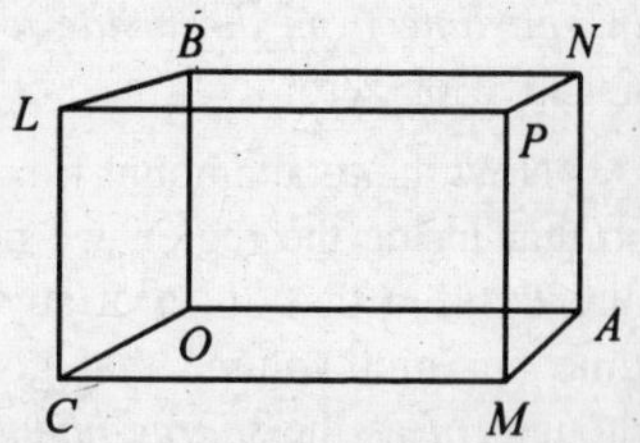

Fig. 7.9.

Over the surface *AOBN*

$$\int_S \mathbf{F}\cdot\mathbf{n}\,dS$$

$$= \int_S \left\{\left(x^3 - yz\right)\mathbf{i} - 2x^2y\mathbf{j} + 2\mathbf{k}\right\}\cdot(-\mathbf{k})\,dS$$

$$= \int_0^a \int_0^a -2dx\,dy = -2a^2.$$

Over the surface *PLCM*,

$$\int_S \mathbf{F}\cdot\mathbf{n}\,dS = \int_S \left\{\left(x^3 - yz\right)\mathbf{i} - 2x^2y\mathbf{j} + 2\mathbf{k}\right\}\cdot\mathbf{k}\,dS$$

$$= \int_0^a \int_0^a 2dx\,dy = 2a^2.$$

Similarly over the faces *NPMA, BLCO, AOCM, NBLP,* the corresponding surface integrals are respectively

$$\int_S \mathbf{F}\cdot\mathbf{i}\,dS = \iint_S \left(x^3 - yz\right) dy\,dz = \int_0^a \int_0^a \left(a^3 - yz\right) dy\,dz$$

$$= a^5 - \frac{1}{4}a^4,$$

$$\int_S \mathbf{F}\cdot(-\mathbf{i})\,dS = -\iint_S \left(x^3 - yz\right) dy\,dz = \int_0^a \int_0^a yz\,dy\,dz = \frac{1}{4}a^4$$

$$\int_S \mathbf{F}\cdot(-\mathbf{j})\,dS = \int_0^a \int_0^a 2x^2 y\,dx\,dy = 0$$

$$\int_S \mathbf{F}\cdot\mathbf{j}\,dS = -\int_0^a \int_0^a 2x^2 y\,dx\,dy = -2a\int_0^a \int_0^a x^2 dx\,dz$$

$$= \frac{-2}{3}a^5.$$

Adding we see that over the whole surface

$$\int_S \mathbf{F}\cdot\mathbf{n}\,dS = -2a^2 + 2a^2 + a^5 - \frac{1}{4}a^4 + \frac{1}{4}a^4 + 0 - \frac{2}{3}a^5 = \frac{1}{3}a^5.$$

Hence the verification.

Example 3. *Evaluate*

$$\int_S -\gamma\frac{\mathbf{r}}{r^3}\cdot d\mathbf{a},$$

where S denotes any closed surface enclosing a region V.

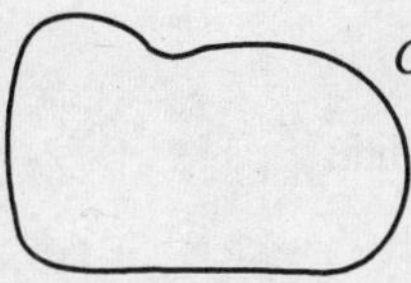

Fig. 7.10.

Solution. Two cases arise according as the origin lies within S or without S.

Suppose first that the origin lies without S. In this case

$$\mathbf{F} = -\gamma\frac{\mathbf{r}}{r^3},$$

is continuously differentiable throughout the region enclosed by S.

$$\therefore \qquad \int_S -\frac{\gamma\mathbf{r}}{r^3}\cdot d\mathbf{a} = \int_V \operatorname{div}\left(-\frac{\gamma\mathbf{r}}{r^3}\right) dv.$$

It may be easily shown that

$$\operatorname{div}\frac{-\gamma\mathbf{r}}{r^3} = 0.$$

Thus, the surface integral is 0.

Suppose now that the origin O lies within S. In this case $\mathbf{F}$ is not defined at the point O within S so that we cannot apply Gauss's theorem as such.

We now enclose O in a small sphere with its centre at O and lying within S. Let its radius be a. Let its surface be denoted by S_1. In relation to the application of Gauss's theorem, the normal at any point of the surface S_1 of the sphere has to be directed inwards towards O.

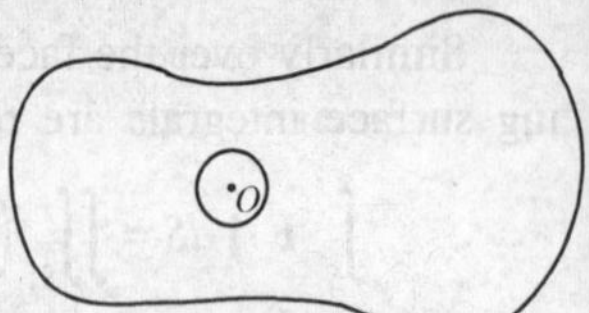

Fig. 7.11.

The function **F** is continuously differentiable at all points of the region V' enclosed between S and S_1.

$\therefore$ By Gauss's theorem

$$\int_S -\frac{\gamma \mathbf{r}}{r^3} \cdot d\mathbf{a} + \int_{S_1} -\frac{\gamma \mathbf{r}}{r^3} \cdot d\mathbf{a} = \int_{V'} \operatorname{div}\left(-\frac{\gamma \mathbf{r}}{r^3}\right) \cdot dv = 0.$$

As in Ex. 6, on Page 334, we have

$$-\gamma \int_{S_1} \frac{\mathbf{r}}{r^3} \cdot d\mathbf{a} = -4\pi\gamma.$$

$$\therefore \qquad \int_S -\gamma \frac{\mathbf{r}}{r^3} \cdot d\mathbf{a} = 4\pi\gamma.$$

Example 4. *Convert*

$$\int_S \mathbf{F} \times d\mathbf{a},$$

taken over a closed surface into a volume integral and verify the result in the case

$$\mathbf{F} = xy\mathbf{k}$$

where S is the surface of a cube whose base is in the XY plane and which is bounded by the planes

$$x = 0, \ x = a; \ y = 0, \ y = a.$$

Solution. In Cor. 1 to § 7.5, it has been shown that

$$\int_S \mathbf{F} \times d\mathbf{a} = -\int_S \operatorname{curl} \mathbf{F} \, dv.$$

Now when $\mathbf{F} = xy\mathbf{k}$, we have

$$\operatorname{curl} \mathbf{F} = \mathbf{i}x - \mathbf{j}y.$$

Now $\displaystyle\int_V \operatorname{curl} \mathbf{F} \, dv = \mathbf{i} \int_V x \, dv - \mathbf{j} \int_V y \, dv$

$$= \mathbf{i} \int_0^a \int_0^a \int_0^a x \, dx \, dy \, dz - \mathbf{j} \int_0^a \int_0^a \int_0^a y \, dx \, dy \, dz$$

$$= \mathbf{i}\frac{a^4}{2} - \mathbf{j}\frac{a^4}{2} = \frac{1}{2} a^4 (\mathbf{i} - \mathbf{j}).$$

[Refer Fig. Ex. 2, Page 345]

We shall now calculate

$$\int_S \mathbf{F} \times d\mathbf{a},$$

over the six faces of the cube.

Over the face *AOCM*,

$$\int \mathbf{F} \times d\mathbf{a} = \iint (xy\mathbf{k}) \times -\mathbf{j} dx\, dz = \mathbf{i} \int_0^a \int_0^a xy\, dx\, dz = 0,$$

for y is zero at every point of this face.

Over the face *PNBL*,

$$\int \mathbf{F} \times d\mathbf{a} = \iint (xy\mathbf{k}) \times \mathbf{j} dx\, dz = -\mathbf{i} \int_0^a \int_0^a xy\, dx\, dz$$

$$= -\mathbf{i}a \int_0^a \int_0^a x\, dx\, dz = -\mathbf{i}\frac{a^4}{2}\cdot$$

Similarly, over the face *NPMA, BLCO, AOBN, MCLP,* the corresponding surface integrals respectively are :

$$\int \mathbf{F} \times d\mathbf{a} = \iint xy\mathbf{k} \times \mathbf{i}\, dy\, dz = \mathbf{j}a \int_0^a \int_0^a y\, dy\, dz = \mathbf{j}\frac{a^4}{2}$$

$$\int \mathbf{F} \times d\mathbf{a} = \iint xy\mathbf{k} \times (-\mathbf{i}\, dy\, dz) = 0, \text{ for } \mathbf{F} = \mathbf{0} \text{ at every point.}$$

$$\int \mathbf{F} \times d\mathbf{a} = \iint xy\mathbf{k} \times -\mathbf{k}\, dx\, dy = 0,$$

$$\int \mathbf{F} \times d\mathbf{a} = \iint xy\mathbf{k} \times \mathbf{k}\, dx\, dy = 0,$$

$\therefore$ Over the six faces,

$$\int \mathbf{F} \times d\mathbf{a} = 0 - \mathbf{i}\frac{a^4}{2} + \mathbf{j}\frac{a^4}{2} + 0 + 0 + 0 = -\frac{1}{2}a^4 (\mathbf{i} - \mathbf{j}).$$

Hence, the verification.

Example 5. *Prove that*

$$\int_S \mathbf{n} \times (\mathbf{a} \times \mathbf{r})\, dS = 2aV$$

where **a** *is a constant vector.*

Solution. By Gauss's divergence theorem

$$\int_S \mathbf{n} \times (\mathbf{a} \times \mathbf{r})\, dS = \int_V \nabla \times (\mathbf{a} \times \mathbf{r})\, dV$$

Now, $$\nabla \times (\mathbf{a} \times \mathbf{r}) = \Sigma \mathbf{i} \times \frac{\partial}{\partial x}(\mathbf{a} \times \mathbf{r})$$

$$= \Sigma \mathbf{i} \times \left(\frac{\partial \mathbf{a}}{\partial x} \times \mathbf{r} + \mathbf{a} \times \frac{\partial \mathbf{r}}{\partial x} \right)$$

$$= \Sigma \mathbf{i} \times (0 + \mathbf{a} \times \mathbf{i})$$

$$= \Sigma \mathbf{i} \times (\mathbf{a} \times \mathbf{i})$$

$$= \Sigma \, (\mathbf{i} \cdot \mathbf{i} \, \mathbf{a} - \mathbf{i} \cdot \mathbf{a} \, \mathbf{i})$$

$$= \Sigma \, (\mathbf{a} - \mathbf{a}_1 \mathbf{i}) = \Sigma \mathbf{a} - \Sigma \mathbf{a}_1 \mathbf{i}$$

$$= 3\mathbf{a} - \mathbf{a} = 2\mathbf{a}$$

$$\therefore \quad \int_S \mathbf{n} \times (\mathbf{a} \times \mathbf{r}) \, dS = \int_V 2a \, dV = 2aV.$$

Example 6. *Prove that*

$$\int_V (\mathbf{g} \cdot curl\ curl\ \mathbf{f} - \mathbf{f} \cdot curl\ curl\ \mathbf{g}) \, dV$$
$$= \int_S \{(\mathbf{f} \times \text{curl } \mathbf{g}) - (\mathbf{g} \times \text{curl } \mathbf{f})\} \cdot d\mathbf{a}$$

Solution. $$\int_S (\mathbf{f} \times \text{curl } \mathbf{g}) \cdot d\mathbf{a} = \int_S (\mathbf{f} \times \text{curl } \mathbf{g}) \cdot \mathbf{n} dS$$

$$= \int_V \text{div} \, [\mathbf{f} \times \text{curl } \mathbf{g}] \cdot dV$$

$$= \int_V [\text{curl } \mathbf{g} \cdot \text{curl } \mathbf{f} - \mathbf{f} \cdot \text{curl curl } \mathbf{g}] \, dV \quad \text{...(1)}$$

$$\int_S (\mathbf{g} \times \text{curl } \mathbf{f}) \cdot d\mathbf{a} = \int_S (\mathbf{g} \times \text{curl } \mathbf{f}) \cdot \mathbf{n} \cdot dS$$

$$= \int_V \text{div} \, (\mathbf{g} \times \text{curl } \mathbf{f}) \, dV$$

$$= \int [\text{curl } \mathbf{f} \cdot \text{curl } \mathbf{g} - \mathbf{g} \cdot \text{curl curl } \mathbf{f}] \, dV \quad \text{...(2)}$$

Subtracting (1) from (2), we get the required result

EXERCISES

1. Evaluate

$$\int_S (\mathbf{i}x + \mathbf{j}y + \mathbf{k}z) \cdot d\mathbf{a}$$

where S denotes the surface of the cube bounded by the planes $x = 0$, $x = a$; $y = 0$, $y = a$; $z = 0$, $z = a$ by the application of Gauss's theorem and verify by direct manipulation.

2. Evaluate

$$\int_S \left(y^2z^2\mathbf{i}+z^2x^2\mathbf{j}+z^2y^2\mathbf{k}\right)\cdot d\mathbf{a},$$

where S is the part of the sphere $x^2 + y^2 + z^2 = 1$ above the xy-plane.

3. Show that

$$\int_S \left(x^2\mathbf{i}+y^2\mathbf{j}+z^2\mathbf{k}\right)\cdot \mathbf{n}\,dS=0,$$

where S denotes the surface of the ellipsoid

$$\frac{x^2}{a^2}+\frac{y^2}{b^2}+\frac{z^2}{c^2}=1.$$

4. Show that

$$\int_S (ax\mathbf{i}+by\mathbf{j}+cz\mathbf{k})\cdot d\mathbf{a}=\frac{4}{3}\pi\,(a+b+c),$$

where S is the surface of the sphere $x^2 + y^2 + z^2 = 1$.

(*Ajmer 98, Rohilkhand 99*)

5. Evaluate

$$\int_S \left(x\mathbf{i}+y\mathbf{j}+z^2\mathbf{k}\right)\cdot d\mathbf{a},$$

where S denotes the closed surface bounded by the cone $x^2 + y^2 = z^2$ and the plane $z = 1$.

6. Show that

$$\int_S (x+y+z)\,(lx+my+nz)\,dS=2a^2h^2\pi,$$

where S is the boundary of the region

$$x^2+y^2\le a^2,\ 0\le z\le h$$

and l, m, n are the direction cosines of the outward drawn normal.

7. Show that

$$\int_S d\mathbf{a}=0,\ \int_S \mathbf{r}\times d\mathbf{a}=0,$$

over any closed surface S.

8. If $\mathbf{F}$ is normal to the surface S at each point, show that

$$\int_V \text{curl}\,\mathbf{F}\,dv=0.$$

9. Show that

$$\int_S \nabla\phi\times\mathbf{n}\,dS=0.$$

10. Prove that

$$\int_V \mathbf{F}\cdot\nabla\phi\,dv=\int_S \phi\,\mathbf{F}\cdot d\mathbf{a}-\int_V \phi\nabla\cdot\mathbf{F}\,dv. \qquad \textit{(Ajmer, 99)}$$

11. Prove that

$$\int_V \nabla\phi \cdot \text{curl } \mathbf{F} dv = \int_S (\mathbf{F} \times \nabla\phi) \cdot d\mathbf{a}.$$

12. Show that

$$\int_S \nabla\phi \times \nabla\psi \cdot d\mathbf{a} = 0.$$

13. If $\mathbf{W} = \frac{1}{2} \nabla \times \mathbf{V}$ and $\mathbf{V} = \nabla \times \mathbf{U}$, show that

$$\frac{1}{2}\int \mathbf{V}^2 dv = \frac{1}{2}\int \mathbf{U} \times \mathbf{V} \cdot d\mathbf{a} + \int \mathbf{U} \cdot \mathbf{W}\, dv.$$

14. Prove that

$$\int \mathbf{f} \cdot \text{curl } \mathbf{F}\, d\mathbf{v} = \int \mathbf{F} \times \mathbf{f} \cdot d\mathbf{a} + \int \mathbf{F} \cdot \text{curl } \mathbf{f}\, dv.$$

(Gorakhpur 2000)

15. **f** and **g** are two continuously differentiable vector point functions in a region V such that

$$\text{curl } \mathbf{f} = 0, \quad \text{div } \mathbf{g} = 0$$

in V and

$$\mathbf{g} \,.\, \mathbf{n} = 0$$

on the boundary S of V where **n** is a unit vector along the normal to S.

16. Show that

$$\int_S \mathbf{f} \,.\, \mathbf{g}\, dv = 0.$$

[Show that there exists a scalar point function ϕ such that $\mathbf{F} = \nabla\phi$.]

REDUCTION OF SURFACE TO LINE INTEGRALS

7.8. STOKE'S THEOREM

If **F** *is any continuously differentiable vector point function and S is a surface bounded by a curve C, then*

$$\int_C \mathbf{F} \cdot d\mathbf{r} = \int_S \text{curl } \mathbf{F} \cdot \mathbf{n}\, dS,$$

where the unit normal vector **n** *at any point of S is drawn in the sense in which a right handed screw would move when rotated in the sense of description of C.* *(Ajmer 99, Gorakhpur 98, Lucknow 97)*

Stoke's Theorem for Plane. Firstly we shall prove the theorem for the case of a plane surface. Stoke's theorem in a plane is also often referred to as Green's theorem. Take a system of cartesian rectangular co-ordinate axes such that the plane of the given surface S is the OXY plane and the z-axis lies along the direction of the normal vector **n**. In the present case of plane the normal vector is constant.

Let

$$\mathbf{F} = \mathbf{i}F_1 + \mathbf{j}F_1 + \mathbf{k}F_3.$$

We have

$$\int_C \mathbf{F}\cdot d\mathbf{r} = \int_C \mathbf{F}\cdot\mathbf{t}\, ds$$

$$= \int_C (\mathbf{i}F_1 + \mathbf{j}F_2 + \mathbf{k}F_3)\cdot\left(\mathbf{i}\frac{dx}{ds} + \mathbf{j}\frac{dy}{ds} + \mathbf{k}\frac{dz}{ds}\right) ds$$

$$= \int_C \left(F_1\frac{dx}{ds} + F_2\frac{dy}{ds}\right) ds = \int_C (F_1 dx + F_2 dy).$$

for $dz/dx = 0$; the tangent at any point lying in the OXY plane.

$$\text{Also} \qquad \int_C \text{curl } \mathbf{F}\,.\,\mathbf{n}\, dS = \int_C \text{curl } \mathbf{F}\,.\,\mathbf{k}\, dS$$

$$= \iint_S \left(\frac{\partial F_2}{\partial x} - \frac{\partial F_1}{\partial y}\right) dx\, dy.$$

Thus, for the case of a plane surface the theorem is equivalent to showing

$$\int_C (F_1 dx + F_2 dy) = \iint_C \left(\frac{\partial F_2}{\partial x} - \frac{\partial F_1}{\partial y}\right) dx\, dy.$$

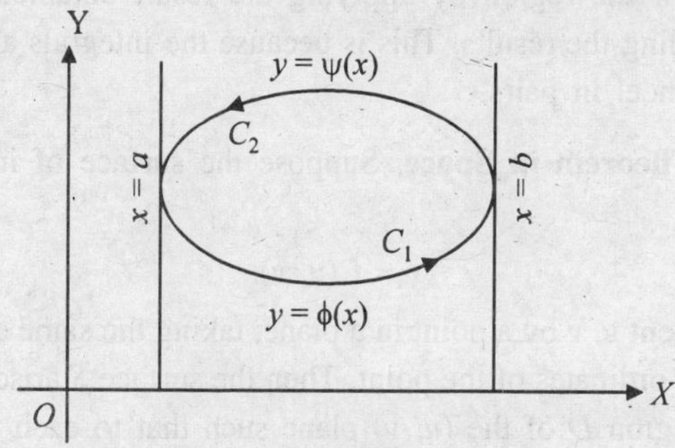

Fig. 7.12.

We shall now prove it. Firstly suppose that the region S is such that any line parallel to either axis meets the boundary C in at the most two points. Let the region be included between the lines $x = a$, $x = b$ and any line parallel to y-axis with abscissa x meet C in the points given by

$$y = \psi(x), \quad y = \phi(x)$$

where $\psi(x) \geq \phi(x)$. Thus, the boundary curve C is split up into two arcs C_1 and C_2 as shown in the figure.

We have

$$\iint_S \frac{\partial F_1}{\partial y} dx\, dy = \int_a^b \left[\int_{\phi(x)}^{\psi(x)} \frac{\partial F_1}{\partial y} dy\right] dx$$

$$= \int_a^b \left[F_1[x, \psi(x)] - F_1[x, \phi(x)]\right] dx$$

$$= -\int_{C_2} F_1(x, y)\, dx - \int_{C_1} F_1(x, y)\, dx$$

$$= -\int_C F_1(x, y)\, dx. \qquad \text{...(1)}$$

Similarly, we may show that

$$\iint_S \frac{\partial F_2}{\partial x} dx\, dy = \int_C F_2(x, y)\, dy. \qquad \text{...(2)}$$

From (1) and (2)ve

$$\int_C (F_2 dy + F_1 dx) = \iint_S \left(\frac{\partial F_2}{\partial x} - \frac{\partial F_1}{\partial y}\right) dx\, dy.$$

Now if a plane region be such that it can be divided into a finite number of sub-regions such that the boundary of each is met in at the most two points by any line parallel to either axis, then we can see the truth of the theorem for the region by applying the result obtained to each sub-region and adding the results. This is because the integrals along boundary curves will cancel in pairs.

Stoke's Theorem in Space. Suppose the surface of integration S is given by

$$\mathbf{r} = \mathbf{f}\,(u, v).$$

We represent u, v by a point in a plane, taking the same as the cartesian rectangular co-ordinates of the point. Then the surface S arises as the image of a certain region D of the (u, v) plane such that to each point (u, v) of D, there corresponds one and only one point of S. Let K be the boundary of D whose sense of description corresponds to that of the curve C bounding the surface S.

We have

$$\int_C \text{curl } \mathbf{F} \,.\, \mathbf{n}\, dS = \iint_S \left(\frac{\partial F_3}{\partial y} - \frac{\partial F_2}{\partial z}\right) dy\, dz + \left(\frac{\partial F_1}{\partial z} - \frac{\partial F_3}{\partial x}\right) dz\, dx + \left(\frac{\partial F_2}{\partial x} - \frac{\partial F_1}{\partial y}\right) dx\, dy$$

$$= \iint_D \left[\left(\frac{\partial F_3}{\partial y} - \frac{\partial F_2}{\partial z}\right)\frac{\partial(y, z)}{\partial(u, v)} + \left(\frac{\partial F_1}{\partial z} - \frac{\partial F_3}{\partial x}\right)\frac{\partial(z, x)}{\partial(u, v)} + \left(\frac{\partial F_2}{\partial x} - \frac{\partial F_1}{\partial y}\right)\frac{\partial(x, y)}{\partial(u, v)}\right] du\, dv.$$

In the integrand corresponding to the double integral over the domain D, in the u, v plane, the terms involving F_1 are

$$\frac{\partial F_1}{\partial z}\frac{\partial(z, x)}{\partial(u, v)} - \frac{\partial F_1}{\partial y}\frac{\partial(x, y)}{\partial(u, v)}$$

$$= -\frac{\partial F_1}{\partial z}\frac{\partial(x, z)}{\partial(u, v)} - \frac{\partial F_1}{\partial y}\frac{\partial(x, y)}{\partial(u, v)} - \frac{\partial F_1}{\partial x}\frac{\partial(x, x)}{\partial(u, v)},$$

the last term introduced being zero. Again these terms are

$$= -\frac{\partial F_1}{\partial z}\left[\frac{\partial x}{\partial u}\frac{\partial z}{\partial v} - \frac{\partial x}{\partial v}\frac{\partial z}{\partial u}\right] - \frac{\partial F_1}{\partial y}\left[\frac{\partial x}{\partial u}\frac{\partial y}{\partial v} - \frac{\partial x}{\partial v}\frac{\partial y}{\partial u}\right] - \frac{\partial F_1}{\partial x}\left[\frac{\partial x}{\partial u}\frac{\partial x}{\partial v} - \frac{\partial x}{\partial v}\frac{\partial x}{\partial u}\right]$$

$$= -\frac{\partial x}{\partial u}\left[\frac{\partial F_1}{\partial z}\frac{\partial z}{\partial v} + \frac{\partial F_1}{\partial y}\frac{\partial y}{\partial v} + \frac{\partial F_1}{\partial x}\frac{\partial x}{\partial v}\right] + \frac{\partial x}{\partial v}\left[\frac{\partial F_1}{\partial z}\frac{\partial z}{\partial u} + \frac{\partial F_1}{\partial y}\frac{\partial y}{\partial u} + \frac{\partial F_1}{\partial x}\frac{\partial x}{\partial u}\right]$$

$$= -\frac{\partial x}{\partial u}\frac{\partial F_1}{\partial v} + \frac{\partial x}{\partial v}\frac{\partial F_1}{\partial u} = \frac{\partial F_1}{\partial u}\frac{\partial x}{\partial v} - \frac{\partial F_1}{\partial v}\frac{\partial x}{\partial u}.$$

Similarly the terms involving F_2, F_3 are

$$\frac{\partial F_2}{\partial u}\frac{\partial y}{\partial v} - \frac{\partial F_2}{\partial v}\frac{\partial y}{\partial u}, \quad \frac{\partial F_3}{\partial u}\frac{\partial z}{\partial v} - \frac{\partial F_3}{\partial v}\frac{\partial z}{\partial u}$$

Thus, the double integral over the domain in the (u, v) plane is

$$= \iint_D \left[\left(\frac{\partial F_1}{\partial u}\frac{\partial x}{\partial v} - \frac{\partial F_1}{\partial v}\frac{\partial x}{\partial u}\right) + \left(\frac{\partial F_2}{\partial u}\frac{\partial y}{\partial v} - \frac{\partial F_2}{\partial v}\frac{\partial y}{\partial u}\right) + \left(\frac{\partial F_3}{\partial u}\frac{\partial z}{\partial v} - \frac{\partial F_3}{\partial v}\frac{\partial z}{\partial u}\right)\right] du\, dv.$$

By Stoke's theorem in plane,

$$\iint_D \left(\frac{\partial F_1}{\partial u}\frac{\partial x}{\partial v} - \frac{\partial F_1}{\partial v}\frac{\partial x}{\partial u}\right) du\, dv$$

$$= \int_C \left(F_1 \frac{\partial x}{\partial u} du + F_1 \frac{\partial x}{\partial v} dv\right)$$

$$= \int_C F_1 dx,$$

and similarly for the other two parts of the double integral over D. Thus,

$$\int_S \text{curl } \mathbf{F} \,.\, \mathbf{n} \, dS = \int_C (F_1 dx + F_2 dy + F_3 dz)$$

$$= \int_C \mathbf{F} \,.\, d\mathbf{r}.$$

Note. The theorem also holds good for a surface S (shaded) such as that shown in the figure. It is enclosed by two curves C_1 and C_2. We have

$$\int_S \text{curl } \mathbf{F} \,.\, \mathbf{n} \, dS = \int_{C_1} \mathbf{F} \,.\, d\mathbf{r} + \int_{C_2} \mathbf{F} \,.\, d\mathbf{r},$$

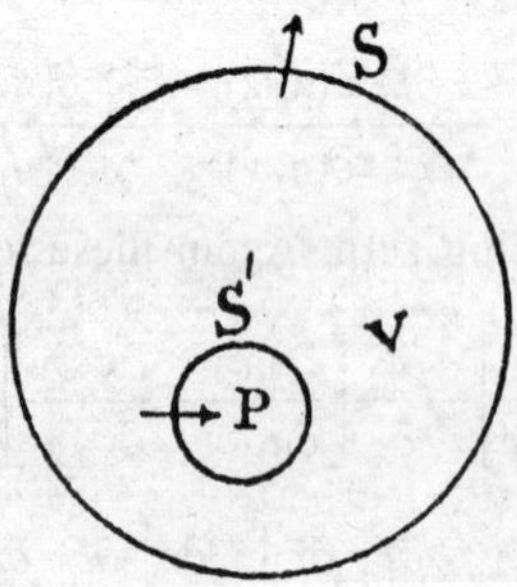

Fig. 7.13.

where the line integrals on the right are taken along the directions of C_1 and C_2 as shown in the figure. The result will follow, if we join the curves by any lines PQ and LM as shown and apply the theorem to the two surfaces thus obtained. The line integrals along PQ and LM will be taken in opposite directions and will consequently vanish on addition.

Cor. (*i*) We write

$$F = \mathbf{a}\phi,$$

where **a** is a constant vector and ϕ, a continuously differentiable scalar point function. Applying Stoke's theorem, we have

$$\int_C \mathbf{a}\phi \cdot d\mathbf{r} = \int_S \text{curl } (\mathbf{a}\phi) \cdot \mathbf{n} \, dS. \qquad ...(1)$$

Also

$$\text{curl } (\mathbf{a}\phi) = \nabla \times (\mathbf{a}\phi)$$

$$= \nabla\phi \times \mathbf{a}. \qquad [\S\ 10.15,\ \text{p. } 295] \qquad ...(2)$$

From (1) and (2),

$$\mathbf{a} \cdot \int_C \phi \, d\mathbf{r} = \int_S \nabla\phi \times \mathbf{a} \cdot \mathbf{n} \, dS$$

$$= -\,\mathbf{a} \cdot \int_S \nabla\phi \times \mathbf{n} \, dS$$

or

$$\mathbf{a} \cdot \left[\int_C \phi \, d\mathbf{r} + \int_S \nabla\phi \times \mathbf{n} \, dS\right] = 0.$$

$$\therefore \qquad \int_C \phi \, d\mathbf{r} + \int_S \nabla\phi \times \mathbf{n} \, dS = 0;$$

a being any arbitrary vector.

$$\therefore \qquad \int_C \phi \, d\mathbf{r} = \int_S \mathbf{n} \times \nabla\phi \, dS.$$

(*ii*) Again, we write

$$\mathbf{f} = \mathbf{a} \times \mathbf{F},$$

where **F** is a constant vector and apply Stoke's theorem to the function **f**.

$$\therefore \qquad \int_C \mathbf{a} \times \mathbf{F} \cdot d\mathbf{r} = \int_S \text{curl}\,(\mathbf{a} \times \mathbf{F}) \cdot \mathbf{n} \, dS.$$

Also

$$\text{curl}\,(\mathbf{a} \times \mathbf{F}) = \nabla \times (\mathbf{a} \times \mathbf{F}) = \mathbf{a}\,(\nabla \cdot \mathbf{F}) - (\mathbf{a} \cdot \nabla)\,\mathbf{F}$$

and

$$[(\mathbf{a} \cdot \nabla)\,\mathbf{F}] \cdot \mathbf{n} = \mathbf{a} \cdot \nabla\,(\mathbf{F} \cdot \mathbf{n}),$$

where it is understood that ∇ does not operate on **n** and for the purpose of ∇, **n** is thought of as constant. Thus,

$$\int_C \mathbf{a} \cdot \mathbf{F} \times d\mathbf{r} = \int_S \left[\mathbf{a} \cdot (\nabla \cdot \mathbf{F})\mathbf{n} - \mathbf{a} \cdot \nabla(\mathbf{F} \cdot \mathbf{n})\right] dS$$

or

$$\mathbf{a} \cdot \int_C \mathbf{F} \times d\mathbf{r} = \mathbf{a} \cdot \int_S \left[(\nabla \cdot \mathbf{F})\,\mathbf{n} - \nabla\,(\mathbf{F} \cdot \mathbf{n})\right] dS$$

so that, as before, **a** being a constant vector,

$$\int_C \mathbf{F} \times d\mathbf{r} = \int_S \left[(\nabla \cdot \mathbf{F})\,\mathbf{n} - \nabla\,(\mathbf{F} \cdot \mathbf{n})\right] dS.$$

Looking upon ∇ as a vector, we may write

$$(\nabla \cdot \mathbf{F})\,\mathbf{n} - \nabla\,(\mathbf{F} \cdot \mathbf{n}) = -\,(\mathbf{n} \times \nabla) \times \mathbf{F}.$$

Hence

$$\int_C d\mathbf{r} \times \mathbf{F} = \int_S (\mathbf{n} \times \nabla) \times \mathbf{F} \, dS.$$

The three results connecting line and surface integrals which we have arrived at may now be restated as follows :

$$\int_C d\mathbf{r} \cdot \mathbf{F} = \int_S (\mathbf{n} \times \nabla) \cdot \mathbf{F} \, dS,$$

$$\int_C d\mathbf{r} \times \mathbf{F} = \int_S (\mathbf{n} \times \nabla) \times \mathbf{F}\, dS,$$

$$\int_C d\mathbf{r}\phi = \int_S (\mathbf{n} \times \nabla)\, \phi dS,$$

so that we have replaced $d\mathbf{r}$ by $\mathbf{n} \times \nabla$. It should also be understood that ∇ on the right does not operate on $\mathbf{n}$.

Note. The truth of Stoke's theorem in three dimensions can also be shown to be a simple *intuitively* perceived consequence of that in plane, as we shall now see.

Let C be any curve and S any surface through the same. Divide S into any finite number of sub-surfaces. The boundary curve of each sub-surface is supposed to be described in the sense of rotation of a right handed screw when it is moved along the direction of the normal.

Let δS_p denote the magnitude of the area of any sub-surface and $\mathbf{n}_p$ the unit normal vector drawn at any point of the sub-surface. Then, by def.

$$\int_S \text{curl } \mathbf{F} \cdot \mathbf{n}\, dS = \lim \Sigma \text{ curl } \mathbf{F} \cdot \mathbf{n}_p\ \delta S_p, \qquad \text{...(1)}$$

where curl $\mathbf{F}$ denotes the value at any point of the sub-surface.

Also by the theorem for the plane surface,

$$\int_{\delta S_p} \text{curl } \mathbf{F} \cdot \mathbf{n}\, dS = \int_{C_p} \mathbf{F} \cdot d\mathbf{r}_p \qquad \text{...(2)}$$

for we can regard each sub-surface as approximately a plane region. As δS_p is small, we can re-write (2) as

$$\text{curl } \mathbf{F} \cdot \mathbf{n}_p\ \delta S_p = \int_{C_p} \mathbf{F} \cdot d\mathbf{r}. \qquad \text{...(3)}$$

On addition, we now get, from (1) and (3)

$$\int_S \text{curl } \mathbf{F} \cdot \mathbf{n}\, dS = \int_C \mathbf{F} \cdot d\mathbf{r},$$

for the line integrals along bounding arcs of sub-surfaces cancel in pairs.

EXAMPLES

Example 1. *Verify Stoke's theorem for the function*

$$\mathbf{F} = x\,(\mathbf{i}x + \mathbf{j}y),$$

integrated round the square in the plane $z = 0$ whose sides are along the lines

$$x = 0,\ \ y = 0,\ \ x = a,\ \ y = a.$$

Solution. We have

curl $x\,(\mathbf{i}x + \mathbf{j}y) = \mathbf{k}y$.

$$\therefore \qquad \int_S \text{curl } x\,(\mathbf{i}x + \mathbf{j}y) \cdot d\mathbf{a}$$

$$= \int_0^a \int_0^a \mathbf{k}y.\mathbf{k}\, dx\, dy$$

$$= \int_0^a \int_0^a y\, dx\, dy = \frac{a^3}{2}$$

Again

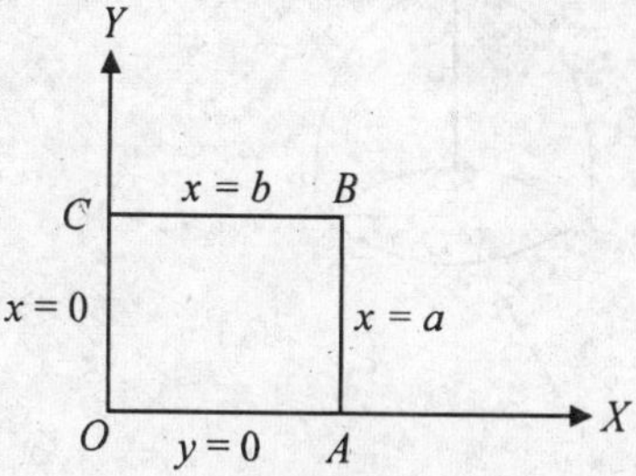

Fig. 7.14.

$$+ \int_{BC} \mathbf{F} \cdot d\mathbf{r} + \int_{CO} \mathbf{F} \cdot d\mathbf{r}.$$

$$+ \int_{BC} \mathbf{F} \cdot d\mathbf{r} + \int_{CO} \mathbf{F} \cdot d\mathbf{r}.$$

Now

$$\int_{OA} \mathbf{F} \cdot d\mathbf{r} = \int_0^a x\,(\mathbf{i}x + \mathbf{j}y) \cdot \mathbf{i}dx = \int_0^a x^2 dx = \frac{1}{3}a^3,$$

$$\int_{AB} \mathbf{F} \cdot d\mathbf{r} = \int_0^a x\,(\mathbf{i}x + \mathbf{j}y) \cdot \mathbf{j}dy = \int_0^a ay\, dy = \frac{1}{2}a^3,$$

$$\int_{BC} \mathbf{F} \cdot d\mathbf{r} = \int_a^0 x\,(\mathbf{i}x + \mathbf{j}y) \cdot \mathbf{i}dx = -\int_0^a x^2 dx = -\frac{1}{3}a^3,$$

$$\int_{CO} \mathbf{F} \cdot d\mathbf{r} = \int_a^0 x\,(\mathbf{i}x + \mathbf{j}y) \cdot \mathbf{j}dy = 0.$$

$$\therefore \quad \int_C \mathbf{F} \cdot d\mathbf{r} = \frac{1}{3}a^3 + \frac{1}{2}a^3 - \frac{1}{3}a^3 + 0 = \frac{1}{2}a^3.$$

Hence the verification.

Example 2. *Find the value of*

$$\int \text{curl } \mathbf{F} \cdot d\mathbf{a},$$

taken over the portion of the surface

$$x^2 + y^2 - 2ax + az = 0,$$

for which $z \geq 0$, when

$$\mathbf{F} = \left(y^2 + z^2 - x^2\right)\mathbf{i} + \left(z^2 + x^2 - y^2\right)\mathbf{j} + \left(x^2 + y^2 - z^2\right)\mathbf{k}.$$

Solution. Rewriting the equation

$$x^2 + y^2 - 2ax + az = 0$$

as

$$(x - a)^2 + y^2 = -a\,(z - a),$$

we see that the surface is a paraboloid with its vertex at (a, o, a) and axis parallel to z-axis and turned towards the negative direction of the same. It meets the plane $z = 0$ in the circle C, given by

$$x^2 + y^2 - 2ax = 0, \quad z = 0.$$

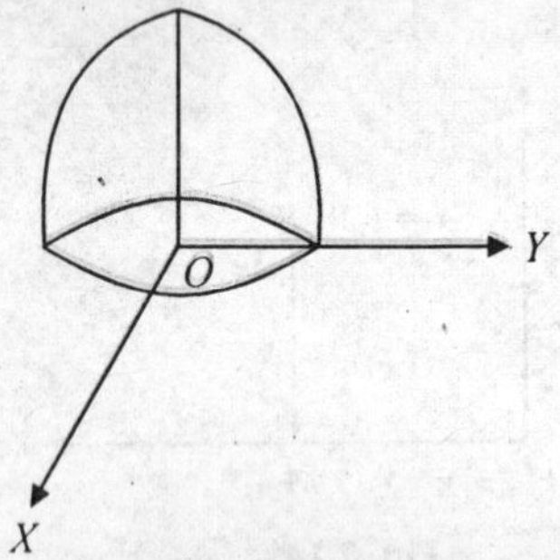

Fig. 7.15.

(In the Fig., O is the point (a, o, o), and ox, oy, oz, are the lines parallel to the co-ordinate axes).

By Stoke's theorem, the given surface integral is equal to the line integral

$$\int_C \mathbf{F} \cdot d\mathbf{r}.$$

The circle C is given by

$$x = a\ (1 + \cos\theta), \quad y = a \sin\theta, \quad z = 0.$$

Along C,

$$\mathbf{F} = \left[a^2 \sin^2\theta - a^2(1+\cos\theta)^2\right]\mathbf{i} + \left[a^2(1+\cos\theta)^2 - a^2 \sin^2\theta\right]\mathbf{j} + \left[a^2(1+\cos\theta)^2 + a^2\sin^2\theta\right]\mathbf{k}$$

$$\therefore \quad \mathbf{F} \cdot \left(\frac{dx}{d\theta}\mathbf{i} + \frac{dy}{d\theta}\mathbf{j} + \frac{dz}{d\theta}\mathbf{k}\right)$$

$$\therefore \quad = [a^2 \sin^2\theta - a^2 (1 + \cos\theta)^2]\ [- a \sin\theta - \text{a} \cos\theta].$$

$$\therefore \quad \int_C \mathbf{F} \cdot d\mathbf{r} = \int_0^{2\pi} \mathbf{F} \cdot \frac{d\mathbf{r}}{d\theta} d\theta = 2a^3\pi.$$

Another method. By a further application of Stoke's theorem, we see that the given integral

$$= \int_{S_1} \text{curl}\ \mathbf{F} \cdot \mathbf{n}\ dS,$$

where S_1 is the plane region bounded by the circle C.

Here

$$\mathbf{n} = \mathbf{k}$$

Thus, $$\text{curl}\ \mathbf{F}\ .\ \mathbf{n} = 2(x - y).$$

$$\therefore \quad \text{The integral } = 2\iint (x - y)\ dx\ dy,$$

taken over S_1. Changing to polar co-ordinates, so that

$$x = a + r\cos\theta, \quad y = r \sin\theta,$$

we see that the integral

$$= 2\int_0^a \int_0^{2\pi} (a + r\cos\theta - r\sin\theta)\ r d\theta\ dr$$

$$= 2a^3\pi.$$

Example 3. *Applying Stoke's theorem to prove that*

$$\int_C (ydx + zdy + xdz) = -2\sqrt{2\pi a^2},$$

where C is the curve given by

$$x^2 + y^2 + z^2 - 2ax - 2ay = 0, \quad x + y = 2a$$

and begins at the point (2a, 0, 0) and goes at first below the z-plane.

Solution. It may be easily seen that the curve C is the circle drawn on AB as diameter and lying in the plane $x + y = 2a$ which passes through A, B and is perpendicular to the z-plane.

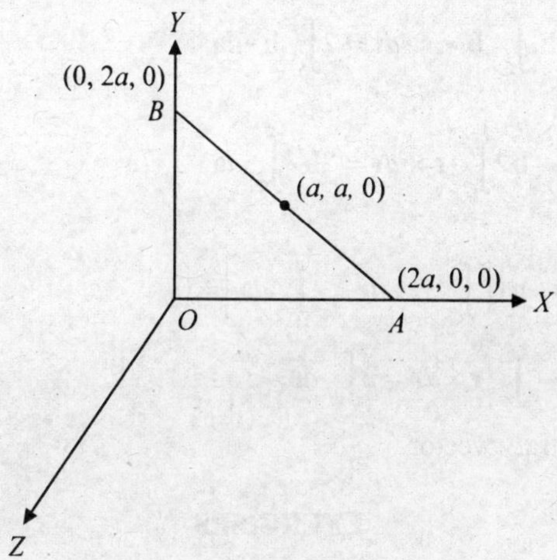

Fig. 7.16.

The given line integral

$$= \int_C (y\mathbf{i} + z\mathbf{j} + x\mathbf{k}) \cdot d\mathbf{r} = \int_S \text{curl}\,(y\mathbf{i} + z\mathbf{j} + x\mathbf{k}) \cdot \mathbf{n}\, dS$$

where S is the surface of the circle referred to above.

We have

$$\text{curl}\,(y\mathbf{i} + z\mathbf{j} + x\mathbf{k}) = -(\mathbf{i} + \mathbf{j} + \mathbf{k}),$$

$$\mathbf{n} = \frac{1}{\sqrt{2}}\mathbf{i} + \frac{1}{\sqrt{2}}\mathbf{j}.$$

$\therefore$ The given line integral

$$= \int_S -(\mathbf{i} + \mathbf{j} + \mathbf{k}) \cdot \left(\frac{1}{\sqrt{2}}\mathbf{i} + \frac{1}{\sqrt{2}}\mathbf{j}\right) dS$$

$$= -\frac{2}{\sqrt{2}}\int_S dS = -\sqrt{2\left(2a^2\pi\right)} = -2\sqrt{2\pi a^2}.$$

Example 4. *Prove that*

$$\int_C \mathbf{r} \times d\mathbf{r} = 2\int_S d\mathbf{a},$$

where S is a diaphragm enclosing a circuit C. *(Lucknow, 98)*

Solution. We apply Stoke's theorem to the function $\mathbf{b} \times \mathbf{r}$ where $\mathbf{b}$ is any constant vector.

Thus,

$$\int_C \mathbf{b} \times \mathbf{r} \cdot d\mathbf{r} = \int_S \text{curl}\,(\mathbf{b} \times \mathbf{r}) \cdot d\mathbf{a}.$$

Now $$\nabla \times (\mathbf{b} \times \mathbf{r}) = \mathbf{b}\nabla \cdot \mathbf{r} - \mathbf{b} \cdot \nabla\mathbf{r} = 3\mathbf{b} - \mathbf{b} = 2\mathbf{b},$$

$$\therefore \quad \int_C \mathbf{b} \times \mathbf{r} \cdot d\mathbf{r} = 2\int_S \mathbf{b} \cdot d\mathbf{a},$$

or

$$\mathbf{b} \cdot \int_C \mathbf{r} \times d\mathbf{r} = 2\mathbf{b} \cdot \int_S d\mathbf{a},$$

or

$$\mathbf{b} \cdot \left[\int_C \mathbf{r} \times d\mathbf{r} - 2\int_S d\mathbf{a}\right] = 0,$$

$$\therefore \quad \int_C \mathbf{r} \times d\mathbf{r} = 2\int_S d\mathbf{a},$$

for $\mathbf{b}$ is an arbitrary vector.

EXERCISES

1. Evaluate

$$\int \left(xy\,dx + xy^2 dy\right)$$

taken round the positively oriented square with vertices (± 1, 0), (0, ± 1) and verify by using Stoke's theorem.

2. Show that

$$\int_S \text{curl}\,\mathbf{f} \cdot \mathbf{n}\,ds = 0,$$

S being a closed surface.

[This question may be done both by Gauss's as well as Stoke's theorem.]

3. If the normal surface integral of a vector point function **G** over every open surface is equal to the tangential line integral of another **F** round its boundary, prove that

$$\mathbf{G} = \text{curl}\,\mathbf{F}.$$

4. Show that

(*i*) $\int_C \mathbf{r} \cdot d\mathbf{r} = 0.$

(ii) $\int_C \phi \nabla \phi \cdot d\mathbf{r} = 0.$

(iii) $\int_C \phi \nabla \psi \cdot d\mathbf{r} = -\int_C \psi \nabla \phi \cdot d\mathbf{r}.$

5. Show that

$$\phi(p) = \int_A^P \mathbf{F} \cdot d\mathbf{r}$$

6. f= $\nabla\phi$, $\mathbf{g} = \nabla\psi$ are two functions such that $\nabla^2\phi = 0$, $\nabla^2\psi = 0$, show that

$$\int_S \phi \text{ curl } \mathbf{f} \cdot d\mathbf{a} = \int_C \phi \mathbf{f}\, d\mathbf{r} - \int_S \text{grad } \phi \times \mathbf{f} \cdot d\mathbf{a}.$$

7.9. CONDITION FOR IRROTATIONAL VECTOR FUNCTIONS

We have already obtained one form of condition for a vector point function to be irrotational and we shall now obtain another with the help of Stoke's theorem.

Theorem. *The necessary and sufficient condition for a vector point function* **F** *to be irrotational in a simply connected region is that curl* **F** = *0 at every point of the region.*

The condition is necessary. Let **F** be irrotational so that, as shown in § 7.2.1, Page 325, there exists a scalar ϕ such that

$$\mathbf{F} = \nabla\phi.$$

$$\therefore \qquad \text{curl } \mathbf{F} = \text{curl } \nabla\phi = \nabla \times \nabla\phi = 0.$$

The condition is sufficient. Let curl **F** = 0. We have, by Stoke's theorem

$$\int_C \mathbf{F} \cdot d\mathbf{r} = \int_S \text{curl } \mathbf{F} \cdot d\mathbf{a} \equiv 0.$$

Hence, **F** is irrotational.

Note. The gradient of a scalar point function is a vector point function. It has now been seen that a given vector function **F** is the gradient of a scalar function if and only if

$$\text{curl } \mathbf{F} = 0$$

7.9.1. *The determination of a scalar function* ϕ *such that*

$$\mathbf{F} = \nabla\phi\,; \qquad \text{...(1)}$$

F *being a given irrotational vector function.*

Firstly we notice that ϕ is not unique. In fact, if ϕ be any solution of (1), then $\phi + \mathbf{c}$ is the *general* solution; c being any arbitrary constant.

[Refer § 10.14, Page 295].

We have seen that [Refer § 7.2.1, Page 325]

$$\phi(p) = \int_A^P \mathbf{F} \cdot d\mathbf{r} \qquad ...(2)$$

A being any fixed point.

In terms of rectangular cartesian axes, let A be (x_0, y_0, z_0) and P be any point (x, y, z). The integral being independent of the curve joining A to P, we choose the same as consisting of 3 portions of straight lines AB, BC, CP respectively parallel to the three co-ordinate axes. Thus, we have from (2),

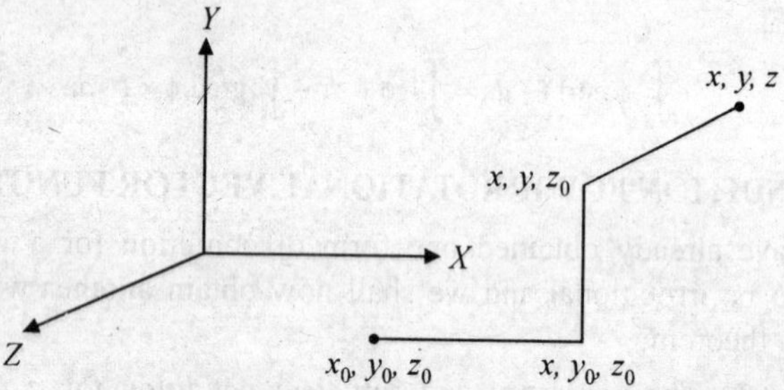

Fig. 7.17.

$$\phi(x, y, z) = \int_{(x_0, y_0, z_0)}^{(x, y, z)} (F_1 dx + F_2 dy + F_3 dz)$$

$$= \int_{(x_0, y_0, z_0)}^{(x, y_0, z_0)} F_1 dx + \int_{(x, y_0, z_0)}^{(x, y, z_0)} F_2 dy + \int_{(x, y, z_0)}^{(x, y, z)} F_3 dz,$$

for $\int F_2 dy$ and $\int F_3 dz$ are zero for lines parallel to x-axis and so on.

For different choices of the point (x_0, y_0, z_0), the values of $\phi(x, y, z)$ will differ from each other by constants.

Illustration. Let

$$\mathbf{F} = (y \sin z - \sin x)\,\mathbf{i} + (x \sin z + 2yz)\,\mathbf{j} + (xy \cos z + y^2)\,\mathbf{k}.$$

We may verify that curl $\mathbf{F} = 0$. We may take $(0, 0, 0)$ for (x_0, y_0, z_0).

$$\therefore \quad \phi = \int_{(0,0,0)}^{(x,0,0)} (y \sin z - \sin x)\, dx + \int_{(x,0,0)}^{(x,y,0)} (x \sin z + 2yz)\, dy$$

$$+ \int_{(x,y,0)}^{(x,y,z)} (xy \cos z + y^2)\, dz$$

$$= \Big| \, xy \sin z + \cos x \, \Big|_{(0,0,0)}^{(x,0,0)} + \Big| \, xy \sin z + y^2 z \, \Big|_{(x,0,0)}^{(x,y,0)}$$

$$+ \Big| \, xy \sin z + y^2 z \, \Big|_{(x,y,0)}^{(x,y,z)}$$

$= -1 + \cos x + xy \sin z + y^2 z.$

$\therefore \quad \mathbf{F} = \nabla (\cos x + xy \sin z + y^2 z)$, omitting the constants -1.

The result may be easily verified.

Ex. 1. Show that the following vector functions **F** are irrotational and find the corresponding scalars ϕ such that $\mathbf{F} = \nabla\phi$

(*i*) $yz\mathbf{i} + zx\mathbf{j} + xy\mathbf{k}$.

(*ii*) $(\sin y + z \cos x)\, \mathbf{i} + (x \cos y + \sin z)\, \mathbf{j} + (y \cos z + \sin x)\, \mathbf{k}$.

(*iii*) $(x^2 - yz)\, \mathbf{i} + (y^2 - zx)\, \mathbf{j} + (z^2 - xy)\, \mathbf{k}$.

7.10. CONDITION FOR SOLENOIDAL VECTOR POINT FUNCTIONS

Theorem. *The necessary and sufficient condition for a vector point function* **F** *to be solenoidal in a simply connected region is that*

$$div\ \mathbf{F} = 0,$$

at every point of the region.

The condition is necessary. We have see in Cor. 2, § 7.6, Page 339, that at any point

$$\text{div}\, \mathbf{F} = \lim_{v \to 0} \frac{\int_S \mathbf{F} \,.\, d\mathbf{a}}{V} = 0,$$

if **F** is solenoidal [§ 7.4, Page 329].

The condition is sufficient. The condition is sufficient, for if div $\mathbf{F} = 0$, we have by Gauss's theorem

$$\int_S \mathbf{F} \cdot d\mathbf{a} = \int_V \text{div}\, \mathbf{F} dv = 0,$$

so that the flux

$$\int_S \mathbf{F} \cdot d\mathbf{a}$$

across every closed surface is zero.

Hence the theorem.

7.10.1. We know that

$$\text{div curl}\, \mathbf{f} = 0,$$

so that the divergence of every curl is zero, *i.e.*, the curl of every function is solenoidal. It will now be shown that every solenoidal vector point function is the curl of some vector function.

Theorem. *If* **F** *is a continuously differentiable vector point function such that div* $\mathbf{F} = 0$, *then there exists a vector point function* **f** *such that*

$$\mathbf{F} = \text{curl}\, \mathbf{f}.$$

Firstly, we shall see *that if* **f** *is any function whose curl is* **F**, *then the general function whose curl is* **F** *is*

$$\mathbf{f} + \nabla\phi,$$

where ϕ is any continuously differentiable scalar point function.

Firstly, if

$$\text{curl } \mathbf{f} = \mathbf{F} = \text{curl } \mathbf{g},$$

so that

$$\text{curl } (\mathbf{g} - \mathbf{f}) = 0,$$

we see that $\mathbf{g} - \mathbf{f}$ is the gradient of some scalar ϕ, [§ 7.9, Page 362].

i.e., $$\mathbf{g} - \mathbf{f} = \nabla\phi \quad \text{or} \quad \mathbf{g} = \mathbf{f} + \nabla\phi.$$

Also if ϕ be *any* scalar point function, then

$$\text{curl } (\mathbf{f} + \nabla\phi) = \text{curl } \mathbf{f} + \text{curl } \nabla\phi$$

$$= \mathbf{F} + \nabla \times \nabla\phi = \mathbf{F}.$$

Hence the result.

Now we attend to the theorem as stated.

Let

$$\mathbf{F} = F_1\mathbf{i} + F_2\mathbf{j} + F_3\mathbf{k}, \quad \mathbf{f} = f_1\mathbf{i} + f_2\mathbf{j} + f_3\mathbf{k}.$$

$$\therefore \qquad \text{curl } \mathbf{f} = \left(\frac{\partial f_3}{\partial y} - \frac{\partial f_2}{\partial z}\right)\mathbf{i} + \left(\frac{\partial f_1}{\partial z} - \frac{\partial f_3}{\partial x}\right)\mathbf{j} + \left(\frac{\partial f_2}{\partial x} - \frac{\partial f_1}{\partial y}\right)\mathbf{k}.$$

Then

$$\mathbf{F} = \text{curl } \mathbf{f}$$

gives

$$F_1 = \frac{\partial f_3}{\partial y} - \frac{\partial f_2}{\partial z}, \quad F_2 = \frac{\partial f_1}{\partial z} - \frac{\partial f_3}{\partial x}, \quad F_3 = \frac{\partial f_2}{\partial x} - \frac{\partial f_1}{\partial y}. \qquad \text{...(1)}$$

Also we have

$$\frac{\partial F_1}{\partial x} + \frac{\partial F_2}{\partial y} + \frac{\partial F_3}{\partial z} = 0. \qquad \text{...(2)}$$

We have to establish the existence of some **f** such that

$$\text{curl } \mathbf{f} = \mathbf{F}.$$

We suppose that $f_1 = 0$.

Then we have, from (1)

$$F_2 = -\frac{\partial f_3}{\partial x} \text{ which gives } f_3 = -\int_{x_0}^{x} F_3 dx + \phi(y, z), \qquad \text{...(3)}$$

$$F_3 = -\frac{\partial f_2}{\partial x} \text{ which gives } f_2 = -\int_{x_0}^{x} F_3 dx + \psi(y, z). \qquad ...(4)$$

In these integrations y and z have to be thought of as parameters. From (3) and (4),

$$\frac{\partial f_3}{\partial x} = -\int_{x_0}^{x} \frac{\partial F_2}{\partial y} dx + \frac{\partial \phi}{\partial y} \qquad ...(5)$$

$$\frac{\partial f_2}{\partial z} = \int_{x_0}^{x} \frac{\partial F_3}{\partial z} dx + \frac{\partial \psi}{\partial z} \qquad ...(6)$$

$$\therefore \quad F_1 = \frac{\partial f_3}{\partial y} - \frac{\partial f_2}{\partial z} = -\int_{x_0}^{x} \left(\frac{\partial F_2}{\partial y} + \frac{\partial F_3}{\partial z} \right) dx + \frac{\partial \phi}{\partial y} - \frac{\partial \psi}{\partial z}$$

$$= \int_{x_0}^{x} \frac{\partial F_1}{\partial x} dx + \frac{\partial \phi}{\partial y} - \frac{\partial \psi}{\partial z} \qquad \text{[By (2)]}$$

$$= F_1(x, y, z) - F_1(x_0, y, z) + \frac{\partial \phi}{\partial y} - \frac{\partial \psi}{\partial z},$$

or

$$F_1(x_0, y, z) = \frac{\partial \phi}{\partial y} - \frac{\partial \psi}{\partial z}.$$

We now *suppose* that $\psi = 0$.

$$\therefore \quad \frac{\partial \phi}{\partial y} = F_1(x_0, y, z) \text{ or } \phi = \int_{y_0}^{y} F_1(x_0, y, z)\, dy.$$

Thus, we have

$$f_1 = 0, \; f_2 = \int_{x_0}^{x} F_3 dx, \; f_3 = -\int_{x_0}^{x} F_2 dx + \phi(y, z),$$

where

$$\phi(y, z) = \int_{y_0}^{y} F_1(x_0, y, z)\, dy.$$

It may now be easily shown that f_1, f_2, f_3 as determined above, are the components of a vector **f** whose curl is the given function **F**.

Illustration. Let

$$\mathbf{F} = (y - z)\,\mathbf{i} + (z - x)\,\mathbf{j} + (x - y)\,\mathbf{k}.$$

We may verify that div **F** = 0. We take $x_0 = 0$, $y_0 = 0$.

Now

$$f_2 = \int_0^x (x - y)\, dx = \frac{1}{2} x^2 - xy,$$

$$f_3 = -\int_0^y (z - x)\, dx + \phi(y, z) = \frac{1}{2}x^2 - xz + \phi(y, z),$$

$$\phi(y, z) = \int_0^y (y - z)\, dy = \frac{1}{2}y^2 - yz.$$

$$\therefore \qquad \mathbf{f} = f_1\mathbf{i} + f_2\mathbf{j} + f_3\mathbf{k}$$

$$= \left(\frac{1}{2}x^2 - xy\right)\mathbf{j} + \left[\frac{1}{2}\left(x^2 + y^2\right) - (x + y)\, z\right]\mathbf{k}.$$

The *general* function **f** whose curl is **F** is

$$\left(\frac{1}{2}x^2 - xy\right)\mathbf{j} + \left[\frac{1}{2}\left(x^2 + y^2\right) - (x + y)\, z\right]\mathbf{k} + \nabla g,$$

where g is any scalar point function.

Note. The curl of a vector point function is a vector point function but every vector point function is not necessarily the curl of a vector function. It follows from above that a vector point function **F** is the curl of a function if and only if

$$\text{div } \mathbf{F} = 0,$$

i.e., every solenoidal vector function is the curl of a function.

Ex. Show that the following vector functions **F** are solenoidal and find the functions **f** such that **F** = curl **f**.

(*i*) $(y + z)\,\mathbf{i} + (z + x)\,\mathbf{j} + (x + y)\,\mathbf{k}$.

(*ii*) $x\,(y - z)\,\mathbf{i} + y\,(z - x)\,\mathbf{j} + z\,(x - y)\,\mathbf{k}$.

(*iii*) $yz\mathbf{i} + zx\mathbf{j} + xy\mathbf{k}$.

7.11. GREEN'S THEOREM

If ϕ and ψ are two continuously differentiable scalar point functions such that $\nabla\phi$ and $\nabla\psi$ are also continuously differentiable, then

$$\int_V \left(\phi\nabla^2\psi - \psi\nabla^2\phi\right) dv = \int_S (\phi\nabla\psi - \psi\nabla\phi)\cdot\mathbf{n}\, dS,$$

where V is the region enclosed by a surface S.

(*Gorakhpur 98, 2000; Lucknow 98*)

Applying Gauss's Theorem to the function $\phi\nabla\psi$, we obtain

$$\int_S (\phi\nabla\psi\cdot\mathbf{n}dS) = \int_V \text{div}\,(\phi\nabla\psi)\, dv.$$

Also

$$\text{div}\,(\phi\nabla\psi) = \nabla\cdot(\phi\nabla\psi)$$

$$= \nabla\phi\cdot\nabla\psi + \phi\nabla^2\psi,$$

$$\therefore \qquad \int_S \phi\nabla\psi\cdot\mathbf{n}\,dS = \int_V \nabla\phi\cdot\nabla\psi dv + \int_V \phi\nabla^2\psi dv, \qquad ...(1)$$

or
$$\int_V \nabla\phi\cdot\nabla\psi dv = \int_S \phi\nabla\psi.\mathbf{n}dS - \int_V \phi\nabla^2\psi dv. \qquad ...(2)$$

The result (2) is known as Green's first identity.

Interchanging ϕ and ψ in (1), we get

$$\int_S \psi\nabla\phi\cdot\mathbf{n}dS = \int_V \nabla\psi.\nabla\phi dv + \int_V \psi\nabla^2\phi dv. \qquad ...(3)$$

Subtracting (3) from (1), we get

$$\int_S (\phi\nabla\psi - \psi\nabla\phi)\cdot\mathbf{n}dS = \int_V \left(\phi\nabla^2\psi - \psi\nabla^2\phi\right) dv.$$

Hence the theorem.

The result of the theorem can also be stated as

$$\int_V \left(\phi\nabla^2\psi - \psi\nabla^2\phi\right) dv = \int_S \left(\phi\frac{\partial\psi}{\partial n} - \psi\frac{\partial\phi}{\partial n}\right) dS,$$

where $\partial\phi / \partial n$ and $\partial\psi / \partial n$ denote the directional derivatives of ϕ and ψ respectively along the outward drawn normal at any point of S.

7.11.1. Green's Formula.

Suppose that ϕ is a scalar point function which is twice continuously differentiable in a region V enclosed by a surface S.

Take any fixed point O, *within* the region and let $\mathbf{r}$ denote the position vector of any point P relative to O and let

$$OP = |\,\mathbf{r}\,| = r.$$

We write

$$\psi = \frac{1}{r}$$

so that, except at O, ψ is a twice continuously differentiable scalar function.

We surround O by a small sphere of radius $\mathbf{r}$. Let S_1 denote the surface of this sphere and V_1 the region bounded by S and S_1.

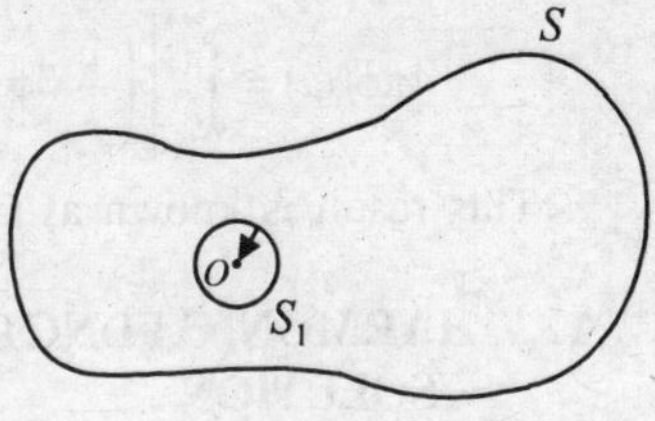

Fig. 7.18.

We have

$$\nabla\psi = -\frac{\mathbf{r}}{r^2},\ \nabla^2\psi = 0.$$

We apply Green's theorem to the region V_1 enclosed by the surfaces S, S_1. Thus

$$\int_V -\frac{1}{r}\nabla^2\phi dv = \int_S \left[\phi\frac{\partial}{\partial n}\left(\frac{1}{r}\right)-\frac{1}{r}\frac{\partial\phi}{\partial n}\right]dS + \int_{S_1}\left[\phi\frac{\partial}{\partial n}\left(\frac{1}{r}\right)-\frac{1}{r}\frac{\partial\phi}{\partial n}\right]dS, \quad ...(1)$$

where the unit normal at any point of the sphere S_1 is directed toward O.

At any point of S_1,

$$\frac{\partial}{\partial n}\left(\frac{1}{r}\right) = \nabla\left(\frac{1}{r}\right)\cdot \mathbf{n}$$

$$= -\frac{\mathbf{r}}{r^3}\cdot -\frac{\mathbf{r}}{r} = \frac{\mathbf{r}^2}{r^4} = \frac{1}{r^2} = \frac{1}{r^2}.$$

Now

$$\int_{S_1} \phi\frac{\partial}{\partial n}\left(\frac{1}{r}\right)dS = \frac{1}{\varepsilon^2}\int_{S_1}\phi dS,$$

$$\lim_{\varepsilon\to 0}\int_{S_1}\phi\frac{\partial}{\partial n}\left(\frac{1}{r}\right)dS = 4\pi\phi(O)$$

$$\lim_{\varepsilon\to 0}\int_{S_1}\frac{1}{r}\frac{\partial\phi}{\partial n}dS = 0$$

Thus, from (1), taking the limits when $\varepsilon \to 0$, we obtain

$$\int_V -\frac{1}{r}\nabla^2\phi dv = \int_S \left[\phi\frac{\partial}{\partial n}\left(\frac{1}{r}\right)-\frac{1}{r}\frac{\partial\phi}{\partial n}\right]dS + 4\pi\phi(O)$$

or $$4\pi\phi(O) = \int_S \left[\frac{1}{r}\frac{\partial\phi}{\partial n}-\phi\frac{\partial}{\partial n}\left(\frac{1}{r}\right)\right]dS - \int_V \frac{1}{r}\nabla^2\phi dv$$

which may be re-written as

$$4\pi\phi(O) = \int_S \left[\left(\frac{1}{r}\nabla\phi - \phi\nabla\frac{1}{r}\right)\right]\cdot d\mathbf{a} - \int_V \frac{1}{r}\nabla^2\phi dv.$$

This result is known as Green's formula.

7.12. HARMONIC FUNCTIONS, LAPLACE'S EQUATION AND ITS SOLUTION

Def. *A function ϕ is said to be harmonic in a region if it is twice continuously differentiable in the region and*

$$\nabla^2\phi = 0$$

at every point of the region.

Also the equation $\nabla^2\phi = 0$ is known as Laplace's equation.

Thus, if ϕ be a harmonic function in a region V enclosed by a surface S then by Green's formula, we see that for any point O within the region

$$4\pi\phi(O) = \int_S \left(\frac{1}{r}\nabla\phi - \phi\nabla\frac{1}{r}\right)\cdot d\mathbf{a}, \qquad ...(1)$$

so that it is proved *that the value of a harmonic function ϕ at any point within a region is known in terms of the values of ϕ and $\partial\phi/\partial n$ at any point of the surface enclosing the region.*

7.13. POISSON'S EQUATION AND SOLUTION

The equation

$$\nabla^2\phi = -4\pi\rho,$$

where ϕ is a scalar point function vanishing outside a finite region is known as *Poisson's equation.*

Consider a region V bounded by a surface S. Applying Green's formula to the region V; ϕ being a scalar point function satisfying (1), we have

$$4\pi\phi(O) = \int_S \left(\frac{1}{r}\nabla\phi - \phi\nabla\frac{1}{r}\right)\cdot d\mathbf{a} + 4\pi\int_V \frac{\rho}{r}\,dv.$$

We now let the region V tend to infinity in all directions so that the surface S also recedes to infinity.

We now assume that for sufficiently large values of r, ϕ is of the form k/r where k remains bounded. Then for sufficiently large values of r, $|\nabla\phi|$ is of the form k/r^2.

$$\therefore \qquad \int_S \left(\frac{1}{r}\nabla\phi - \phi\nabla\frac{1}{r}\right)\cdot d\mathbf{a} \to 0,$$

as S recedes to infinity in all directions.

Hence, we see that

$$\phi(O) = \int_V \frac{\rho}{r}\,dv,$$

where the volume integral is carried over the whole space. The volume integral carried over the whole space is the same as the volume integral carried over the region outside which ρ is zero.

Cor. Consider

$$\nabla^2\mathbf{F} = -4\pi\mathbf{f}. \qquad ...(1)$$

Let

$$\mathbf{f} = f_1\mathbf{i} + f_2\mathbf{j} + f_3\mathbf{k}.$$

We write

$$\mathbf{F} = F_1\mathbf{i} + F_2\,\mathbf{j} + F_3\mathbf{k}.$$

The equation (1) is equivalent to

$$\nabla^2 F_1 = -\,4\pi f_1,\ \ \nabla^2 F_2 = -\,4\pi f_2,\ \ \nabla^2 F_3 = -\,4\pi f_3.$$

Under suitable conditions imposed on f_1, f_2, f_3, we see that for any point O,

$$F_1(O) = \int \frac{f_1}{r}\,dv,\ F_2(O) = \int \frac{f_2}{r}\,dv,\ F_3(O) = \int \frac{f_3}{r}\,dv.$$

These give

$$\mathbf{F}(O) = \int \frac{\mathbf{f}}{r}\,dv.$$

EXAMPLES

Example 1. *Show that two functions which are harmonic in a region enclosed by a surface and take on the same values at any point of the surface are identical.*

Solution. Let ϕ and ψ be two functions harmonic in a region bounded by a surface S and let $\phi = \psi$ at every point of S.

We have

$$\int_V \theta\nabla^2\theta\,dv + \int_V \nabla\theta\cdot\nabla\theta\,dv = \int_S \theta\nabla\theta\cdot d\mathbf{a}.$$

Take $\theta = \phi - \psi$ so that $\nabla^2\theta = \nabla^2\phi - \nabla^2\psi = 0$ at every point of V and $\theta = 0$ at every point of S.

$$\therefore \qquad \int_V (\nabla\theta)^2\,dv = 0.$$

$$\therefore \qquad \nabla\theta = 0,$$

i.e., θ is a constant.

Also θ is zero at every point of S, we see that θ is everywhere zero.

Hence the result.

Example 2. *Evaluate by Green's theorem in the plane*

$$\oint_C \left[\left(x^2 - \cosh y\right)dx + \left(y + \sin x\right)dy\right],$$

where C is the rectangle with vertices (0, 0) (π, 0), (π, 1), (0, 1).

Solution. The path of integration C is the rectangle OA, BC, as shown in figure. Let R be the plane region bounded by C.

Using Green's theorem in plane, we have

$$\oint_C \left[\left(x^2 - \cosh y\right) dx + \left(y + \sin x\right) dy\right]$$

$$= \iint_R \left[\frac{\partial}{\partial x}(y + \sin x) - \frac{\partial}{\partial y}\left(x^2 - \cosh y\right)\right] dx\, dy$$

$$= \int_{x=0}^{\pi} \int_{y=0}^{1} (\cos x + \sinh y)\, dx\, dy$$

$$= \int_0^{\pi} \left[y \cos x + \cosh y\right]_0^1 dx$$

$$= \int_0^{\pi} (\cos x + \cosh 1 - 1)\, dx$$

$$= \pi \cosh 1 - \pi = \pi\left(\cosh 1 - 1\right).$$

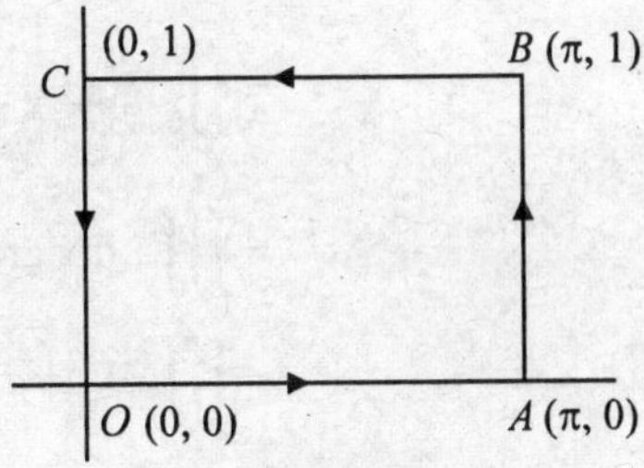

Fig. 7.19.

Example 3. *Show that the area bounded by a simple closed curve C is given by* $\frac{1}{2}\oint_C (x\, dy - y\, dx)$. *Hence, find the area of the ellipse* $x = a \cos \theta$, $y = b \sin \theta$. (*Garhwal, 92*)

Solution. If R is a closed region in the xy-plane bounded by a simple closed curve C, then by Green's theorem

$$\oint_C (\psi dx + \phi dy) = \iint_R \left(\frac{\partial \phi}{\partial x} - \frac{\partial \psi}{\partial y}\right) dx\, dy \qquad ...(1)$$

Putting $\psi = -y$ and $\phi = x$, we get

$$\oint_C (x\, dy - y\, dx) = \iint_R \left[\frac{\partial x}{\partial x} - \frac{\partial(-y)}{\partial y}\right] dx\, dy$$

$$\Rightarrow \qquad \frac{1}{2}\oint_C (x\, dy - y\, dx) = \iint_R dx\, dy = A.$$

where A is the area of the region R bounded by C.

Parametric equations of ellipse are

$$x = a \cos \theta, \quad y = b \sin \theta,$$

$$\Rightarrow \qquad dx = -a \sin \theta\, d\theta, \quad dy = b \cos \theta\, d\theta.$$

$\therefore$ Required area of ellipse

$$= \frac{1}{2}\int_0^{2\pi}\left(ab \cos^2 \theta + ab \sin^2 \theta\right) d\theta$$

$$= \frac{1}{2} ab \left[\theta\right]_0^{2\pi} = \pi\, ab.$$

Example 4. *If* ϕ *is harmonic in V, then show that* $\iint_S (\partial\phi / \partial n)\, dS = 0$, *where S is the surface enclosing V.*

Solution. $$\iint_S \frac{\partial\phi}{\partial n} dS = \iint_S \left(\frac{\partial\phi}{\partial n}\mathbf{n}\right) \cdot \mathbf{n}\, dS$$

$$= \iint_S \text{grad } \phi \cdot \mathbf{n}\, dS \qquad \left[\because \text{ grad } \phi = \frac{\partial\phi}{\partial n}\mathbf{n}\right]$$

$$= \iiint_V \text{div grad } \phi\, dV, \text{ by Gauss's divergence theorem.}$$

$$= \iiint_V \nabla^2\phi\, dV$$

$$= 0 \qquad \left[\because \phi \text{ is harmonic in } V \Rightarrow \nabla^2\phi = 0\right]$$

Example 5. *Show that*

$$\mathbf{F} = (\sin y + z)\,\mathbf{i} + (x \cos y - z)\,\mathbf{j} + (x - y)\,\mathbf{k}$$

is irrotational and find a function ϕ *such that* $\mathbf{F} = \nabla\phi$.

Solution. $$\text{curl } \mathbf{F} = \begin{vmatrix} \mathbf{i} & \mathbf{j} & \mathbf{k} \\ \partial/\partial x & \partial/\partial y & \partial/\partial z \\ \sin y + z & x\cos y - z & x - y \end{vmatrix}$$

$$= \mathbf{i}(-1+1) - \mathbf{j}(1-1) + \mathbf{k}(\cos y - \cos y) = 0.$$

$\therefore$ The given vector is irrotational and so $\mathbf{F} = \nabla\phi$. [§ 7.9]

Hence, $$(\sin y + z)\,\mathbf{i} + (x \cos y - z)\,\mathbf{j} + (x - y)\,\mathbf{k}$$

$$= \frac{\partial\phi}{\partial x}\mathbf{i} + \frac{\partial\phi}{\partial y}\mathbf{j} + \frac{\partial\phi}{\partial z}\mathbf{k}$$

$$\therefore \quad \frac{\partial\phi}{\partial x} = \sin y + z, \text{ hence, } \phi = x \sin y + xz + f_1(y, z) \qquad ...(1)$$

$$\frac{\partial\phi}{\partial y} = x \cos y - z, \text{ hence, } \phi = x \sin y - yz + f_2(x, z) \qquad ...(2)$$

$$\frac{\partial\phi}{\partial z} = x - y, \text{ hence, } \phi = xz - yz + f_3(x, y) \qquad ...(3)$$

(1), (2) and (3) each represents ϕ. These agree if we choose $f_1(y, z) = -yz$, $f_2(x, z) = xz$ and $f_3(x, y) = x \sin y$. Hence, the required ϕ is given by

$$\phi = x \sin y + xz - yz + c,$$

c being a constant.

EXERCISES

1. Verify Green's theorem in the plane for

$$\oint_C \left[\left(x^2 + y^2\right) dx - 2xy\, dy\right],$$

where C is the rectangle bounded by $y = 0$, $x = 0$, $y = b$, $x = a$.

2. Verify Green's theorem in the plane for

$$\oint_C \left[\left(xy + y^2\right) dx + x^2 dy\right],$$

where C is the closed curve of the region bounded by $y = x$ and $y = x^2$.

3. Evaluate by Green's theorem

$$\oint_C \left[(\cos x \sin y - xy)\, dx + \sin x \cos y\, dy\right],$$

where C is the circle $x^2 + y^2 = 1$.

4. If ϕ is harmonic in V, then prove that

$$\iint_S \phi \frac{\partial \phi}{\partial n} dS = \iiint_V |\nabla \phi|^2\, dV.$$

5. If ϕ and ψ are harmonic in V and $\partial\phi / \partial n = \partial\psi / \partial n$ on S, then show that $\phi = \psi + c$, where c is a constant.

6. If ϕ is harmonic in V and $\partial\phi / \partial n = 0$ on S, then prove that ϕ is a constant in V.

7. Show that the vector

$$\mathbf{F} = (x^2 - yz)\, \mathbf{i} + (y^2 - zx)\, \mathbf{j} + (z^2 - xy)\, \mathbf{k}$$

is irrotational. Find a scalar ϕ such that $\mathbf{F} = \nabla\phi$.

OBJECTIVE QUESTIONS

For each of the following questions, four alternatives are given for the answer. Only one of them is correct. Choose the correct alternative.

1. A force **F** acts upon a particle which is displaced along a given curve c in space from the point P whose position vector is **r**, the workdone by the force is

(*a*) $\mathbf{F} \cdot d\mathbf{r}$ (*b*) $\mathbf{F} \cdot \dfrac{d\mathbf{r}}{ds}$

(*c*) $\int_O^P \mathbf{F} \cdot d\mathbf{r}$ (*d*) $\int_C \mathbf{F} \cdot d\mathbf{r}$

2. If $\mathbf{F} = 3xy\mathbf{i} - y^2\mathbf{j}$, then value of $\int_C \mathbf{F} \cdot d\mathbf{r}$, where c is the curve $y = 2x^2$ in the xy-plane, from $(0, 0)$ to $(1, 2)$ is

(*a*) $-6/7$ (*b*) $7/6$ (*c*) $-7/6$ (*d*) $6/7$

3. If $\mathbf{t}$ is the unit tangent vector and c is the unit circle in xy-plane with centre at origin, then value of $\int_C \mathbf{t} \cdot d\mathbf{r}$

(a) 2π (b) π (c) $\pi/2$ (d) 0

4. Value of $\int (x\,dy - y\,dx)$ around the circle $x^2 + y^2 = 1$ is

(a) 0 (b) $\pi/2$ (c) π (d) 2π

5. If c be a closed curve, then $\oint_C \mathbf{r} \cdot d\mathbf{r} =$

(a) r (b) r^2 (c) $1/r$ (d) 0

6. Area bounded by a simple closed curve c is

(a) $\oint_C (x\,dy - y\,dx)$

(b) $\frac{1}{2}\oint_C (x\,dy - y\,dx)$

(c) $\oint_C (x\,dy + y\,dx)$

(d) $\frac{1}{2}\oint_C (x\,dy + y\,dx)$

7. Value of $\oint_C \mathbf{r} \times d\mathbf{r} =$

(a) $\iint_S \mathbf{n}\,ds$ (b) 0

(c) $2\iint_S \mathbf{n}\,ds$ (d) None of these

8. Value of $\iint_S \mathbf{n} \times \mathbf{F}\,ds$ is

(a) $\iiint_V \text{div}\,\mathbf{F}\,dV$ (b) $\iiint_V \text{curl}\,\mathbf{F}\,dV$

(c) $\iiint_V \mathbf{F}\,dV$ (d) zero

9. Value of $\iint_S \nabla r^2 \cdot d\mathbf{s}$ is

(a) V (b) $3V$ (c) $6V$ (d) $12V$

10. For any closed surface S, prove that $\iint_S \text{curl}\,\mathbf{F}\,.\,\mathbf{n}\,dS$ is

(a) 0 (b) V (c) $2V$ (d) $3V$

11. For any closed surface S, which of the following is true :

(*a*) $\iint_S \mathbf{r} \times \mathbf{n}\, dS = 0$ (*b*) $\iiint_V \frac{dV}{r^2} = 0$

(*c*) $\iint_S \mathbf{r} \cdot \mathbf{n}\, dS = 0$ (*d*) $\iint_S \nabla r^2 \cdot dS = 0$

12. The necessary and sufficient condition that $\int_C \mathbf{F} \cdot d\mathbf{r}$ should be independent of the path **c** joining A and B is

(*a*) $\mathbf{F} \times d\mathbf{F} = 0$ (*b*) div $\mathbf{F} = 0$

(*c*) curl $\mathbf{F} = 0$ (*d*) Nones of these

ANSWERS

1. (*d*) **2.** (*c*) **3.** (*a*) **4.** (*d*) **5.** (*d*)
6. (*b*) **7.** (*c*) **8.** (*b*) **9.** (*c*) **10.** (*a*)
11. (*a*) **12.** (*c*)

8

Hydrodynamics

8.1. STRESSES IN FLUIDS

We start considering the case of any arbitrary fluid.

Consider any point P, in a fluid and imagine a small surface drawn througl the same. The portions of the fluid on the two sides of the surface act anc react on each other with equal and opposite forces.

Let, **n**, be the unit vector normal to the surface at P drawn on either sid of the same. Then, $-$**n**, is the unit vector normal to the surface drawn on th other side. Thus we distinguish between the two sides of a surface and th two sides corresponding to the vectors **n** and $-$ **n** may be referred to as the r and $-$ **n** sides respectively.

Let, A, be the area of the small surface through P and let, **F** be the actio due to the fluid on $-$**n** side on that at the **n** side. The direction of, **F**, may no necessarily be perpendicular to the surface. We shall suppose that, **F**, acts a P and say that

F *acts at* P *across the surface perpendicular to the vector* **n.** Then th stress at P across the surface perpendicular to **n** is defined as

$$\lim_{A \to O} \frac{\mathbf{F}}{A} = \mathbf{P}.$$

The stress at the point across the surface perpendicular to $-$**n** is then, $-$**p**

The stress being force per unit area, the dimensions of the stress are

$$MLT^{-2}L^{-2} = ML^{-1}T^{-3}.$$

Thus we have a stress at a point, P, associated with every vector **n**.

We shall now show that *the stress at a point across any plane can b expressed as a linear combination of the stresses across any three mutuall perpendicular planes through the point.*

Through the point P, draw any three mutually perpendicular lines PA PB, PC and regard the same as the co-ordinate axes of a rectangular cartesia system. Let, as usual, **i**, **j**, **k**, denote unit vectors along the axes.

Let $\mathbf{P}_1, \mathbf{P}_2, \mathbf{P}_3$ denote the stresses at P across the co-ordinate planes perpendicular to the vectors **i**, **j**, **k**.

Consider now, any other plane through P and let **n** be the unit vector normal to the same. Also draw any plane perpendicular to **n** so drawn that **n** is the outward drawn normal to the plane in relation to the tetrahedron formed by the plane and the co-ordinate planes. Let his plane meet the axes at A, B, C.

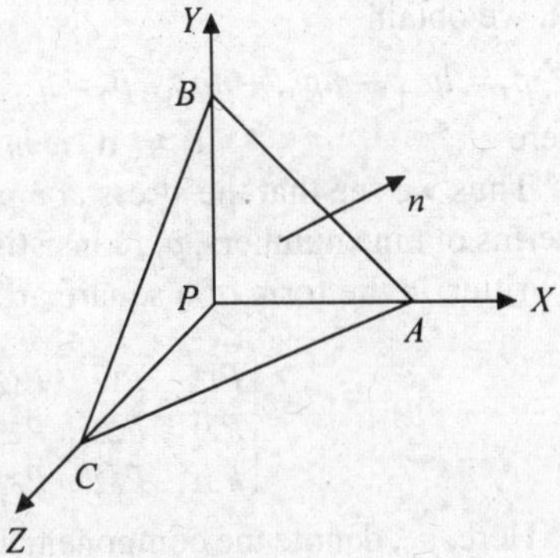

Fig. 8.1

Let Δ be the area of the triangle ABC. The triangles PBC, PCA, PAB being the projections of the ΔABC an the co-ordinate planes, we have

$$\Delta PBC = (\mathbf{n} : \mathbf{i})\,\Delta,$$
$$\Delta PCA = (\mathbf{n} \,.\, \mathbf{j})\,\Delta,$$
$$\Delta PAB = (\mathbf{n} \,.\, \mathbf{k})\,\Delta.$$

The volume of the tetrahedron $PABC$ is $\frac{1}{3}\,h\,\Delta$, where h, denotes the perpendicular distance of the plane ABC from the point P.

Let $\overline{\mathbf{P}}_1, \overline{\mathbf{P}}_2, \overline{\mathbf{P}}_3, \overline{\mathbf{P}}$ be the average stresses across the faces of the tetrahedron perpendicular to the vectors **i**, **j**, **k**, **n**.

Also let, **F**, be the force per unit mass acting on the fluid and $\overline{\rho}$, the average density of the fluid within the tetrahedron.

If, **a**, denote the acceleration of the centroid of the fluid within the tetrahedron, we have,

$$\frac{1}{3}\,h\,\Delta\,\overline{\rho}\,\mathbf{a} = \frac{1}{3}\,h\,\Delta\,\overline{\rho}\,\mathbf{F} + (\mathbf{n.i})\,\Delta\,\overline{\mathbf{P}}_1 + (\mathbf{n.j})\,\Delta\,\overline{\mathbf{P}}_2 + (\mathbf{n.k})\,\Delta\,\overline{\mathbf{P}}_3 - \Delta\overline{\mathbf{P}}$$

or

$$\frac{1}{3}\,h\,\overline{\rho}\,(\mathbf{a}-\mathbf{F}) = (\mathbf{n.i})\,\overline{\mathbf{P}}_1 + (\mathbf{n.j})\,\overline{\mathbf{P}}_2 + (\mathbf{n.k})\,\overline{\mathbf{P}}_3 - \overline{\mathbf{P}}. \qquad ...(1)$$

Let, now, $h \to 0$ so that the limiting position of the plane ABC is the plane through P normal the vector **n**.

If $\mathbf{P}_1, \mathbf{P}_2, \mathbf{P}_3, \mathbf{P}$, denote the stresses at P across planes perpendicular to **i**, **j**, **k**, **n**, we have, from (1), in the limit

$$\mathbf{P} = (\mathbf{n}\,.\,\mathbf{i})\,\mathbf{P}_1 + (\mathbf{n}\,.\,\mathbf{j})\,\mathbf{P}_2 + (\mathbf{n}\,.\,\mathbf{k})\,\mathbf{P}_3. \qquad ...(2)$$

The equation (2) expresses **P** as a linear combination of $\mathbf{P}_1, \mathbf{P}_2, \mathbf{P}_3$.

Note. Resolving $\mathbf{P}, \mathbf{P}_1, \mathbf{P}_2, \mathbf{P}_3$, along the co-ordinate axes, we write

$$\mathbf{P} = p_1\mathbf{i} + p_2\mathbf{j} + p_3\mathbf{k},$$

$$\mathbf{P}_1 = p_{11}\mathbf{i} + p_{12}\mathbf{j} + p_{13}\mathbf{k},\ \mathbf{P}_2 = p_{21}\mathbf{i} + p_{22}\mathbf{j} + p_{23}\mathbf{k},\ \mathbf{P}_3 = p_{31}\mathbf{i} + p_{32}\mathbf{j} + p_{33}\mathbf{k}.$$

Substituting these expressions in (2) and equating the co-efficients of **i**, **j**, **k**, we obtain

$$p_1 = lp_{11} + mp_{21} + np_{31},\ p_2 = lp_{12} + mp_{22} + np_{32},\ p_3 = lp_{13} + mp_{23} + np_{33}$$

where $\quad l = \mathbf{n}\,.\,\mathbf{i},\ m = \mathbf{n}\,.\,\mathbf{j},\ n = \mathbf{n}\,.\,\mathbf{k}.$

Thus we see that the stress at a point across any plane can be described in terms of nine numbers, p_{ij}, consisting of three triads of numbers which can be written in the form of a square array as follows:

$$\left\{\begin{matrix} p_{11} & p_{12} & p_{13} \\ p_{21} & p_{22} & p_{23} \\ p_{31} & p_{32} & p_{33} \end{matrix}\right\} \begin{matrix} i = 1, 2, 3 \\ j = 1, 2, 3 \end{matrix}$$

Here, p_{ij}, denote the component along the jth axis of the stress across the plane normal to the ith axis. Here the x-axis, y-axis and z-axis are respectively denoted as the first, second and third axes.

8.2. PERFECT FLUIDS

The stress at a point across any plane can be resolved into two components perpendicular to the plane and along the plane. The component along the plane is known as **shearing stress** and is of the nature of friction. In the case of *viscous* fluids, these shearing stresses do appear but in many problems these are so small that they may be neglected. We shall now idealise and define *a Perfect fluid as one for which shearing stress at every point across every plane is zero* and develop the subject for perfect fluids only.

Thus for a perfect fluid the stress at every point across every plane is normal to the same.

A perfect fluid is also said to be Frietionless, non-viscous or inviscid.

For perfect fluids we have the following important theorem which introduces the notion of *pressure at a point* as a scalar point function.

Theorem. *The stress at a point of a perfect fluid is the same across every plane, i.e.,* it is the same in all directions.

Let **P**, $\mathbf{P}_1$, $\mathbf{P}_2$, $\mathbf{P}_3$ denote the stresses at a point across planes normal to the vectors **n, i, j, k,** so that we have

$$\mathbf{P} = (\mathbf{n.i})\,\mathbf{P}_1 + (\mathbf{n.j})\,\mathbf{P}_2 + (\mathbf{n.k})\,\mathbf{P}_3. \qquad ...(1)$$

If p, p_1, p_2, p_3 denote the magnitudes of these stresses across the four planes, we have, the fluid being perfect.

$$\mathbf{P} = p\mathbf{n},\ \mathbf{P}_1 = p_1\mathbf{i},\ \mathbf{P}_2 = p_2\mathbf{j},\ \mathbf{P}_3 = p_3\mathbf{k}.$$

Substituting in (1), we get

$$p\mathbf{n} = (\mathbf{n.i})p_1\mathbf{i} + (\mathbf{n.j})p_2\mathbf{j} + (\mathbf{n.k})p_2\mathbf{k}.$$

Multiplying scalarly by **i, j, k** respectively, we obtain

$$p = p_1,\ p = p_2,\ p = p_3.$$

Thus $\quad p = p_1 = p_2 = p_3.$

Hence the theorem.

8.3. PRESSURE AND DENSITY

8.3.1. Pressure at a point of a perfect fluid

By experience, the stress at any point of a fluid across any plane is in the nature of pressure and not tension. On this account, the stress at any point of a perfect fluid which is the same for all directions is called **Pressure**. Thus for a perfect fluid we have evolved an important scalar point function, *viz.*, Pressure.

If, p, denote the pressure at a point and A, be the area of a small surface through the point and $\mathbf{n}$ the unit vector normal to the surface drawn on either side, then the force acting on the fluid on the, $-\mathbf{n}$, side due to that on the $\mathbf{n}$ is $-pA\mathbf{n}$.

For a perfect fluid in motion, the pressure at a point may also depend upon time so that we write

$$p = f(x, y, z, t) = f(\mathbf{r}, t),$$

where, $\mathbf{r}$, denotes the position vector of the point.

8.3.2. Density

In addition to pressure, we need to deal with another scalar point function also, namely *density*, ρ, which may also depend upon time.

If δm, be the mass of a fluid of volume δv around a point P at time t, then

$$\lim_{\delta v \to 0} \frac{\delta m}{\delta v},$$

is the density of the fluid at the point P at time t. We may write

$$\rho = \phi(x, y, z, t) = \phi(\mathbf{r}, t).$$

8.3.3. Relation between pressure and density

For every fluid, there exists a relation between pressure and density which is known from considerations other than purely mechanical and which may also depend upon other circumstances such as temperature. We shall, however, in genera deal with fluids for which p is a function of ρ only. For incompressible fluid ρ, is constant for any fluid particle.

8.4. LOCAL AND INDIVIDUAL TIME-RATES OF CHANGE OF POINT FUNCTIONS

Associated with a fluid in motion, we have several point functions, such as,

pressure, density, velocity

which depend upon time also. For these point functions, both scalar and vector, it is important to distinguish between two-time-rates of changes, known as *Local* and *Individual*. These will now be introduced.

Firstly, we shall consider the case of any scalar function

$$\phi(x, y, z, t) = \phi(\mathbf{r}, t).$$

The value $\phi(x, y, z, t)$ of the function is associated with, (*i.e.*, it describes) the character of the fluid particle at the point (x, y, z) at time t.

8.4.1. Local time-rate of change

Let $P(x, y, z)$ be any point. We now keep the *point fixed* and find the change

$$\phi(x, y, z, t+\delta t) - \phi(x, y, z, t),$$

i.e.,

$$\phi(\mathbf{r}, t+\delta t) - \phi(\mathbf{r}, t)$$

in ϕ, associated with the same point at time t and $t + \delta t$. Then

$$\lim_{\delta t \to 0} \frac{\phi(\mathbf{r}, t+\delta t) - \phi(\mathbf{r}, t)}{\delta t},$$

is called the local time-rate of change and according to the usual convention, is denoted by the symbol

$$\frac{\partial \phi}{\partial t}.$$

8.4.2. Individual time-rate of change

We now attend to a fluid particle and keep the *particle fixed* and find the change

$$\phi(x+\delta x, y+\delta y, z+\delta z, t+\delta t) - \phi(x, y, z, t)$$

i.e.,

$$\phi(\mathbf{r}+\delta\mathbf{r}, t+\delta t) - \phi(\mathbf{r}, t)$$

in, ϕ, associated with the same particle at time t and $t + \delta t$, $\mathbf{r} + \delta\mathbf{r}$ denoting the position of the particle, in question, at time $t + \delta t$.

The change, $\delta\mathbf{r}$, in the position of the particle during time δt depends upon the velocity of the particle at time t. Thus if the vector, $\mathbf{q}$, denote the velocity of the particle at time t, we have

$$\delta\mathbf{r} = \mathbf{q}\delta t.$$

Then

$$\lim_{\delta t \to 0} \frac{\phi(\mathbf{r}+\mathbf{q}\,\delta t, t+\delta t) - \phi(\mathbf{r}, t)}{\delta t}$$

is called the individual time-rate of change and is denoted by the symbol

$$\frac{d\phi}{dt}.$$

8.4.3. Relation between the local and individual rates

We shall prove that

$$\frac{d\phi}{dt} = \frac{\partial \phi}{\partial t} + \mathbf{q} \cdot \nabla\phi,$$

where the vector, $\mathbf{q}$, denotes the velocity of the particle at, $\mathbf{r}$, at time t.

We write

$$\begin{aligned}\phi(\mathbf{r}+\delta\mathbf{r}, t+\delta t) - \phi(\mathbf{r}, t) &= \phi(\mathbf{r}+\delta\mathbf{r}, t+\delta t) - \phi(\mathbf{r}, t+\delta t) \\ &\quad + \phi(\mathbf{r}, t+\delta t) - \phi(\mathbf{r}, t).\end{aligned}$$

As seen in, § 6.7.3

$$\phi(\mathbf{r}+\delta\mathbf{r}, t+\delta t) - \phi(\mathbf{r}, t+\delta t) = \delta\mathbf{r}.\nabla\phi(\mathbf{r}, t+\delta t) + \varepsilon\,|\delta\mathbf{r}|;$$

where $\varepsilon \to 0$ as $\delta\mathbf{r} \to \mathbf{0}$.

Thus we have

$$\frac{\phi(\mathbf{r}+\delta\mathbf{r}, t+\delta t) - \phi(\mathbf{r}, t)}{\delta t} = \frac{\delta\mathbf{r}}{\delta t} \,.\, \nabla\phi(\mathbf{r}, t+\delta t) + \varepsilon\frac{|\delta\mathbf{r}|}{\delta t} + \frac{\phi(\mathbf{r}, t+\delta t) - \phi(\mathbf{r}, t)}{\delta t}$$

Proceeding to the limit when $\delta t \to 0$, we obtain

$$\frac{d\phi}{dt} = \frac{d\mathbf{r}}{dt} \,.\, \nabla\phi(\mathbf{r}, t) + \frac{\partial\phi}{\partial t},$$

i.e.,
$$\frac{d\phi}{dt} = \frac{\partial\phi}{\partial t} + \mathbf{q} \,.\, \nabla\phi,$$

which is the relation sought.

The same result can also be obtained in *another* manner as follows:

Resolving the velocity $\mathbf{q}$ into its components along the co-ordinate axes, we write

$$\mathbf{q} = u\mathbf{i} + v\mathbf{j} + w\mathbf{k},$$

so that
$$u = \frac{dx}{dt},\ v = \frac{dy}{dt},\ w = \frac{dz}{dt},$$

Now $d\phi/dt$ denotes the total derivative of, ϕ, *w.r.* to t, allowing for the variation of x, y, z, as a result of the change in t and $\partial\phi/\partial t$ denotes the partial derivative of, ϕ, *w.r.* to t.

Thus we have,

$$\frac{d\phi}{dt} = \frac{\partial\phi}{\partial x}\frac{dx}{dt} + \frac{\partial\phi}{\partial y}\frac{dy}{dt} + \frac{\partial\phi}{\partial z}\frac{dz}{dt} + \frac{\partial\phi}{\partial t}$$

$$= u\frac{\partial\phi}{\partial x} + v\frac{\partial\phi}{\partial y} + w\frac{\partial\phi}{\partial z} + \frac{\partial\phi}{\partial t}$$

$$= \mathbf{q} \,.\, \nabla\phi + \frac{\partial\phi}{\partial t}.$$

If, instead of a scalar point function, we consider a vector point function $\mathbf{f}(\mathbf{r}, t)$, then as above, we can show that

$$\frac{d\mathbf{f}}{dt} = \frac{\partial\mathbf{f}}{\partial t} + \mathbf{q}.\ \nabla\mathbf{f},$$

where $\mathbf{q}.\nabla\,\mathbf{f}$ means $(\mathbf{q}\,.\,\nabla)\,\mathbf{f}$.

It may again be noted here that the symbol $\mathbf{q}(\nabla\mathbf{f})$ has not been given any meaning.

8.5. STEADY MOTION

The motion of a fluid is said to be steady when the conditions at every point fixed in space remain unchanged with time *i.e.,* the local time rates

$$\frac{\partial \mathbf{f}}{\partial t}, \frac{\partial \rho}{\partial t}, \frac{\partial p}{\partial t}$$

are all zero.

8.6: EQUATION OF CONTINUITY

We shall now obtain an equation which holds for a fluid in motion irrespective of the forces acting on the same and of the nature of the fluid, perfect or otherwise.

Consider any closed surface drawn in the region occupied by moving fluid and *fixed* in space. The increase in the mass of fluid contained within the surface during any interval of time is equal to the mass of fluid that enters the surface during the same time, on the supposition that no fluid is being created or destroyed within the surface, *i.e.,* there are no sources or sinks within the surface.

Proceeding to the limit when the interval of time tends to zero, we see that the time rate of change in the mass of the fluid within the fixed surface at any time must equal the mass of fluid that enters the surface per unit time. The analytical equivalent of this physical fact is known as the equation of continuity and we, now, proceed to obtain the same.

Let, dv, be an element of volume within the fixed surface and ρ, the density of the fluid element at time t. Then the mass of the fluid within the surface at time t

$$= \int_V \rho dv,$$

where, V, denotes the volume enclosed within the surface. Its time rate of change

$$= \frac{\partial}{\partial t}\int_V r dv = \int_V \frac{\partial \rho}{\partial t} dv$$

The time rate of change considered here is local, for the surface is being kept fixed. The element of volume dv is associated with a fixed point thereof and, ρ, is the density of the fluid at that point.

Consider, now, any element, dS, of the surface and let, $\mathbf{n}$, be the unit outward drawn normal at any point of dS and let, $\mathbf{q}$, be the velocity of the fluid particle at time, t, at the point P.

The resolved part of, $\mathbf{q}$, along the outward drawn normal at $\mathbf{P}$

$$= \mathbf{q.n}$$

so that the inward normal velocity

$$= -\mathbf{q.n}.$$

The mass of fluid entering per unit time across the element dS

$$= -\rho\mathbf{q.n}dS.$$

Therefore the mass of fluid entering per unit time across the whole surface

$$= -\int_S \rho\mathbf{q}.\mathbf{n}dS$$

$$= -\int_V \text{div}(\rho\mathbf{q})\, dv,$$

by Gauss's theorem.

Thus we have

$$\int_V \frac{\partial\rho}{\partial t} = -\int_V \text{div}(\rho\mathbf{q})\, dv$$

or $$\int_V \left[\frac{\partial\rho}{\partial t} + \text{div}(\rho\mathbf{q})\right] dv = 0.$$

As the surface S is arbitrary, we deduce that

$$\frac{\partial\rho}{\partial t} + \text{div}(\rho\mathbf{q}) = 0, \qquad ...(1)$$

which is the required *equation of continuity.*

The equation of continuity can also be written down in another form so as to involve individual rate of change $d\rho/dt$ instead of the local rate of change $\partial\rho/\partial t$.

By § 6.1.5,

$$\text{div}(\rho\mathbf{q}) = \nabla.(\rho\theta) = \rho\nabla.\mathbf{q} + \mathbf{q}.\nabla\rho. \qquad ...(2)$$

From (1) and (2), we have

$$\frac{\partial\rho}{\partial t} + \mathbf{q}.\nabla\rho + \rho\nabla.\mathbf{q} = 0$$

i.e.,

$$\frac{\partial\rho}{\partial t} + \rho\nabla.\mathbf{q} = 0 \qquad (\S\ 8.4)$$

or

$$\frac{\partial\rho}{\partial t} + \rho\ \text{div}\ \mathbf{q} = 0.$$

Cor. Equation of continuity for incompressible fluids. For an incompressible fluid, $d\rho/dt = 0$, so that the equation of continuity reduces to

$$\text{div}\ \mathbf{q} = 0,$$

i.e.,

$$\frac{\partial u}{\partial x} + \frac{\partial v}{\partial u} + \frac{\partial w}{\partial z} = 0;$$

u, v, w being the components of **q** along the co-ordinate axes. Thus *in the case of incompressible fluid, the velocity vector* **q** *of an incompressible fluid is solenoidal.*

8.7. EULER'S EQUATION OF MOTION FOR A PERFECT FLUID

Two types of forces act on the fluid contained inside any closed surface S:

Body forces and surface forces.

Body forces are of the nature of gravitational forces and are described as forces per unit mass so that we may have a body force **F** per unit mass acting at each point of the fluid. Thus if we consider any element of mass, m, around a point **P** and if a force, **P**, acts on the same, then,

$$\lim_{m \to 0} \frac{\mathbf{P}}{m} = \mathbf{F}$$

is the force per unit mass at P.

Surface Forces. If, p, be the pressure at any point P, of a surface, S, then a force, $p\,A$, known as surface force acts at an elementary area, A, of the surface around P along the inward drawn normal to the surface at P.

We shall now obtain the fundamental equation of motion, due to Euler, for a perfect fluid in motion under the action of any given system of body forces.

Consider now any *arbitrary* closed surface drawn in the region occupied by the fluid and moving with the fluid so that it contains the same fluid particles at every time.

Let, S, be the position of the moving surface at time t.

Let, dv, be an element of volume enclosed within the surface S around a point P and let, ρ, be the density of the fluid particle at P. Then, ρdv, is the mass of the element.

Let, **F**, be the body force per unit mass. Then the body force acting on the fluid within the surface at time t

$$= \int_V \mathbf{F}\,\rho dv,$$

where, V, denotes the region enclosed by S.

Also the total surface force

$$= \int_S -p\mathbf{n}dS.$$

Here we have taken negative sign. This is so for, by assumption, **n** is the *outward* drawn normal vector and the surface force acts inwards; p being necessarily positive.

Also since

$$\mathbf{a} = \frac{d\mathbf{q}}{dt},$$

denotes the acceleration of the fluid particle of mass ρdv at P, we have by § 4.15.1,

$$\int_V \frac{d\mathbf{q}}{dt}\rho dv = \int_V \mathbf{F}\rho dv - \int p\mathbf{n}dS \qquad ...(1)$$

Also, by § 7.6 **Cor.** 1, (*ii*),

$$\int_S p\mathbf{n}dS = \int_V \nabla p dv. \qquad ...(2)$$

From (1) and (2),

$$\int_V \frac{d\mathbf{q}}{dt}\rho dv = \int_V \mathbf{F}\rho dv - \int_V \nabla p dv$$

or $$\int_V \left(\frac{d\mathbf{q}}{dt}\rho - \mathbf{F}r + \nabla p\right) dv = \mathbf{0}.$$

As the result holds for the region enclosed by any arbitrary closed surface S, we must have

$$\frac{d\mathbf{q}}{dt}\rho - \mathbf{F}\rho + \nabla p = \mathbf{0},$$

i.e., $$\frac{d\mathbf{q}}{dt} = \mathbf{F} - \frac{1}{\rho}\nabla p,$$

which is the required Euler's equation of motion.

If we write

$$\frac{d\mathbf{q}}{dt} = \frac{\partial \mathbf{q}}{\partial t} + \mathbf{q}.\nabla\mathbf{q},$$

we can re-write the equation of motion in the form

$$\frac{\partial \mathbf{q}}{\partial t} + \mathbf{q}.\nabla\mathbf{q} = \mathbf{F} - \frac{1}{\rho}\nabla p, \qquad ...(3)$$

involving local rate $\partial\mathbf{q}/\partial t$ instead of individual rate $d\mathbf{q}/dt$.

8.7.1. Cartesian equations.

We have

$$\frac{d\mathbf{q}}{dt} = \frac{d^2x}{dt^2}\mathbf{i} + \frac{d^2y}{dt^2}\mathbf{j} + \frac{d^2z}{dt^2}\mathbf{k},$$

$$\nabla p = \frac{\partial p}{\partial x}\mathbf{i} + \frac{\partial p}{\partial y}\mathbf{j} + \frac{\partial p}{\partial z}\mathbf{k}.$$

Also let $$\mathbf{F} = F_1\mathbf{i} + F_2\mathbf{j} + F_3\mathbf{k}.$$

$$\frac{d^2x}{dt^2} = \mathbf{F}_1 - \frac{1}{\rho}\frac{\partial p}{\partial x};$$

$$\frac{d^2y}{dt^2} = \mathbf{F}_2 - \frac{1}{\rho}\frac{\partial p}{\partial y};$$

$$\frac{d^2z}{dt^2} = \mathbf{F}_3 - \frac{1}{\rho}\frac{\partial p}{\partial z};$$

as three scalar cartesian equations of motion equivalent to the single vector equation (3).

Note. From the Eulerian point of view, the problem of Hydrodynamics consists in the determination of the velocity **q**, the pressure p and the density ρ as functions of position and time, *i.e., x, y, z, t.*

Also we have obtained two equations, *viz.*

$$\frac{\partial \rho}{\partial t} + \text{div}(\rho\mathbf{q}) = 0. \quad \textit{Equation of Continuity.}$$

$$\frac{d\mathbf{q}}{dt} = \mathbf{F} - \frac{1}{\rho} \nabla p. \textit{ Equation of Motion.}$$

In addition to these, we have a relation connecting ρ and p depending upon the nature of the fluid. Thus we have three equations, two scalar and one vector to determine three unknowns p, ρ, **q** two of which are scalar and one vector. From the scalar point of view, we have five scalar unknowns, and five scalar equations. These equations together with the *initial and boundary conditions* of a problem enable us to determine the solution of any given problem, apart from the difficulties of mathematical nature.

8.8. EQUATION OF MOTION WHEN THE BODY FORCES ARE CONSERVATIVE AND, P, IS A FUNCTION OF P, ONLY

We shall now transform the Euler's equation of motion on the basis of the following two suppositions.

I. *Body forces form a conservative system, i.e.,* possess a potential. Then there exists a scalar point function, say, Ω such that

$$\mathbf{F} = -\nabla\Omega.$$

II. *The density ρ, is a function of the pressure p, only.* In this case we introduce a point function, P, defined as

$$P = \int_c^p \frac{dp}{\rho},$$

c, being any constant.

Clearly, P, is a function of, p, such that

$$\frac{dP}{dp} = \frac{1}{\rho}.$$

Also, we have

$$\nabla p = \Sigma \mathbf{i} \frac{\partial p}{\partial x}$$

$$= \Sigma \mathbf{i} \frac{dp}{dP} \frac{\partial p}{\partial x}$$

$$= \rho \Sigma \mathbf{i} \frac{\partial P}{\partial x} = \rho \nabla P.$$

Thus the equation of motion takes the form

$$\frac{d\mathbf{q}}{dt} = \mathbf{F} - \frac{1}{\rho} \nabla p = -\nabla\Omega - \nabla P = -\nabla(\Omega + P)$$

or $$\frac{\partial \mathbf{P}}{\partial t} + \mathbf{q}.\nabla\mathbf{q} = -\nabla(\Omega + P). \quad ...(1)$$

Also, by § 6.15, p.

$$\nabla(\mathbf{q}.\mathbf{q}) = \mathbf{q} \times \text{curl } \mathbf{q} + \mathbf{q} \times \text{curl } \mathbf{q} + \mathbf{q}.\nabla\mathbf{q} + \mathbf{q}.\nabla\mathbf{q}$$

or $$\frac{1}{2}\nabla\mathbf{q}^2 = \mathbf{q}\times\text{curl}\,\mathbf{q}+\mathbf{q}.\nabla\mathbf{q},$$

or $$\mathbf{q}.\nabla\mathbf{q} = \frac{1}{2}\nabla\mathbf{q}^2-\mathbf{q}\times\text{curl}\,\mathbf{q}. \quad ...(2)$$

From (1) and (2), we have

$$\frac{\partial\mathbf{q}}{\partial t}+\frac{1}{2}\nabla\mathbf{q}^2-\mathbf{q}\times\text{curl}\,\mathbf{q}-\nabla(\Omega+P),$$

or $$\frac{\partial\mathbf{q}}{\partial t}-\mathbf{q}\times\text{curl}\,\mathbf{q} = -\nabla\left(\Omega+P+\frac{1}{2}\mathbf{q}^2\right),$$

which is the equation of motion on the two suppositions made above.

8.9. VORTICITY

The vector ω defined by the equality

$$\omega = \frac{1}{2}\text{curl}\,\mathbf{q}=\frac{1}{2}\nabla\times\mathbf{q}$$

is called *vorticity.* The vector ω is also a function of **r** and *t*.

8.9.1. Irrotational motion. Scalar velocity potential

The motion is said to be irrotational at any instant if the vorticity ω at the instant is zero. Also a motion is said to be permanently irrotational if the vorticity is zero at every instant.

In the case of permanent irrotational motion,

$$\omega = \frac{1}{2}\text{curl}\,\mathbf{q}=\mathbf{0},$$

i.e., $$\text{curl}\,\mathbf{q}=\mathbf{0},$$

so that there exists, by § 7.9, a scalar point function, say $\psi(x, y, z, t)$, such that

$$\mathbf{q} = -\nabla\psi.$$

The function, ψ, is called *scalar velocity potential.*

We can write down the equation of continuity of an incompressible fluid for irrotational motion in terms of the scalar velocity potential.

By § 8.6 Cor. the equation of continuity for an incompressible fluid is

$$\nabla.\mathbf{q} = 0$$

Also $$\mathbf{q} = -\nabla\psi.$$

∴ we have $\nabla.\nabla\psi = \mathbf{0}$, *i.e.,* $\nabla^2\psi = 0$.

This is equivalent to

$$\frac{\partial^2\psi}{\partial x^2}+\frac{\partial^2\psi}{\partial y^2}+\frac{\partial^2\psi}{\partial z^2} = 0,$$

in terms of fixed rectangular cartesian axes so that ψ satisfied the Laplacian equation.

Note. In the general case of motion, irrotational or not, the velocity vector $\mathbf{q}$ satisfies the equation $\nabla.\mathbf{q} = 0$ if fluid is incompressible. There exists, therefore, by § 7.10.1 p., a vector A such that

$$\mathbf{q} = \text{curl}\, \mathbf{A}.$$

This vector, **A**, is called **velocity vector potential**.

Thus the flow velocity of an incompressible fluid in general motion can be derived from a vector potential.

8.10. STREAM LINES AND VORTEX LINES. DEF.

A curve drawn in the fluid at any given instant such that the direction of the tangent at any point of the curve coincides with that of the velocity of the fluid particle at the point and at the instant is called a stream line.

Again, a **vortex line** *is a curve drawn in the fluid at any given instant such that the direction of the tangent at any point of the curve coincides with that of the* **vorticity** *of the fluid particle at the point at the instant.*

Thus if, **t**, denote the unit tangent to a stream line, we have

$$\mathbf{t} \times \mathbf{q} = \mathbf{0},$$

for any point of the stream line, and if, **t**, denote the unit tangent to a vortex line, we have

$$\mathbf{t} \times \omega = \mathbf{0}.$$

Thus at any point of a stream line as well as vortex line, the scalar triple product

$$[\mathbf{t}\ \mathbf{q}\ \omega] = 0$$

For non-steady motion, the patterns of stream lines and vortex lines are different at different times.

The path followed by a fluid particle is called a *line of motion* and should be distinguished from stream lines.

8.11. BERNOUILLI'S THEOREM. STEADY MOTION

If the body force form a conservative system and, ρ, *is a function of,* p, *only and the flow is steady, then*

$$\Omega + P + \frac{1}{2}\mathbf{q}^2,$$

is constant along every stream line and every vortex line.

In § 8.8 we have seen that when the body forces form a conservative system and ρ is a function of *p* only, the equation of motion takes the form

$$\frac{\partial \mathbf{q}}{\partial t} - \mathbf{q} \times \text{curl}\, \mathbf{q} = \nabla\left(\Omega + P + \frac{1}{2}\mathbf{q}^2\right). \quad ...(1)$$

The flow being also assumed steady so that $\partial\mathbf{q}/\partial t = 0$, we rewrite (1) as

$$\mathbf{q} \times \text{curl}\, \mathbf{q} = \nabla\left(\Omega + P + \frac{1}{2}\mathbf{q}^2\right)$$

or

$$2\mathbf{q} \times \omega = \nabla\left(\Omega + P + \frac{1}{2}\mathbf{q}^2\right) \quad ...(2)$$

ω being the vorticity.

Multiplying (2) scalarly by **t**, which denotes a unit tagent vector to any given stream line or vortex line, we obtain

$$\mathbf{t} \,.\, \nabla\left(\Omega + P + \frac{1}{2}\mathbf{q}^2\right) = 0, \text{ for } [\mathbf{t}\ \mathbf{q}\ \omega] = 0.$$

Thus the directional derivative of

$$\left(\Omega + P + \frac{1}{2}\mathbf{q}^2\right)$$

also every stream line and vortex line is zero.

Hence the theorem.

The constant will, in general, be different for different stream line and vortex line.

The theorem states that at any given instant, $\Omega + P + \frac{1}{2}\mathbf{q}^2$, has the same value at every point of a given stream line (or a vortex line).

Col. 1. The case of Steady as well as irrotational motion.

If the motion is also irrotational in addition to being steady so that we have

$$\omega = \mathbf{0} \text{ and } \partial\mathbf{q}/\partial t = 0,$$

then the equation of motion gives

$$\nabla\left(\Omega + P + \frac{1}{2}\mathbf{q}^2\right)$$

and accordingly $\Omega + P = \frac{1}{2}\mathbf{q}^2$ *is constant having the same value at any point.*

Cor. 2. The case of irrotational motion. In this case curl $\mathbf{q} = 0$ and there exists velocity potential ψ.

We have

$$\mathbf{q} = -\nabla\psi.$$

$$\therefore \quad \frac{\partial \mathbf{q}}{\partial t} = -\frac{\partial}{\partial t}\nabla\psi = -\nabla\frac{\partial\psi}{\partial t}.$$

The equation of motion, therefore, takes the form

$$-\nabla\frac{\partial\psi}{\partial t} = -\nabla\left(\Omega + P + \frac{1}{2}\mathbf{q}^2\right)$$

i.e.,
$$\nabla\left(\Omega + P + \frac{1}{2}\mathbf{q}^2 - \frac{\partial\psi}{\partial t}\right) = 0$$

Thus

$$\Omega + P + \frac{1}{2}\mathbf{q}^2 - \frac{\partial\psi}{\partial t} = \text{constant}$$

where the constant may depend upon time and not on position. Calling it $f(t)$, we obtain

$$\Omega + P + \frac{1}{2}\mathbf{q}^2 - \frac{\partial \psi}{\partial t} = f(t).$$

Note. Under the conditions stated in cor. 1, the conclusion

$$\Omega + P + \frac{1}{2}\mathbf{q}^2 = \text{constant}.$$

provides a solution of the equation of motion.

8.12. STEADY IRROTATIONAL MOTION OF AN INCOMPRESSIBLE FLUID

In this case, we have the two equations

(1) $\dfrac{\partial^2 \psi}{\partial x^2} + \dfrac{\partial^2 \psi}{\partial y^2} + \dfrac{\partial^2 \psi}{\partial z^2} = 0.$ Equation of Continuity.

(2) $\Omega + P + \dfrac{1}{2}\mathbf{q}^2 = \text{constant}.$ Equation of motion.

The solution of the problem, now, consists in determining the solution (1) consistents with the initial and boundary conditions of the problem.

This then leads to

$$\mathbf{q} = -\nabla\psi.$$

The case of steady irrotational motion of an incompressible fluid is very important.

8.13. CIRCULATION ALONG A CLOSED CURVE

Def. *If, Γ, be any closed curve drawn in a fluid at time t, then*

$$\mathbf{c} = \int_\Gamma \mathbf{q}.d\mathbf{r} = \int_\Gamma \mathbf{q}.\mathbf{t}\, ds$$

is called the circulation round the curve at the time; **q** denoting the velocity of the fluid particle at the point **r** at time *t*.

Theorem. *If the motion in a simply connected region is irrotational at any instant, then the circulation around any closed curve at the instant is also zero.*

We have by Stoke's theorem,

$$c = \int_\Gamma \mathbf{q}.d\mathbf{r} = \int_\Gamma \text{curl}\ \mathbf{q}.\mathbf{n}dS,$$

where, S, denotes **a** surface enclosed by Γ.

The motion being irrotational at the instant, curl $\mathbf{q} = \mathbf{0}$ at the instant at every point of the fluid. Hence

$$c = 0.$$

8.13.1. Time-Rate of change of circulation round a closed curve

Kelvin's Theorem. *The circulation round any closed curve moving with the fluid does not change with time, if the field of force is conservative and, ρ, is a function of p only.*

Let, Γ, be the position of a moving curve at any time t. We have the circulation

$$c = \int_\Gamma \mathbf{q}.d\mathbf{r}.$$

The curve, **r**, moves so as to consist always of the same fluid particles. In fact the position of the moving curve at any time is determined by the positions at any time of the fluid particles lying on the same at any given time. The curve Γ and the circulation are both functions of time t.

We have $$c = \int_\Gamma \frac{\mathbf{q}.d\mathbf{r}}{ds}\,ds.$$

$$\therefore \quad \frac{dc}{dt} = \int_\Gamma \frac{d}{dt}\left(\frac{\mathbf{q}.d\mathbf{r}}{ds}\right) ds, \qquad ...(1)$$

Also

$$\frac{d}{dt}\left(\mathbf{q}\cdot\frac{d\mathbf{r}}{ds}\right) = \frac{d\mathbf{q}}{dt}\cdot\frac{d\mathbf{r}}{ds} + \mathbf{q}\cdot\frac{d}{dt}\left(\frac{d\mathbf{r}}{ds}\right)$$

$$= \frac{d\mathbf{q}}{dt}\cdot\frac{d\mathbf{r}}{ds} + \mathbf{q}\cdot\frac{d}{ds}\left(\frac{d\mathbf{r}}{dt}\right)$$

$$= \frac{d\mathbf{q}}{dt}\cdot\frac{d\mathbf{r}}{ds} + \mathbf{q}\cdot\frac{d\mathbf{q}}{ds}$$

$$= \frac{d\mathbf{q}}{dt}\cdot\frac{d\mathbf{r}}{ds} + \frac{1}{2}\frac{d\mathbf{q}^2}{ds}. \qquad ...(2)$$

Also as the body forces form a conservative system and ρ is a function of p only, the equation of motion is

$$\frac{d\mathbf{q}}{dt} = -\nabla(\Omega + P) = -\nabla V, \text{ say}$$

$$\therefore \quad \frac{d}{dt}\left(\mathbf{q}\cdot\frac{d\mathbf{r}}{ds}\right) = -\nabla V.\ \frac{d\mathbf{r}}{ds} + \frac{1}{2}\frac{d\mathbf{q}^2}{ds}$$

$$= -\nabla V.\,\mathbf{t} + \frac{1}{2}\frac{d\mathbf{q}^2}{ds}$$

$$= -\frac{dV}{ds} + \frac{1}{2}\frac{d\mathbf{q}^2}{ds}$$

$$= \frac{d}{dt}\left(\frac{1}{2}\mathbf{q}^2 - V\right). \qquad ...(3)$$

From (1), (2) and (3),

$$\frac{dc}{dt} = \int_\Gamma \frac{d}{ds}\left(\frac{1}{2}\mathbf{q}^2 - V\right) ds$$

$$= \text{the difference in the values of } \frac{1}{2}\mathbf{q}^2 - V$$

at the two ends of the curve. As however the curve is closed, the end points coincide and we have

$$\frac{dc}{dt} = 0.$$

Hence the circulation, c, is constant.

If the motion of a non-viscous fluid is initially irrotational, then it remains irrotational, if the body forces are conservative and pressure is a function of density only.

If follows from the theorems in § 8.13, 8.13.1.

8.14. KELVIN'S MINIMUM ENERGY THEOREM

The irrotational motion of a liquid occupying a simply connected region has less kinetic energy than any other motion consistent with the same normal velocity of the boundary.

Since the actual motion is set up by giving impulses on the boundaries, the motion must be irrotational and acyclic, so that a single valued velocity potential ϕ exists such that

$$\mathbf{q} = -\nabla\phi \qquad ...(1)$$

Let $\mathbf{q}_0$ be the fluid velocity at any other motion consistent with the same normal velocity of the boundary whose kinetic energy be to.

Then $\mathbf{n} \cdot \mathbf{q}_0$

$= \mathbf{n} \cdot \mathbf{q} \;\Rightarrow\; \mathbf{n}.(\mathbf{q}_0 - \mathbf{q}) = 0$

$$\Rightarrow \quad \mathbf{n} \cdot \mathbf{u} = 0, \qquad \mathbf{u} = \mathbf{q}_0 - \mathbf{q} \qquad ...(2)$$

Also from the equation of continuity we have

$$\nabla \cdot \mathbf{q}_0 = 0 = \nabla \cdot \mathbf{q} \;\Rightarrow\; \nabla \cdot (\mathbf{q}_0 - \mathbf{q}) = 0$$

$$\Rightarrow \quad \nabla \cdot \mathbf{u} = 0 \qquad ...(3)$$

Now
$$T_0 = \frac{1}{2}\int_V \rho\, \mathbf{q}_0{}^2\, dv = \frac{1}{2}\int_V \rho\, (\mathbf{q} + \mathbf{u})^2\, dv$$

$$= \frac{1}{2}\int_V \rho\, (\mathbf{q}^2 + \mathbf{u}^2 + 2\mathbf{q} \cdot \mathbf{u})\, dv$$

$$= \frac{1}{2}\int_V \rho\, \mathbf{q}^2\, dv + \frac{1}{2}\int_V \rho\, \mathbf{u}^2\, dv + \rho\int_V \mathbf{q} \cdot \mathbf{u}\, dv$$

or
$$T_0 = T + \frac{1}{2}\int_V \rho\, \mathbf{u}^2\, dv - \rho\int_V (\nabla\phi \cdot \mathbf{u})\, dv,$$

where T is the kinetic energy of the irrotational motion of which ϕ is the velocity potential and $\mathbf{q}$ is the fluid velocity. The third integral is zero by Ganss Theorem.

$$\therefore \quad T_0 - T = \frac{1}{2}\int_V \rho\, \mathbf{u}^2\, dv$$

But T_0, T are essentially positive qualities so that

$$T_0 - T > 0 \quad \Rightarrow \quad T < T_0.$$

This theorem is due to Lord Kelvin and was subsequently generalised by him so as to apply to all dynamical systems started impulsively from rest.

8.15. HELMHOLTZ'S VORTICITY EQUATION

If the body forces are conservative and ρ is a function of p only, then

$$\frac{d}{dt}\left(\frac{\omega}{\rho}\right) = \left(\frac{\omega}{\rho} \cdot \nabla\right) \mathbf{q},$$

where $\omega = \dfrac{1}{2}$ *curl* $\mathbf{q}$ *is the vorticity.*

Consider the equation no motion

$$\frac{\partial \mathbf{q}}{\partial t} - \mathbf{q} \times \text{curl } \mathbf{q} = -\nabla\left(\Omega + P + \frac{1}{2}\mathbf{q}^2\right) \qquad ...(1)$$

and the equation of continuity

$$\frac{d\rho}{dt} + \rho \text{div } \mathbf{q} = 0. \qquad ...(2)$$

Taking curl of both sides of (1), we obtain

$$\nabla \times \left(\frac{\partial \mathbf{q}}{\partial t} - \mathbf{q} \times \text{curl } \mathbf{q}\right) = \mathbf{0}. \quad \text{for} \quad \nabla \times \nabla\phi = \mathbf{0};$$

$$\therefore \quad \frac{\partial}{\partial t}(\nabla \times \mathbf{q}) - \nabla \times (\mathbf{q} \times \text{curl } \mathbf{q}) = \mathbf{0}.$$

or

$$\frac{\partial \omega}{\partial t} - \nabla \times (\mathbf{q} \times \omega) = \mathbf{0}. \qquad ...(3)$$

By § 6.15

$$\begin{aligned} \nabla \times (\mathbf{q} \times \omega) &= (\nabla \cdot \omega)\,\mathbf{q} - (\nabla \cdot \mathbf{q})\,\omega + (\omega \cdot \nabla)\mathbf{q} - (\mathbf{q} \cdot \nabla)\omega \\ &= -(\nabla \cdot \mathbf{q})\omega + (\omega \cdot \nabla)\mathbf{q} - (\mathbf{q} \cdot \nabla)\omega, \qquad ...(4) \end{aligned}$$

$$\text{for } \nabla \cdot \omega = \frac{1}{2} \nabla \cdot \omega \times \mathbf{q} = 0.$$

Also, by § 8.4.3,

$$\frac{\partial \omega}{\partial t} = \frac{d\omega}{dt} - \mathbf{q} \cdot \nabla\omega.$$

From (3), (4) and (5), we have

$$\frac{d\omega}{dt} + (\nabla \cdot \mathbf{q})\omega - (\omega \cdot \nabla)\mathbf{q} = \mathbf{0}. \qquad ...(6)$$

From (2) and (6), we obtain

$$\frac{d\omega}{dt} - \frac{\omega}{\rho}\frac{d\rho}{dt} - (\omega \cdot \nabla)\,\mathbf{q} = \mathbf{0},$$

or

$$\frac{1}{\rho}\frac{d\omega}{dt} - \frac{\omega}{\rho^2}\frac{d\rho}{dt} - \left(\frac{\omega}{\rho} \cdot \nabla\right) \mathbf{q} = \mathbf{0}.$$

or $$\frac{d}{dt}\left(\frac{\omega}{\rho}\right) = \left(\frac{\omega}{\rho}.\nabla\right)\mathbf{q},$$

which is the required vorticity equation due to Helmholtz.

8.16. BOUNDARY SURFACE

Physical conditions that should be satisfied on given boundaries of the fluid are called as **boundary conditions.**

When the fluid is in contact with an impermeable bounding surface the velocity of fluid particle at any point of the boundary relative to the surface must be tangential to the boundary. Thus at a fixed boundary, the velocity of the fluid perpendicular to the surface must vanish and the normal component of the velocity of the fluid must be equal to the normal component of the velocity of the surface. Thus, at the boundary of the fluid, the equation of continuity is replaced by a special surface condition.

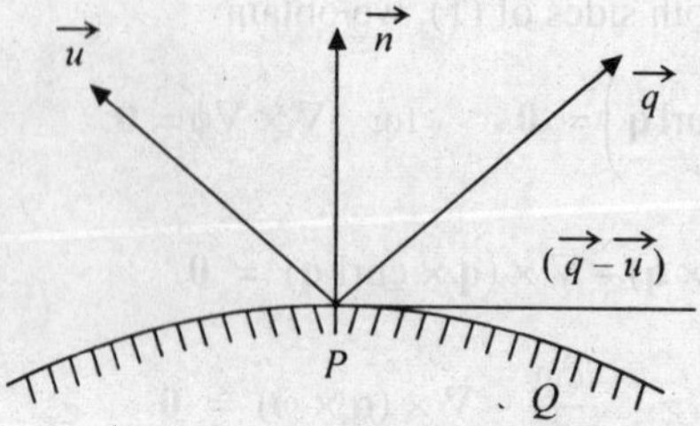

Fig. 8.2

Let **q** be the velocity of the fluid and **u** be the velocity of the surface at the point P. Let **n** be the unit normal vector drawn at the point P on the boundary surface $F(\mathbf{r}, t) = 0$. Since there must be no relative normal velocity at P between boundary and fluid so we must have the two normal components equal, i.e.,

$$\mathbf{q}\,.\,\mathbf{n} = \mathbf{u}\,.\,\mathbf{n} \Rightarrow (\mathbf{q}-\mathbf{u})\,.\,\mathbf{n} = 0 \qquad ...(1)$$

For two fluids in contact, a dynamical boundary is required, *i.e.,* the pressure must be continuous across the interface.

$$(\mathbf{q}-\mathbf{u})\,.\,\nabla F = 0 \qquad [\because \mathbf{n} = \nabla F] \quad ...(2)$$

The position of the point P on the moving surface at any instant of time $t + \delta t$ is given by

$$F(\mathbf{r}+\delta\mathbf{r},\ t+\delta t) = 0$$

Expanding by Taylor's theorem, we have

$$F(\mathbf{r}, t) + \delta\mathbf{r}\,.\,\nabla F + \delta t.\ \frac{\partial F}{\partial t} = 0$$

$$\Rightarrow \qquad \frac{\partial F}{\partial t} + \frac{\partial \mathbf{r}}{\partial t}\,.\,\nabla F = 0.$$

The above equation reduces to

$$\frac{\partial \mathbf{F}}{\partial t} + \mathbf{u} \cdot \nabla F = 0; \quad \delta\mathbf{r} \to 0,\ \delta t \to 0,$$

$$\mathbf{u} = \frac{d\mathbf{r}}{dt} \qquad ...(3)$$

$$\frac{\partial \mathbf{F}}{\partial t} + \mathbf{q} \cdot \nabla F = 0 \Rightarrow \frac{\partial F}{\partial t} + u\frac{\partial F}{\partial x} + v\frac{\partial F}{\partial y} + w\frac{\partial F}{\partial z} = 0 \qquad ...(4)$$

This the equation of boundary surface.

If the surface is at rest then $\frac{\partial \mathbf{F}}{\partial t} = 0$, then

$$\mathbf{q} \cdot \nabla F = 0.$$

This is the condition when the liquid is in contact with a rigid surface.

In order that contact is maintained, the fluid and the surface must have the same velocity normal to the surface.

Also, the normal velocity of the boundary is given by

$$\mathbf{u} \cdot \mathbf{n} = \frac{u \cdot \nabla \mathrm{F}}{|\nabla \mathrm{F}|} = \frac{\frac{\partial \mathrm{F}}{\partial t}}{\sqrt{\left(\frac{\partial \mathrm{F}}{\partial x}\right)^2 + \left(\frac{\partial \mathrm{F}}{\partial y}\right)^2 + \left(\frac{\partial \mathrm{F}}{\partial z}\right)^2}}.$$

EXAMPLES

1. *Show that for a fluid in equilibrium*

$$\mathbf{F} \cdot \text{curl}\, \mathbf{F} = 0$$

The equation of motion is

$$\rho\frac{d\mathbf{q}}{dt} - \mathbf{F}\rho + \nabla p = \mathbf{0}.$$

For equilibrium $d\mathbf{q}/dt = \mathbf{0}$ so that we have

$$\rho\mathbf{F} = \nabla p.$$

Taking curl of both members, we have

$$\nabla \times (\rho\mathbf{F}) = \nabla \times \nabla p = \mathbf{0}.$$

$$\therefore\ \rho\nabla \times \mathbf{F} + \nabla\rho \times \mathbf{F} = \mathbf{0}. \qquad [\S\ 6.15,]$$

Multiplying scalarly with **F**, we get

$$\nabla \times \mathbf{F} \cdot \mathbf{F} = 0,$$

$$\mathbf{F} \cdot \text{curl}\, \mathbf{F} = 0.$$

2. *Determine the equations of the vortex lines when the velocity vector of the flow field is given by*

$$\mathbf{q} = \mathbf{i}(Az - By) + \mathbf{j}\,(Bx - Cz) + \mathbf{k}(Cy - Ax)$$

where A, B, C are constants.

The vorticity components are given by

$$\xi = \frac{\partial w}{\partial y} - \frac{\partial v}{\partial z} = C + C = 2C,$$

$$\eta = \frac{\partial u}{\partial z} - \frac{\partial w}{\partial x} = A + A = 2A,$$

$$\zeta = \frac{\partial v}{\partial x} - \frac{\partial u}{\partial y} = B + B = 2B.$$

The equations of the vortex lines are

$$\frac{dx}{\xi} = \frac{dx}{\eta} = \frac{dz}{\zeta}$$

$$\Rightarrow \quad \frac{dx}{2C} = \frac{dy}{2A} = \frac{dz}{2B}$$

which yield vortex lines as

$Ax - Cy = k_1, \qquad By - Az = k_2.$

3. *Show that* $\phi = xf(r)$ *is a possible form for the vorticity potential of an incompressible liquid motion. Given that the liquid speed* $q \to 0$ *as* $r \to \infty$*, deduce that the surfaces of constant speed are* $(r^2 + 3x^2)\, r^{-8} =$ *constant.*

Since $\quad \phi = x f(r)$

$$\mathbf{q} = -\nabla\phi = -\nabla\{x f(r)\}$$

$$\Rightarrow \quad \mathbf{q} = -f(r)\,\mathbf{i} - \frac{x f'(r)}{r}\,\mathbf{r} \qquad \text{...(1)}$$

It will be a possible liquid motion if $\nabla . \mathbf{q} = 0$.

$$\therefore \quad \nabla . \mathbf{q} = -\nabla . \{f(r)\,\mathbf{i}\} - \left\{\frac{x f'(r)}{r}\right\} (\nabla . \mathbf{r})$$

$$- \mathbf{r} . \nabla \left\{\frac{x f'(r)}{r}\right\}$$

$$= -\frac{x}{r}\frac{\partial f}{\partial x} - \frac{\partial}{\partial x}\left\{\frac{x^2 f'(r)}{r}\right\}$$

$$- \frac{\partial}{\partial x}\left\{\frac{xy f'(r)}{r}\right\} - \frac{\partial}{\partial z}\left\{\frac{xz f'(r)}{r}\right\}$$

$$= -\frac{x}{r}\frac{\partial f}{\partial x} - \frac{4x}{r} f'(r) - \frac{x}{r^2}\{(x^2 + y^2 + z^2) f''(r)$$

$$+ \frac{x}{r^3}\{(x^2 + y^2 + z^2) f'(r)\}$$

$$= -\frac{x}{r}\frac{\partial f}{\partial x} - \frac{4x}{r} f'(r) - x f''(r) + \frac{x}{r^2} f'(r)$$

$$= -x\left(f''+\frac{4f'}{r}\right).$$

For an incompressible liquid $\nabla \cdot \mathbf{q} = 0$, hence

$$f'+\frac{4f'}{r} = 0 \Rightarrow \frac{f''}{f'}+\frac{4}{r}=0$$

$$\Rightarrow \quad \log f' + \log r = \log c \Rightarrow f' = Cr^{-4} \quad ...(2)$$

$$\Rightarrow \quad f = -\frac{1}{3}Cr^{-3} + D, \text{ where D is a constant.}$$

From (1)

$$\mathbf{q} = \left(\frac{C}{3r^3} - D\right).\mathbf{i} - \frac{Cx}{r^5}\mathbf{r} \quad ...(3)$$

When $r \to \infty$, $q \to -Di \Rightarrow D = 0$.

Thus, $$\mathbf{q} = \frac{C}{3r^3}\left(\mathbf{i} - \frac{3x\,\mathbf{r}}{r^2}\right)$$

and $$\mathbf{q}^2 = \mathbf{q}\cdot\mathbf{q} = \frac{C^2}{9r^6}\left(1-\frac{6x\,\mathbf{r}.\mathbf{i}}{r^2}+\frac{9x^2r^2}{r^4}\right)$$

Hence q^2 = constant gives the surface of constant speed as

$$(r^2+3x^2)r^{-8} = \text{constant.}$$

4. *Determine the different types of motion involved with velocity field*

$$\mathbf{q} = (Ax+2By+u_0)\mathbf{i}+(Ay+v_0)\mathbf{j}+(-2Az+w_0)\mathbf{k},$$

where A, B, u_0, v_0 and w_0 are constant.

We know that the general motion of a fluid element is made up of three parts, viz., translation, rotation and deformation.

(*i*) *Translational velocity*

$$\mathbf{q} = (Ax+2By+u_0)\mathbf{i}+(Ay+v_0)\mathbf{j}+(-2Az+w_0)\mathbf{k}$$

(*ii*) *Rotational velocity*

$$\vec{\omega} = \frac{1}{2}\text{ curl }\mathbf{q} = \frac{1}{2}\begin{vmatrix}\mathbf{i} & \mathbf{j} & \mathbf{k}\\ \partial/\partial x & \partial/\partial y & \partial/\partial z\\ Ax+2By+u_0 & Ay+v_0 & -2Az+w_0\end{vmatrix}$$

$$= \frac{1}{2}(-2Bk) = -Bk.$$

(*iii*) *Deformation*

Here $$u = Ax+2By+u_0,\ v = Ay+v_0,$$
$$w = -2Az+w_0$$

The components of rate of normal strain are

$$\epsilon_{xx} = \frac{\partial u}{\partial x} = A,$$

$$\in_{yy} = \frac{\partial v}{\partial y} = A,$$

$$\in_{zz} = \frac{\partial w}{\partial z} = -2A.$$

The components of rate of shear strain are

$$\in_{zz} = \in_{yx} \Rightarrow \frac{1}{2}\left(\frac{\partial u}{\partial y} + \frac{\partial v}{\partial x}\right) = \frac{1}{2}\left(\frac{\partial v}{\partial x} + \frac{\partial u}{\partial y}\right)$$

$$\in_{yz} = \in_{zy} \Rightarrow \frac{1}{2}\left(\frac{\partial v}{\partial z} + \frac{\partial w}{\partial y}\right) = \frac{1}{2}\left(\frac{\partial w}{\partial y} + \frac{\partial v}{\partial z}\right) = 0$$

$$\in_{zx} = \in_{xz} \Rightarrow \frac{1}{2}\left(\frac{\partial w}{\partial x} + \frac{\partial u}{\partial z}\right) = \frac{1}{2}\left(\frac{\partial u}{\partial z} + \frac{\partial w}{\partial x}\right) = 0$$

$$\therefore \quad \mathbf{D} = \mathbf{i}\,(\in_{xx}\, dx + \in_{xy}\, dy) + \mathbf{j}\,(\in_{yx}\, dx + \in_{yy}\, dy) + \mathbf{k}\,(\in_{zz}\, dz)$$

$$= \mathbf{i}\,(\text{A}dx + \text{B}dy) + \mathbf{j}\,(\text{B}dx + \text{A}dy) + \mathbf{k}\,(-2\text{A}\, dz).$$

5. *A body of liquid with constant density rotates with constant angular velocity w about a vertical axis under the action of gravity only. Show that the free surface is a paraboloid of revolution with axis vertical and that the vorticity at any point of the fluid is equal to w.*

Take a point on the axis of revolution as origin and the axis directed upwards as z-axis.

The velocity at a point whose position vector relative to the origin is $\mathbf{r}$, is given by

$$\mathbf{q} = \omega \times \mathbf{r}.$$

$$\therefore \quad \frac{d\mathbf{q}}{dt} = \omega \times \frac{d\mathbf{r}}{dt}, \text{ for } w \text{ is constant.}$$

$$= \omega \times (\omega \times \mathbf{r}) = (\omega \,.\, \mathbf{r})\, \omega - (\omega \,.\, \omega)\mathbf{r}.$$

Let $\omega = |\omega|\, \mathbf{k},\ \mathbf{r} = \mathbf{i}x + \mathbf{j}y + \mathbf{k}z.$

$$\therefore \quad \frac{d\mathbf{q}}{dt} = -\omega^2\,(\mathbf{i}x + \mathbf{j}y) = -\frac{1}{2}\,\omega^2\, \nabla(x^2 + y^2).$$

Also $\mathbf{F} = -g\mathbf{k} = -\nabla(gz),$

$$P = \int_c^p \frac{dp}{\rho} = \frac{dp}{\rho}\int_c^p dp = \frac{1}{\rho}\,(p - c).$$

$\therefore$ the equation of motion is

$$-\frac{1}{3}\,\omega^2\, \nabla(x^2 + y^2) = -\nabla\left(gz + \frac{p}{\rho} - \frac{c}{\rho}\right)$$

or $$-\frac{1}{2}\omega^2(x^2+y^2)+gz+\frac{p}{\rho}-\frac{c}{\rho} = \text{constant}$$

or $$gz+\frac{p}{\rho}-\frac{1}{2}\omega^2(x^2+y^2) = \text{constant.}$$

Thus the free surface where p is constant is a paraboloid of evolution with z-axis as its axis.

$$\text{Again the vorticity} = \frac{1}{2}\,\text{curl}\,\mathbf{q}$$

$$= \frac{1}{2}\nabla\times(w\times\mathbf{r})$$

$$= \omega. \qquad \text{[Refer § 6.15]}$$

6. *A body of liquid is revolving about a vertical axis with angular speed which is a function of the perpendicular distance, r, from the axis. Show that, if the motion is irrotational,*

$$\mathbf{q} = \frac{\mu}{r}\,\mathbf{b}, \quad \omega = \frac{\mu}{r^2}\,\mathbf{c},$$

where **a, b, c** *are the unit vectors in the directions of r, θ, z with cylindrical co-ordinates; the upwardly directed axis being the z-axis and w denotes angular velocity.*

The angular velocity w at any point r, θ, z is given by

$$\omega = f(r)\,\mathbf{c}.$$

Also the position vector of the point = $r\mathbf{a}$.

$$\therefore \quad \text{velocity} = f(r)\,\mathbf{c}\times r\mathbf{a} = rf(r)\,\mathbf{b}.$$

We shall now calculate

$$\nabla\times\mathbf{q}.$$

We have

$$\nabla\times\mathbf{q} = \nabla\times[rf(r)\mathbf{b}]$$

$$= \nabla[rf(r)]\times\mathbf{b}+rf(r)\,\nabla\times\mathbf{b},$$

$$\nabla[rf(r)] = [f(r)+rf'(r)]\mathbf{a},$$

$$\nabla\times\mathbf{b} = \frac{1}{r}\,\mathbf{c}.$$

$$\therefore \quad \nabla\times\mathbf{q} = [f(r)+rf'(r)]\mathbf{c}+f(r)\mathbf{c}$$

$$= [2f(r)+rf'(r)]\mathbf{c} = \frac{1}{r}\frac{d}{dr}[r^2f(r)]\mathbf{c}.$$

For irrotational motion, we have $\nabla\times\mathbf{q} = \mathbf{0}$ so that

$$r^2f(r) = \text{constant} = \mu, \text{ say.}$$

$$\therefore \quad \text{angular velocity, } \omega = f(r)\mathbf{c} = \frac{\mu}{r^2}\,\mathbf{c}$$

Also velocity, $\mathbf{q} = rf(r)\mathbf{b} = \dfrac{\mu}{r}\mathbf{b}.$

7. *Find the equation of continuity in spherical polar co-ordinates.*

Let **a, b, c** denote unit vectors tangent to the co-ordinate curves at any point.

Let $\mathbf{q} = u\mathbf{a} + v\mathbf{b} + w\mathbf{c}$,

so that, u, v, w, denote components of the velocity along **a, b, c** respectively.

The equation of continuity is

$$\frac{\partial \rho}{\partial t} + \nabla . (\rho \mathbf{q}) = 0.$$

For spherical polar co-ordinates $h_1 = 1,\ h_2 = r,\ h_3 = r\sin\theta$ [Refer § 6.23, p.205].

$$\therefore \quad \nabla . (\rho \mathbf{q}) = \nabla . (\rho u\mathbf{a} + \rho v\mathbf{b} + \rho w\mathbf{c})$$

$$= \frac{1}{h_1 h_2 h_3}\left[\frac{\partial(h_2 h_3 \rho u)}{\partial r} + \frac{\partial(h_3 h_1 \rho v)}{\partial \theta} + \frac{\partial(h_2 h_3 \rho \omega)}{\partial \phi}\right]$$

$$= \frac{1}{r^2 \sin\theta}\left[\frac{\partial(r^2 \sin\theta \rho u)}{\partial r} + \frac{\partial(r \sin \rho v\theta)}{\partial \theta} + \frac{\partial(r\rho\omega)}{\partial \phi}\right]$$

$$= \frac{1}{r^2 \sin\theta}\left[2r\sin\theta\rho u + r^2 \sin\theta\frac{\partial(u\rho)}{\partial\theta} + \right.$$

$$\left. r\cos\theta\rho v + r\sin\theta\frac{\partial(\rho v)}{\partial\theta} + r\frac{\partial(\rho\omega)}{\partial\theta}\right]$$

$$= \frac{\partial(\rho u)}{\partial r} + \frac{1}{r}\frac{\partial(\rho v)}{\partial\theta} + \frac{\operatorname{cosec}\theta}{r}\frac{\partial(\rho\omega)}{\partial\phi} + \frac{2\rho u}{r} + \frac{\cot\theta}{r}\rho v.$$

Hence the equation of continuity in spherical polar co-ordinates is

$$\frac{\partial\rho}{\partial t} + \frac{\partial(\rho u)}{\partial r} + \frac{1}{r}\frac{\partial(\rho v)}{\partial\theta} + \frac{\operatorname{cosec}\theta}{r}\frac{\partial(\rho\omega)}{\partial\phi} + \frac{2\rho u}{r} + \frac{\cot\theta}{r}\rho v = 0.$$

8. *If a moving fluid is bounded by a fixed surface* $\phi(\mathbf{r}) = 0$, *then the boundary condition is*

$$\mathbf{q} . \nabla\phi = 0,$$

and if the bounding surface is itself moving and given by the equation $\phi(\mathbf{r}, t) = 0$, *then the boundary condition is*

$$\frac{\partial\phi}{\partial t} + \mathbf{q} . \nabla\phi = 0.$$

The normal component of the velocity of the fluid particle at any point of the fixed bounding surface must be zero.

The normal at any point of the surface

$$\phi(\mathbf{r}) = 0$$

being parallel to the vector $\nabla\phi$, the required condition is

$$\mathbf{q} \cdot \nabla\phi = 0.$$

In the case of a moving bounding surface, the velocity of a fluid particle at any point of the boundary, relative to the point of the boundary must be tangential to the boundary. This means that the normal component of the velocity of the particle relative to the boundary is zero,

Let, **u**, be the velocity of a point of the boundary. Thus the boundary condition is

$$(\mathbf{q} - \mathbf{u}) \cdot \nabla\phi = 0. \qquad ...(1)$$

Also the point of the boundary moves so that the individual sate of change of ϕ is zero so that

$$\frac{\partial\phi}{\partial t} + \mathbf{u} \cdot \nabla\phi = 0. \qquad ...(2)$$

From (1) and (2), we have

$$\frac{\partial\phi}{\partial t} + \mathbf{q} \cdot \nabla\phi = 0,$$

as the required boundary condition.

9. *A body of liquid is moving irrotationally in a simply connected region bounded by fixed surface; show that the velocity potential ϕ is constant throughout the interior of the region.*

As the motion is irrotational curl $\mathbf{q} = \mathbf{0}$. Thus there exists a scalar point function, ϕ, called velocity potential such that

$$\mathbf{q} = -\nabla\phi.$$

By virtue of the equation of continuity

$$\nabla^2 \phi = 0$$

so that ϕ is harmonic in the region.

Also the normal component of the velocity at any point of the boundary is zero. Hence the result.

10. *In the steady motion of an incompressible homogeneous fluid under no force, the velocity at any point is*

$$ax\mathbf{i} + ay\mathbf{j} - 2ax\mathbf{k}.$$

Find the surfaces of equal pressure.

The equation of motion is

$$\frac{d\mathbf{q}}{dt} = \mathbf{F} - \frac{1}{\rho}\nabla p,$$

or

$$\frac{\partial\mathbf{q}}{\partial t} + \mathbf{q} \cdot \nabla\mathbf{q} = \mathbf{F} - \frac{1}{\rho}\nabla p.$$

We have $\mathbf{F} = \mathbf{0}$, $\dfrac{\partial\mathbf{q}}{\partial t} = 0$, $p = a$ constant,

$$\therefore \quad \mathbf{q} \cdot \nabla\mathbf{q} = \frac{1}{\rho}\nabla p.$$

Also $$\mathbf{q}\cdot\nabla\mathbf{q} = \left(ax\frac{\partial}{\partial x} + ay\frac{\partial}{dy} - 2az\frac{\partial}{\partial z}\right)(ax\mathbf{i} + ay\mathbf{j} - 2az\mathbf{k})$$
$$= a^2 x\mathbf{i} + a^2y\mathbf{j} + 4a^2 z\mathbf{k}.$$

$$\therefore \; a^2(x\mathbf{i} + y\mathbf{j} + 4z\mathbf{k}) = -\frac{1}{\rho}\nabla p,$$

or $$\nabla\rho = -a^2p(x\mathbf{i} + y\mathbf{j} + 4z\mathbf{k}).$$

$$\therefore \; p = -a_2\rho\left(\frac{1}{2}x^2 + \frac{1}{2}y^2 + 2z^2\right) + \text{constant},$$

$\therefore$ the surfaces of equal pressure are given by

$$x^2 + y^2 + 4z^2 = \text{constant},$$

which are spheroids.

The student may verify that div $\mathbf{q} = 0$.

11. *Show that the kinetic energy of a body of homogeneous liquid within a closed surface S moving irrotationally is*

$$\frac{1}{2}\rho\int_S \psi\nabla\psi\,.\,\mathbf{n}\,dS;$$

ψ being the velocity potential.

We have, by Gauss's Theorem,

$$\frac{1}{2}\rho\int_S \psi\nabla\psi\,.\,\mathbf{n}\,dS = \frac{1}{2}\rho\int_V \nabla.(\psi\nabla\psi)\,dv$$
$$= \frac{1}{2}\rho\int_V [\nabla\psi\,.\,\nabla\psi + \psi(\nabla\,.\,\nabla\psi)]\,dv$$
$$= \frac{1}{2}\rho\int_V (\nabla\psi)^2\,dv + \frac{1}{2}\rho\int_V \psi\nabla^2\psi\,dv$$
$$= \int_V \frac{1}{2}\rho\mathbf{q}^2\,dv, \text{ for } \nabla^2\psi = 0$$
$$= \text{kinetic energy.}$$

12. *If* $\mathbf{q}$ *and* $\mathbf{q}'$ *are the, velocities of some fluid particle relative to two frames of reference S, S′ moving in any manner, the angular velocity of S′ relative to S being ω, show that if the vorticities at P relative to the two frames are ξ and ξ′, then*

$$\xi = \xi' + \omega.$$

Show also that if c, c′ be the circulations in the same closed contour, Γ, relative to the frames S and S′, then

$$c = c' + 2|\omega|\,A,$$

where A is the area enclosed by the projection of Γ on plane perpendicular to ω.

We have $$\mathbf{q} = \mathbf{q}' + \omega\times\mathbf{r}$$ [Refer § 5.1.3, p. 149]

$$\therefore \; \text{curl}\,\mathbf{q} = \text{curl}\,\mathbf{q}' + \text{curl}\,(\omega\times\mathbf{r})$$

or $$2\xi = 2\xi' + 2\omega$$ [Refer § 6.15, p.184]

or $$\xi = \xi' + \omega.$$

Again $$c = \int_\Gamma \mathbf{q} \, . \, d\mathbf{r}, \; c' = \int_\Gamma \mathbf{q}'.d\mathbf{r}.$$

$$\therefore \quad c = \int_\Gamma (\mathbf{q}' + \omega \times \mathbf{r}) \, . \, d\mathbf{r}$$

$$= \int_\Gamma \mathbf{q}' \, . \, d\mathbf{r} + \int_\Gamma (\omega \times \mathbf{r}).d\mathbf{r}$$

$$= c' + \int_S \text{curl} \, (\omega \times \mathbf{r}) \, . \, d\mathbf{a},$$

S being any surface through Γ.

$$\therefore \quad c = c' + 2\int_S \omega \, . \, d\mathbf{a} = c' + 2|\omega| \, A.$$

EXERCISES

1. A fluid of constant density ρ is at rest under the action of conservative body forces derivable from a potential Ω. Show that the pressure p is given by

$$p + \rho\Omega = \text{constant}.$$

2. The vorticity of an incompressible fluid is constant throughout, prove that

$$\nabla^2 \mathbf{q} = \mathbf{0}.$$

3. The motion of an incompressible fluid under the action of conservative forces is steady; show that

$$(\mathbf{q} \, . \, \nabla)\omega = (\omega \, . \, \nabla)\mathbf{q}.$$

4. If c, is the circulation around any closed curve moving with the fluid, prove that

$$\frac{dc}{dt} = \int pd\left(\frac{1}{\rho}\right),$$

if the field is conservative and pressure p depends only on the density ρ.

5. If the motion of a fluid is referred to a moving frame of reference which rotates with angular velocity ω and has translational velocity $\mathbf{u}$, show that the equation of motion is

$$\mathbf{F} - \frac{1}{\rho} \mathbf{v} p = \frac{d\mathbf{u}}{dt} + \frac{d\omega}{dt} \times \mathbf{r} + \omega \times (\omega \times \mathbf{r}) + 2\omega \times \frac{D\mathbf{u}}{dt} + \frac{D^2\mathbf{u}}{dt^2},$$

and the equation of continuity is

$$\frac{\partial \rho}{\partial t} + \mathbf{v} \, . \left(\rho \frac{D\mathbf{r}}{dt}\right) = 0;$$

where $\mathbf{r}$, $D\mathbf{r}/dt$ and $D^2\mathbf{r}/dt^2$ denote the position vector, velocity and acceleration relative to the moving axes.

6. Obtain the equation of continuity in cylindring co-ordinates.

7. If the lines of motion are curves on the surfaces of cones having their vertices at the origin and Z-axis as the common axis, show that the equation of continuity is

$$\frac{\partial \rho}{\partial t} + \frac{\partial(\rho u)}{\partial r} + \frac{\operatorname{cosec}\theta}{r}\frac{\partial(\rho v)}{\partial \phi} + \frac{2\rho u}{r} = 0,$$

where u, v are the components of velocity in the direction of r increasing and ϕ increasing respectively.

8. The lines of motion for a liquid are curves on the surfaces of spheres all touching the OXY plane at the origin O; prove that the equation of continuity is

$$r \sin\theta \frac{\partial \rho}{\partial t} + \frac{\partial(\rho v)}{\partial r} + \sin\theta \frac{\partial(\rho u)}{\partial \theta} + \rho u\,(1 + 2\cos\theta) = 0.$$

9. A sphere of constant radius is moving in a fluid with constant velocity u; show that at any point, $\mathbf{r}$, of the sphere at time t, the velocity $\mathbf{q}$ of the fluid satisfies the condition

$$(\mathbf{q} - \mathbf{u}), (\mathbf{r} - ut) = 0,$$

the centre of the sphere being at the origin at time $t = 0$.

10. If $\phi\,(\mathbf{r}, t) = 0$ is the equation of a moving surface, show that the resolved part of the velocity of a fluid particle at any point of the surface along the normal to the surface is

$$\frac{\partial \rho}{\partial t} \Big/ |\nabla\phi|.$$

11. A liquid of constant density is in equilibrium and the action of an external force $\mathbf{F}$ given as follows: find the surfaces of equal pressure.

(*i*) $\mu yzi + \mu xx\mathbf{j} + \mu xy\mathbf{k}$.

(*ii*) $\mu(y + z)\mathbf{i} + \mu(z + x)\mathbf{j} + \mu(x + y)\mathbf{k}$.

$$\left[\text{Find } p \text{ from } \mathbf{F} = \frac{1}{\rho}\nabla p.\right]$$

12. A body of perfect fluid is rotating about a vertical axis with constant angular velocity in a uniform gravitational field and the density and pressure are connected by the adiabatic law. Obtain the equation which gives pressure.

13. A body of liquid is moving irrotationally in a simple connected region bounded by a closed surface such that the velocity potential at any point of the surface is constant; show that the velocity potential has the same value throughout the interior.

14. Prove that the motion of an incompressible perfect fluid is possible when the velocity $\mathbf{q}$ is given by

(*i*) $\mathbf{q} = x(x - z)\mathbf{i} + y(z - x)\mathbf{j} + z(x - y)\mathbf{k}$.

(*ii*) $\mathbf{q} = (z - y)\phi\mathbf{i} + (x - z)\phi\mathbf{j} + (y - x)\phi\mathbf{k}$.

where $$\phi = \frac{(x+y+z)}{(x^2+y^2+z^2)^2}.$$

(*iii*) $\mathbf{q} = -\dfrac{2y^2}{(x^2+y^2)}\mathbf{i} + \dfrac{(x+y)^2}{(x^2+y^2)^2}\mathbf{j} + \dfrac{2(xy+yz+zx)}{(x^2+y^2)^2}\mathbf{k}$.

(*iv*) $\mathbf{q} = (bz - cy)\mathbf{i} + (cx - az)\mathbf{j} + (ay - bx)\mathbf{k}$.

Examine if the motion in either case is irrotational or not. Show also that in case (*iv*) the fluid moves like a rigid body.

15. A closed surface S is drawn in an inviscid liquid in steady motion under no body force; show that

$$\int_S \{p\mathbf{n} + \rho\mathbf{q}(\mathbf{q.n})\}\, dS = 0.$$

16. Show that in irrotational motion, the acceleration of a fluid particle of velocity $\mathbf{q}$ is

$$\frac{\partial \mathbf{q}}{\partial t} + \frac{1}{2}\,\text{grad}\, \mathbf{q}^2.$$

OBJECTIVE QUESTIONS

For each of the following questions, four alternatives are given for the answer. Only one of them is correct. Choose the correct alternative.

1. For a perfect fluid shearing stress at every point is
(*a*) positive (*b*) negative (*c*) zero (*d*) infinite

2. The motion of a fluid is said to be steady, when
(*a*) Local time-rates of change are zero
(*b*) Individual time-rates of change are zero
(*c*) None of the two is zero
(*d*) Both (*a*) and (*b*) are true simultaneously

3. A flow is defined by $u = 2(1 + t)$, $v = 3(1 + t)$ where t is the time. The velocity at $t = 2$ is
(*a*) 6 (*b*) 9 (*c*) 10.82 (*d*) 6.7

4. The continuity equation is satisfied by
(*a*) $u = A \sin xy,\ v = -A \sin xy$
(*b*) $u = x + y,\ v = x - y$
(*c*) $u = 2x^2 + cy,\ v = 3y^2$
(*d*) $u = x + 2y,\ v = 2x + y$

5. If $\psi = 3x^2 y - y^3$, values of u and v are
(*a*) $6xy,\ 3x^2 - 3y^2$ (*b*) $3x^2 - 3y^2,\ 6xy$
(*c*) $3x^2 - 3y^2,\ -6xy$ (*d*) $3y^2 - 3x^2,\ 6xy$

6. Valid potential function is
(*a*) $\phi = c(x^2 + y^2)$ (*b*) $\phi = c \cos x$
(*c*) $\phi = 3xy$ (*d*) $\phi = c \log x$

7. The equation of a stream line in two dimensional steady flow can be expressed as:

(a) $\frac{dx}{u} = \frac{dy}{v}$ (b) $-\frac{dy}{u} = \frac{dx}{v}$

(c) $\frac{dy}{u} = \frac{dx}{v}$ (d) $\frac{u}{dx} = \frac{dy}{v}$

8. The continuity equation for incompressible two dimensional steady flow is

(a) $\frac{\partial u}{\partial x} + \frac{\partial v}{\partial t} = 0$ (b) $\frac{\partial u}{\partial x} + \frac{\partial v}{\partial y} = 0$

(c) $\frac{\partial u}{\partial y} + \frac{\partial v}{\partial t} = 0$ (d) $\frac{\partial u}{\partial y} + \frac{\partial v}{\partial x} = 0$

9. The following represent steady incompressible flow:

(a) $u = x^2 + y^2,\ v = -2xy + 7$ (b) $u = x/y,\ v = y/x$

(c) $u = 2x + y,\ v = 4y + x$ (d) $u = 4xy + 2y^2,\ v = 6xy + 3x^2$

10. The flow rate between stream lines with values ψ_1 and ψ_2 is given by

(a) $c\,\psi_1 + \psi_2$ (b) $\psi_1 + c\psi_2$ (c) $\psi_1 + \psi_2$ (d) $\psi_2 - \psi_1$

11. The velocity components in the x and y directions in terms of stream function ψ is given by

(a) $u = \frac{\partial \psi}{\partial y},\ v = -\frac{\partial \psi}{\partial x}$ (b) $u = \frac{\partial \psi}{\partial x},\ v = \frac{\partial \psi}{\partial y}$

(c) $u = -\frac{\partial \psi}{\partial y},\ v = \frac{\partial \psi}{\partial x}$ (d) $u = -\frac{\partial \psi}{\partial x},\ v = \frac{\partial \psi}{\partial y}$

12. The condition for irrotational flow is

(a) $\frac{\partial v}{\partial x} = -\frac{\partial v}{\partial y}$ (b) $\frac{\partial v}{\partial x} = \frac{\partial u}{\partial y}$

(c) $\frac{\partial u}{\partial x} = -\frac{\partial v}{\partial y}$ (d) $\frac{\partial u}{\partial x} = \frac{\partial v}{\partial y}$

13. The velocity components in the x and y directions in terms of potential function ϕ are given by

(a) $u = \frac{\partial \phi}{\partial x},\ v = \frac{\partial \phi}{\partial y}$ (b) $u = -\frac{\partial \phi}{\partial x},\ v = -\frac{\partial \phi}{\partial y}$

(c) $u = \frac{\partial \phi}{\partial y},\ v = -\frac{\partial \phi}{\partial x}$ (d) $u = -\frac{\partial \phi}{\partial y},\ v = \frac{\partial \phi}{\partial x}$

14. A path line describes

(a) The velocity direction at all points on the line

(b) The instantaneous position of all particles that have passed a point

(c) The path followed by particles in a flow

(*d*) The path over a period of times of a single particle that has passed out at a point

15. The relationship between stream and potential function ψ and ϕ is

(*a*) $\frac{\partial \phi}{\partial y} = \frac{\partial \phi}{\partial x}$ (*b*) $\frac{\partial \phi}{\partial x} = -\frac{\partial \psi}{\partial y}$

(*c*) $\frac{\partial \phi}{\partial y} = \frac{\partial \psi}{\partial x}$ (*d*) $\frac{\partial^2 \phi}{\partial x^2} = \frac{\partial^2 \psi}{\partial y^2}$

16. The continuity equation is the result of application of the following law to the flow field

(*a*) First law of thermodynamics

(*b*) Newton's second law of motion

(*c*) conservation of energy (*d*) Conservation of mass

17. The value that will satisfy potential function ϕ is

(*a*) $\phi = \sin x$ (*b*) $\phi = x + y$ (*c*) $\phi = x^2 + y^2$ (*d*) $\phi = \log(x + y)$

18. A potential function

(*a*) describes the flow if it is rotational

(*b*) describes the flow if it is irrotational

(*c*) is constant along a stream line

(*d*) is definable if a stream function is available for the flow

19. The stream function is

(*a*) constant along an equipotential line

(*b*) defined only in irrotational flow

(*c*) defined only for incompressible flow

(*d*) along a stream line

20. Bernoulli equation is applicable for

(*a*) steady rotational flow

(*b*) steady rotational compressible flow

(*c*) steady irrotatational incompressible flow

(*d*) unsteady irrotational incompressible flow

ANSWERS

1. (*c*)	**2.** (*a*)	**3.** (*c*)	**4.** (*b*)	**5.** (*c*)
6. (*c*)	**7.** (*a*)	**8.** (*b*)	**9.** (*a*)	**10.** (*d*)
11. (*a*)	**12.** (*b*)	**13.** (*b*)	**14.** (*d*)	**15.** (*c*)
16. (*d*)	**17.** (*b*)	**18.** (*b*)	**19.** (*d*)	**20.** (*c*)

9

Electromagnetic Theory

9.1. ELECTROSTATICS. THE INVERSE SQUARE LAW.

Electrostatics is concerned with electricity at rest. This electricity may be in the form of discrete point charges or in the form of a continuous charge distribution with volume density ρ or surface density σ or even in the form of electric dipoles, where a dipole may be regarded as the limiting case of two large equal and opposite point charges at a small distance apart [*cf* § 9.6]. Volume and surface distributions of dipoles also play an important part in the theory.

The study of Electrostatics is based primarily on Columb's Inverse Square Law *viz.*, "*If two electric charges e_1, e_2 are placed in vacuo at a distance r apart, the force between them is proportional to*

$$\frac{e_1 e_2}{r^2}.$$

Further if the charges are of the same sign, the force is repulsive if they are of opposite signs, the force as attractive."

Thus the force **F** on a charge e_2 at Q due to another charge e_1 at P is given by

$$\mathbf{F} = c\,\frac{e_1 e_2}{r^2}\,\frac{\mathbf{r}}{r} = c\,\frac{e_1 e_2}{r^3}\,\mathbf{r},$$

where c is a constant and **r** is the position-vector of Q relative to P. In C.G.S. electrostatic unit system, we define a unit charge as one which, in vacuo, repels an exactly similar charge at a distance of one cm. with a force of one dyne. In this system $c = 1$, and

$$\mathbf{F} = \frac{e_1 e_2}{r^3}\mathbf{r}, \qquad ...(1)$$

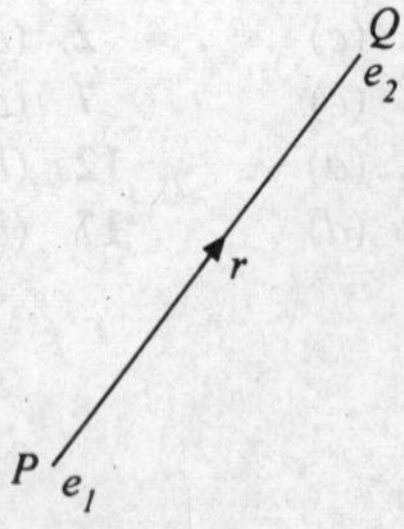

Fig. 9.1

This gives both magnitude and direction of the force; for if e_1, e_2 are of like signs, the force on Q is repulsive; if of opposite signs, the force is attractive.

We define the *electrostatic intentsity* or the *electrostatic field vector* at a point as the force that would act in vacuo on a unit charge placed at that point, due to the rest of the charges, provided the unit test charge is assumed not to disturb the field due to the other charges. This intensity has at each point a definite magnitude and a definite direction and is thus a vector. We denote it by **E**. Thus for n charges

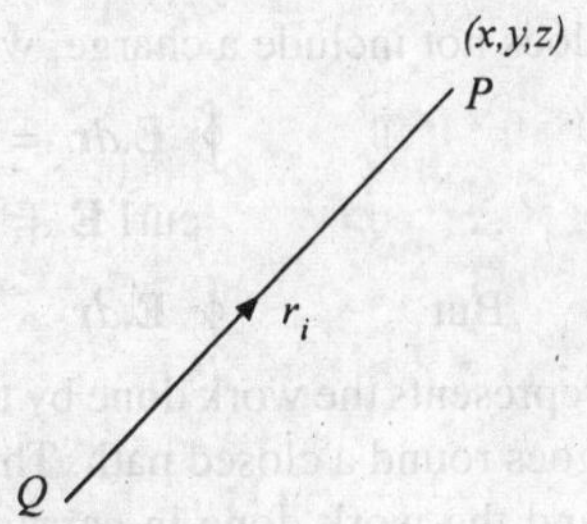

Fig. 9.2

$e_1, e_2, \ldots, e_n$,

$$\mathbf{E} = \sum_{i=1}^{n} \frac{e_i}{r_i^3}\,\mathbf{r}_i, \qquad \ldots(2)$$

where $\mathbf{r}_i$ denotes the position vector of P relative to the ith charge e_i.

Now $$r_i^2 = (x-\xi_i)^2 + (y-\nu_i)^2 + (z-\zeta_i)^2$$

$$\therefore \quad \nabla\left(\frac{1}{r_i}\right) = -\frac{1}{r_i^2}\left(\frac{x-\xi_i}{r_i}\mathbf{i} + \frac{y-\eta_i}{r_i}\mathbf{j} + \frac{z-\zeta_i}{r_i}\mathbf{k}\right)$$

$$= -\frac{1}{r_i^3}\,\mathbf{r}_i.$$

If we define $$\phi = \sum_{i=1}^{n} \frac{e_i}{r_i}, \qquad \ldots(3)$$

then $$\nabla\phi = \sum_{i=1}^{n} \nabla\left(\frac{e_i}{r_i}\right) = -\sum_{i=1}^{n} \frac{e_i}{r_i^3}\,\mathbf{r}_i = -\mathbf{E}$$

Thus $$\mathbf{E} = -\nabla\phi = -\text{grad}\,\phi. \qquad \ldots(4)$$

Further from (2)

$$\text{div}\,\mathbf{E} = \nabla\cdot\left[\sum_{i=1}^{n} \frac{e_i}{r_i^3}\,\mathbf{r}_i\right]$$

$$= \sum_{i=1}^{n} e_i\,\nabla\cdot\left(\frac{\mathbf{r}_i}{r_i^3}\right) = 0 \qquad \ldots(5)$$

$\therefore$ **E** *is a solenoidal vector.*

Also $$\text{curl}\,\mathbf{E} = -\Delta\times(\Delta\phi) = \mathbf{0}. \qquad \ldots(6)$$

$\therefore$ **E** *is also an irrotational vector.*

Combining (4) and (5), we have

$$\nabla . (\nabla \phi) = 0$$

or $$\nabla^2 \phi = 0 \quad ...(7)$$

Therefore, at a point not occupied by charges, ϕ satisfies Laplace's equation (7).

We can verify (6) in an alternative manner. For any closed path, that does not include a charge, we have

$$\oint \mathbf{E}.d\mathbf{r} = -\oint \nabla\phi . d\mathbf{r} = -\oint d\phi = 0$$

$$\therefore \quad \text{curl } \mathbf{E} = 0 \quad ...(8)$$

But $\oint \mathbf{E}.d\mathbf{r}$

represents the work done by the electric intensity, when a unit positive charge goes round a closed path. Thus electrostatic intensity is a conservative force and the work done in carrying a unit positive charge from one position to another does not depend on the path, but only on the position-vector of the end point.

Now

$$\int_{P}^{\infty} \mathbf{E}.d\mathbf{r} = -\int_{P}^{\infty} d\phi = \phi(P) - \phi(\infty) = \phi(P),$$

as from (3)

$$\phi(\infty) = 0$$

Thus ϕ can now be interpreted as the work done by the electrostatic intensity in carrying unit positive charge from P to infinity along any path whatsoever or as the work done against the electrostatic *force* in bringing the unit positive charge from infinity to P.

9.2. SOME FUNDAMENTAL DEFINITIONS

Definition 1. The scalar point function ϕ defined in (3) is called the *electrostatic potential* of the field due to the given system of charges. The surfaces

$$\phi(x, y, z) = \text{constant}$$

for various values of the constant, constitute a family of *equipotential surfaces*. The vector equation of the family is

$$d\phi = d\mathbf{r} . \nabla\phi = 0$$

or $$d\mathbf{r} . \mathbf{E} = 0 \quad ...(1)$$

which shows that no work is done in going from one point to another along an equipotential surface.

Illustrations. For a single point charge, the equipotential surfaces are concentric spheres. For two point charges with opposite signs, the equipotential surface $\phi = 0$ will, in general be a sphere.

Definition 2. A *line of force* is a curve such that the tangent to it at any point gives the direction of **E** at that point.

The vector differential equation of the lines of force is

$$d\mathbf{r} \times \mathbf{E} = 0 \quad ...(2)$$

Through every point in space, in general, only one line of force will pass, but when $\mathbf{E} = \mathbf{0}$, the direction of the line of force becomes indeterminate. Such a point is called a *Neutral point* or a *Point of equilibrium.*

Illustration. The lines of force due to a single point charge are concurrent straight lines passing through the charge. The middle point of the line joining two equal point charges is a neutral point.

Also from (1) and (2), we deduce that at all points of intersection except when $\mathbf{E} = \mathbf{0}$, the lines of force and equipotential surfaces intersect at right angles.

Definition 3. The *Flux* of a vector **E** across the surface element $d\mathbf{a}$ is defined as the scalar $\mathbf{E} \cdot d\mathbf{a}$. If S is any surface, closed or open, the flux of **E** *across* **S** is

$$\int_S \mathbf{E} \cdot d\mathbf{a}$$

If the surface is closed, the flux of **E** *out of* S is

$$\int_S \mathbf{E} \cdot d\mathbf{a}$$

where the direction of $d\mathbf{a}$ is that of the outward drawn normal.

Illustration. The flux of **E** out of the surface of a sphere of radius a, at the centre of which is placed a charge e, is

$$\int_S \mathbf{E} \cdot d\mathbf{a} = \int_S \frac{e}{a^2} \, dS = \frac{e}{a^2} \, 4\pi a^2 = 4\pi e$$

Definition 4. The *solid angle* $d\omega$ subtended at a point O by a surface element of area dS is the area cut off on the unit sphere with centre O, by a cone whose vertex is O and whose base is the surface element.

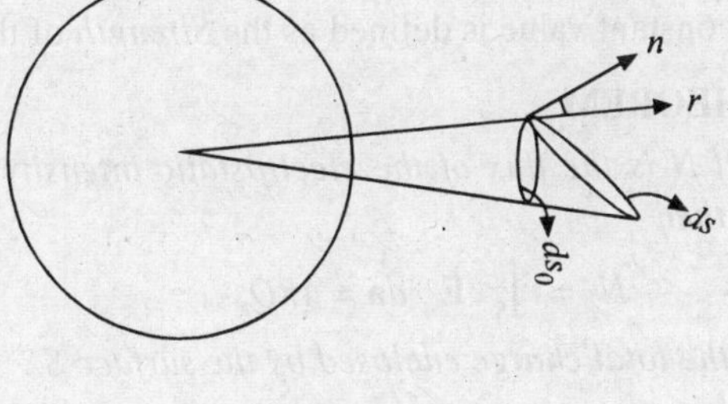

Fig. 9.3

From the figure.

$$\begin{aligned} \mathbf{r}.d\mathbf{a} &= \mathbf{r}.\mathbf{n}dS \\ &= rdS \cos\theta \\ &= rdS_0 = r\frac{dS_0}{d\omega}\, d\omega = r(r^2)d\,\omega = r^3 d\omega \end{aligned}$$

$$\therefore \quad d\omega = \frac{\mathbf{r}.d\mathbf{a}}{r^3}$$

The total solid angle subtended by a surface S at a point 0 is, therefore, given by

$$\omega = \int_S \frac{\mathbf{r}.d\mathbf{a}}{r^3} \qquad \text{...(3)}$$

Illustrations:

(*a*) The total solid angle subtended by a closed surface is

(*i*) at an internal point : the whole area of the unit sphere = 4π. ...(4*a*)

(*ii*) at a point on the surface : surface area of half the unit sphere = 2π. ...(4*b*)

(*iii*) at an external point : solid angle subtended is zero. ...(4*c*)

(*b*) In spherical polar co-ordiantes, $d\omega$ is the element of area on a sphere of radius unity. Thus $d\omega = \sin\theta \, d\theta \, d\phi$

(*c*) The solid angle subtended by a circular disc at a point on its axis

$$= \int_{0=0}^{\alpha} \int_{\phi=0}^{2\pi} \sin\theta \, d\theta \, d\phi$$

$$= 2\pi(1 - \cos\alpha), \qquad \text{...(5)}$$

and that subtended by a lune on a sphere of dihedral angle α at its centre

$$= \int_{0=0}^{\pi} \int_{\phi=0}^{\alpha} \sin\theta \, d\theta \, d\phi = 2\alpha \qquad \text{...(6)}$$

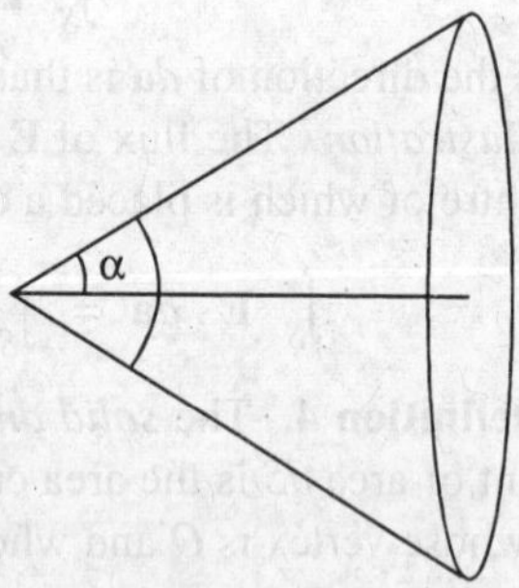

Fig. 9.4

Definition 5. The lines of force passing through all the points of a closed contour C, form a tubular surface called a *Tube of Force*. It will be shown in § 104 that the flux across any section of a tube of force is constant. This constant value is defined as the *Strength* of the tube.

9.3. GAUSS'S THEOREM

Statement. If N is the flux of the electrostatic intensity **E** *out of any closed surface S, then*

$$N = \int_S \mathbf{E} \,.\, d\mathbf{a} = 4\pi Q, \qquad \text{...(1)}$$

where Q denotes the total charge enclosed by the surface S.

Proof :

$$\int_S \mathbf{E} \,.\, d\mathbf{a} = \int_S \left[\sum_{i=1}^{n} \frac{e_i \mathbf{r}_i}{r_i^3} \right] . \, d\mathbf{a}$$

$$= \sum_{i=1}^{n} e_i \int_S \frac{\mathbf{r}_i}{r_i^3} \,.\, d\mathbf{a}$$

$$= \sum_{i=1}^{n} e_i \int d\omega_i = \sum_{i=1}^{n} e_i \omega_i,$$

where ω_i is the solid angle subtended by the closed surface at the ith charge. Now from (4) of § 9.2, $\omega_i = 4\pi$ or 0 according as e_i is inside or outside the surface.

Hence the theorem follows.

Note **1.** For a point charge, if any, on the surface S itself, $\omega_i = 2\pi$, and in this case, the theorem becomes

$$N = 4\pi Q + 2\pi Q', \qquad ...(2)$$

where Q' denotes the total charge on the surface.

Note **2.** We have proved Gauss's theorem on the hypothesis that all the charges are discrete. But if the charge distribution is continuous the contribution, to N by an element dv of charge is $4\pi\ \rho dv$ or zero according as the charge is inside or outside S. Integration through all the space occupied by the charge distribution proves the theorem.

9.4. APPLICATIONS OF GAUSS'S THEOREM

9.4.1. Poisson's and Laplace's equations

If there is a volume distribution of density ρ, the theorem becomes

$$\int_S \mathbf{E}.d\mathbf{a} = 4\pi \int_V \rho dv$$

Applying divergence theorem,

$$\int_V div\, \mathbf{E}\, dv = 4\pi \int_V \rho dv$$

or

$$\int_V (div\, \mathbf{E} - 4\pi\rho)\, dv = 0$$

Since this is applicable to all volumes however small,

$$div\, \mathbf{E} = 4\pi\rho \qquad ...(1)$$

or

$$\nabla.\mathbf{E} = 4\pi\rho$$

or

$$\nabla.(\nabla\phi) = -4\pi\rho \qquad ...(2)$$

or

$$\nabla^2\phi = -4\pi\rho$$

(2) is known as Poisson's equation. In free space, where $\rho = 0$, (2) reduces to the Laplace's equation

$$\nabla^2\phi = 0 \qquad ...(3)$$

9.4.2. Theorem

The potential cannot have a maximum or minimum at a point unoccupied by charge.

For, if ϕ is maximum at a point P, the potential decreases in all directions from P and, therefore, the Flux out of a small sphere with P as centre must be positive. By Gauss's theorem, the total charge inside the sphere must be positive, however small the radius of the sphere must be positive, however small the radius of the sphere may be. Similarly if ϕ is minimum at P, the charge or the charge density at P must be negative.

Further as $\quad \mathbf{E} = -\text{grad}\ \phi,$

$$|\mathbf{E}| = E = -\frac{\partial\phi}{\partial s},$$

where s is the length of the arc of the line of force measured from a fixed point. As $E \geq 0$, ϕ decreases as s increases. Thus *lines of force go from higher potentials to lower potentials* and from the above theorems, we deduce that *lines of force can begin only on positive charges and end on negative charges; they may go to infinity from a positive charge and come from infinity and end on a negative charge.*

9.4.3. Earnshaw's Theorem

A freely movable charge cannot be in stable equilibrium at a point unoccupied by charge; for at this point potential and, therefore, potential energy cannot have a maximum or a minimum; while for stable equilibrium, the potential energy must be minimum.

9.4.4. Theorem

The strength of a tube of force (or *the flux across any section of it*) *remains constant along its path.*

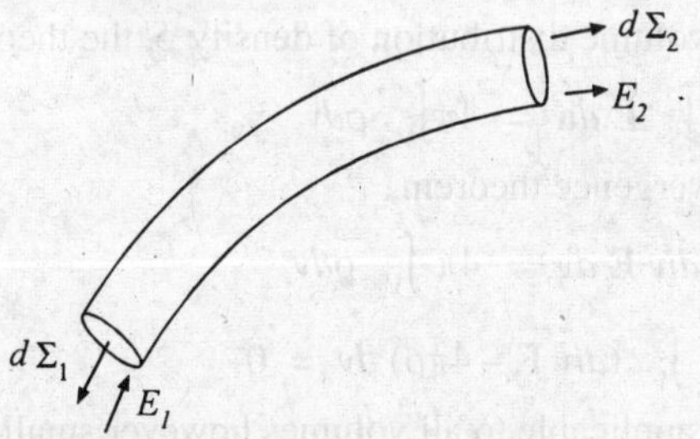

Fig. 9.5

As there is no flux out of a tube of force, applying Gauss's theorem to the closed surface shown in the figure.

$$\mathbf{E}_1 \cdot d\mathbf{a} + \mathbf{E}_2 \cdot d\mathbf{a}_2 = 0$$

or

$$E_1 dS_1 = E_2 dS_2$$

9.4.5. *The equations of lines of force due to a system of collinear point charges.*

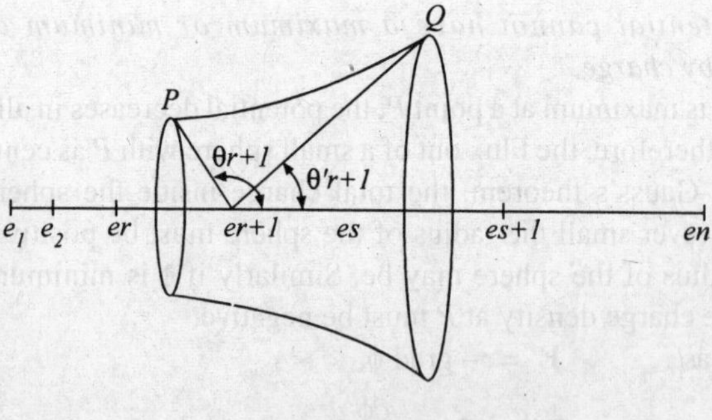

Fig. 9.6

Let PQ be a portion of a line of force. The field has cylindrical symmetry about the line of charges. Applying Gauss's theorem to the closed surface of revolution obtained by rotating PQ about the line of charges, we find that the flux out of the curved surface is zero. Therefore, the total flux out of the two circular plane sections should be equal to 4π times the total charge inside the surface. Using (5) of § 9.2, we get

$$\begin{aligned}
&[-2\pi(1-\cos\theta_1)e_1 - 2\pi(1-\cos\theta_2)\,e_2 \ldots\ldots\ldots\ldots\ldots\ldots - 2\pi(1-\cos\theta_r)e_r]\\
&+ [2\pi(1+\cos\theta_{r+1})\,e_{r+1} + \ldots\ldots\ldots\ldots\ldots\ldots + 2\pi(1+\cos\theta_s)e_s]\\
&+ [2\pi(1+\cos\theta_{s+1})\,e_{s+1} + \ldots\ldots\ldots\ldots\ldots\ldots + 2\pi(1+\cos\theta_n)e_n]\\
&+ [2\pi(1-\cos\theta_1')\,e_1 + 2\pi(1-\cos\theta_2')\,e_2 \ldots\ldots\ldots\ldots + 2\pi(1-\cos\theta_r')e_r]\\
&+ [2\pi(1-\cos\theta_{r+1})\,e_{r+1} + \ldots\ldots\ldots\ldots\ldots\ldots + 2\pi(1-\cos\theta_s')e_s]\\
&+ [-2\pi(1+\cos\theta_{s+1}')\,e_{s+1} - \ldots\ldots\ldots\ldots\ldots\ldots - 2\pi(1+\cos\theta_n')e_n]\\
&= 4\pi\,(e_{r+1} + e_{r+2}) + \ldots\ldots\ldots\ldots\ldots\ldots\ldots\ldots + e_n]
\end{aligned}$$

or
$$\begin{aligned}
e_1\cos\theta_1 + e_2\cos\theta_2 + \ldots\ldots\ldots\ldots &+ e_n\cos\theta_n\\
= e_1\cos\theta_1' + e_2\cos\theta_2' + \ldots\ldots\ldots &+ e_n\cos\theta_n'
\end{aligned}$$

$\therefore$ equation to a line of force is

$$\sum_{i=1}^{n} e_i\cos\theta_i = \text{constant.} \qquad \ldots(4)$$

9.4.6. Conductors

A conductor is a substance which offers no resistance to the motion of charges. In any static situation, *there can be no field inside a conductor*; for, otherwise, the 'free' electrons would move and the field would no longer be static. *The charge on a conductor must reside on its surface* : for from (1) of § 9.4, since $\mathbf{E} = \mathbf{0}$ at all internal points, ρ is also zero at every such point. *The surface of a conductor must be an equipotential,* for any difference in potential would cause the charge to flow and the field would again be static no longer. *The field is everywhere normal to the conductor*; for $\mathbf{E} = -\nabla\phi$ is normal to the equipotential

$$\phi(x, y, z) = \text{constant.}$$

We can easily obtain $\mathbf{E}$ at the surface of a conductor in terms of the surface density σ by applying Gauss's theorem *to the closed cylindrical surface* shown in the figure with its generators normal to the surface. At all points of the curved surface, $\mathbf{E}.d\mathbf{a} = 0$ as $\mathbf{E}$ and $d\mathbf{a}$ are at right angles. At all points inside the conductor $\mathbf{E} = \mathbf{0}$. Thus by Gauss's theorem,

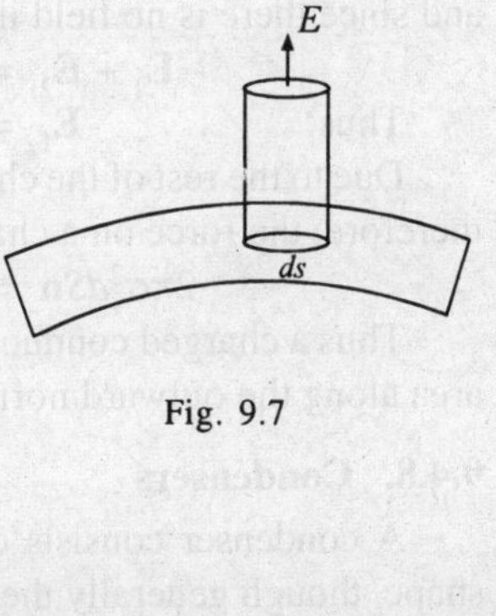

Fig. 9.7

$$\mathbf{E}.d\mathbf{a} = 4\pi\sigma dS$$

or
$$EdS = 4\pi\sigma dS$$

or $\qquad E = 4\pi\sigma$

or $\qquad \mathbf{E} = 4\pi\,\sigma\,\mathbf{n},$ ...(5)

where **n** is the unit normal to the surface. This is called *Coloumb's Law* and gives the field at a point just outside the surface of the conductor.

The *capacity* of a conductor is defined as the charge required to raise the same to unit potential, when it is alone in the field. Thus let a conducting sphere of radius a have charge Q. Applying Gauss's theorem to an imaginary concentric sphere of radius r $(r > a)$, we have by spherical symmetry,

$$\mathbf{E}\,4\pi r^2 = 4\pi Q$$

or $$E = \frac{Q}{r^2},$$

or $$\mathbf{E} = \frac{Q}{r^2}\mathbf{r}$$

$\therefore$ $$\phi(P) = \int_P^{\infty} \frac{Q}{r^3}\mathbf{r}.d\mathbf{r} = \int_a^{\infty} \frac{Q}{r^3}\,r\,dr = \left[\frac{Q}{r}\right]_a^{\infty} = \frac{Q}{a}$$

$\therefore$ capacity $= \dfrac{Q}{Q/a} = a =$ radius of the sphere. ...(6)

9.4.7. Mechanical force on a Conductor

Consider a small area dS on the surface of a conductor. The charge on its is σdS. The field at a point just outside is

(*i*) E_1 due to the charge σdS.

(*ii*) E_2 due to the charge on the rest of the surface of the conductor.

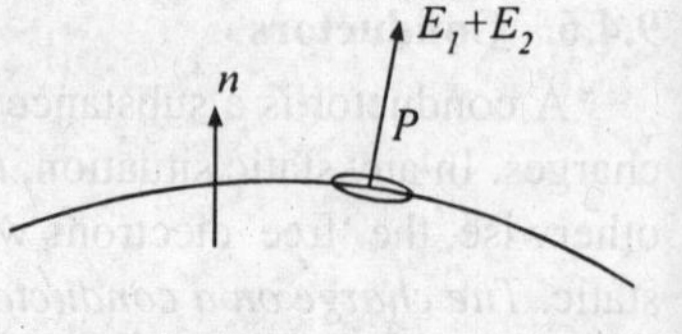

Fig. 9.8

The field at a point just inside is

(*i*) $-\mathbf{E}_1$ due to the charge σdS,

(*ii*) $\mathbf{E}_2$ due to the rest of the surface of the conductor.

Both $\mathbf{E}_1$ and $\mathbf{E}_1 + \mathbf{E}_2$ are normal to the surface and, therefore, $\mathbf{E}_2$ is also normal to the surface. From (5)

$$\mathbf{E}_1 + \mathbf{E}_2 = 4\pi\sigma\,\mathbf{n},$$

and since there is no field inside the conductor

$$-\mathbf{E}_1 + \mathbf{E}_2 = \mathbf{0}$$

Thus $$\mathbf{E}_1 = 2\pi\sigma\mathbf{n},\quad \mathbf{E}_2 = 2\pi\rho\mathbf{n},$$ ...(7)

Due to the rest of the charge, the force on a unit charge at P is $2\pi\sigma\mathbf{n}$, and, therefore, the force on a charge σdS at P is

$$2\pi\sigma^2 dS\mathbf{n} = 2\pi\sigma^2 d\mathbf{a}$$ ...(8)

Thus a charged conductor experiences a mechanical force $2\pi\sigma^2$ per unit area along the outward normal to the surface.

9.4.8. Condensers

A condenser consists of two plates which are conductors of arbitrary shape, though generally the surfaces of the two conductors are parallel. One conductor carries charge + Q and the other – Q and if ϕ_1, ϕ_2 are the potentials of the two plates, the capacity of the condenser is defined as

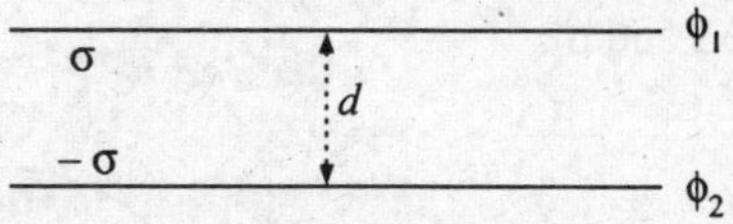

$$Q/(\phi_1 - \phi_2).$$

Fig. 9.9

Illustration. A *parallel-plate condenser* consists of two plane parallel plates, each of area A, at a distance d apart. Neglecting external fields and the irregularities due to edges, we have in the space between the plates, div $\mathbf{E}$ = 0.

$\therefore$ E = constant and by Coloumb's Law = $4\pi\sigma$

Now $\phi_1 - \phi_2$ = work done on a unit charge as it moves from one plate to the other

$$= 4\pi\sigma d.$$

$$\therefore \qquad C = \frac{A\sigma}{4\pi\sigma d} = \frac{A}{4\pi d}. \qquad ...(9)$$

Illustration. A *spherical condenser* consists of two concentric spheres of radii a and b. Applying Gauss's Theorem to imaginary spherical surface of radius r ($a < r < b$), we have, by spherical symmetry.

$$E 4\pi r^2 = 4\pi Q \quad \text{or} \quad E = \frac{Q}{r^2}$$

$$\therefore \qquad \phi_1 - \phi_2 = \int_a^b \frac{Q}{r^2}\, dr = Q\left(\frac{1}{a} - \frac{1}{b}\right)$$

$$\therefore \qquad \text{capacity} = \frac{ab}{b-a}. \qquad ...(10)$$

9.5. GREEN'S THEOREM AND ITS APPLICATIONS

In Divergence theorem,

$$\int_S \mathbf{E}.d\mathbf{a} = \int_V \text{div}\, \mathbf{E}\, dv,$$

by putting

$$\mathbf{E} = \phi_1 \nabla\phi_2 - \phi_2 \nabla \phi_1,$$

$$div\ \mathbf{E} = \phi_1 \nabla^2\phi_2 - \phi_2\nabla^2\phi_2,$$

we get the *Green's Theorem, viz.*,

$$\int_S (\phi_1\nabla \phi_2 - \phi_2\nabla\phi_1) \,.\, d\mathbf{a} = \int_V (\phi_1\nabla^2\phi_2 - \phi_2\nabla^2\phi_1)\, dv \qquad ...(1)$$

9.5.1. Integration of Poisson's equation

Let ϕ be the electrostatic potential in the space V enclosed by a surface S. In (1), put $\phi_1 = \phi_2$, $\phi_3 = \frac{1}{r}$ where r is the distance measured from a point P inside S. This introduces a singularity in ϕ_2 at P, to avoid which we enclose P in a sphere S' of radius ε and apply Green's theorem to the space V bounded by S and S'.

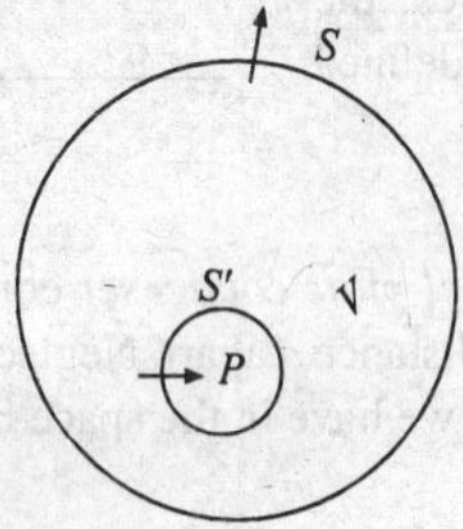

Fig. 9.10

$$\int_S \left[\phi\nabla\left(\frac{1}{r}\right)-\frac{1}{r}\nabla\phi\right].\,d\mathbf{a}+\int_{S'}\left[\phi\nabla\left(\frac{1}{r}\right)-\frac{1}{r}\nabla\phi\right].\,d\mathbf{a}$$

$$= \int_V \left[\phi\nabla^2\left(\frac{1}{r}\right)-\frac{1}{r}\nabla^2\phi\right]dv = 4\pi\int_V \frac{\rho}{r}\,dv, \qquad ...(2)$$

as $$\nabla^2\left(\frac{1}{r}\right) = 0 \text{ and } \nabla^2\phi = -4\pi\rho \text{ in } V$$

Let us evaluate the limits of the two terms in the integral over S' as $\varepsilon \to 0$.

$$\int_{S'} \phi\nabla\left(\frac{1}{r}\right).\,d\mathbf{a} = \int_{S'} \phi\left(-\frac{\mathbf{r}}{r^3}\right).\,d\mathbf{a}$$

$$= \int_{S'} \phi\left(-\frac{\mathbf{r}}{r^3}\right).\left(-r^2 d\omega\,\frac{\mathbf{r}}{r}\right)$$

$$= \int_{S'} \phi\, d\omega \to 4\pi\phi(P) \text{ as } \varepsilon \to 0.$$

$$-\int_{S'} \frac{1}{r}\nabla\phi\, d.\mathbf{a} = -\int_{S'} \frac{1}{r}\nabla\phi.\left(-r^2 d\omega\,\frac{\mathbf{r}}{r}\right)$$

$$= \int_S \mathbf{r}.\nabla\phi\, d\omega$$

$$= \varepsilon\int \mathbf{n}.\nabla\phi\, d\omega \to 0 \text{ as } \varepsilon \to 0$$

Thus, in the limit, (2) gives

$$\phi(P) = 4\pi\int_V \frac{\rho}{r}dv + \frac{1}{4\pi}\int_S\left[\frac{1}{r}\nabla\phi-\phi\nabla\left(\frac{1}{r}\right)\right].\,d\mathbf{a} \qquad ...(3)$$

(3) May be regarded as a formal solution of Poisson's equation. If further we assume the charge distribution to be bounded by a finite sphere,

then, for large distances ϕ is of the order of $\frac{1}{r}$ and $|\nabla\phi|$ is of the order of $\frac{1}{r^2}$ and $d\mathbf{a} = r^2 d\omega\mathbf{n}$. Thus from consideration of orders of magnitude, if S is taken to be the sphere at infinity, we find that the surface integral in (3) vanishes, and we get

$$\phi(P) = \int \frac{\rho}{r}\, dv \qquad \text{...(4)}$$

where the integration extends over all space.

9.5.2. Integration of Laplace's equation

Putting $\rho = 0$ in (3), we get a formal solution of Laplace's equation as

$$\phi(P) = \frac{1}{4\pi}\int_S \left[\frac{1}{r}\nabla\phi - \phi\nabla\left(\frac{1}{r}\right)\right] . d\mathbf{a} \qquad \text{...(5)}$$

This important result shows that *the value at a point P of a function ϕ which satisfies $\nabla^2\phi = 0$ everywhere within a closed surface S enclosing P, may be expressed in terms of the values of ϕ and of $\nabla\phi$ on the surface S.*

9.5.3. Integration of Poisson's equations when the field contains a number of charged conductors.

In (3), let V be the space enclosed by the sphere at infinity S and a system of charged conductors $S_1, S_2, \ldots S_n$ with surface density distribution σ.

As above, the surface integral over the sphere at infinity vanishes and over the ith conductor S_i.

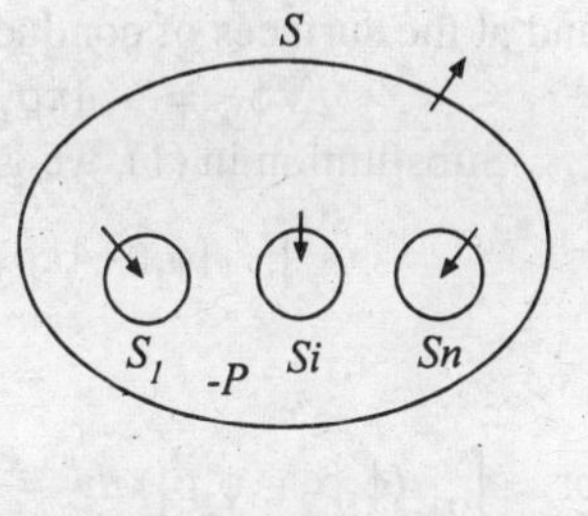

Fig. 9.11

$$\int_{S_i} \phi\nabla\left(\frac{1}{r}\right) . d\mathbf{a} = \int_{S_i} \phi\left(-\frac{1}{r^3}\mathbf{r}\right) d\mathbf{a} = -\phi_i \int_{S_i} d\omega = \mathbf{0}$$

Since ϕ is constant over the ith conductor and the solid angle subtended by a closed surface at an external point is zero,

∴ (3) becomes

$$\phi(P) = \int_V \frac{\rho}{r}\, dv + \frac{1}{4\pi}\sum_i \int_{S_i} \frac{1}{r}\nabla\phi . d\mathbf{a}$$

Remembering that $d\mathbf{a}$ is on the surface of each conductor directed inwards, we can write the general solution as

$$\phi(P) = \int_V \frac{\rho}{r}\, dv - \frac{1}{4\pi}\int \frac{1}{r}\nabla\phi . \mathbf{n}dS, \qquad \text{...(6)}$$

where the surface integral is carried over the surfaces of all the conductors and the flux of $\frac{1}{r}\ \nabla\phi$ is taken out of a conductor.

9.5.4. Proof of Coloumb's Law

Comparing (6) with the usual expression for the potential

$$\phi(P) = \int \frac{\rho}{r}\, dv + \int \frac{\sigma d S}{r} \qquad ...(6)$$

we get $\qquad \nabla\phi, \mathbf{n} = -4\pi\sigma, \qquad ...(8)$

or $\qquad \nabla\phi = -4\pi\sigma\mathbf{n},$

which gives Coloumb's Law

$$\mathbf{E} = 4\pi\sigma\mathbf{n} \qquad ...(9)$$

Alternatively assuming Coloumb's Law or (8), we can deduce (7) from (6).

9.5.5. Green's Reciprocal Theorem

We are given a fixed system of conductors. The potential function is ϕ_1, volume distribution is ρ_1 and the surface distribution is σ_1. A new distribution of charges in the same system leads to a potential function ϕ_2, volume density ρ_2 and surface density σ_2.

Then $\qquad \nabla^2\phi_1 = -4\pi\rho_1, \quad \nabla^2\phi_2 = -4\pi\rho_2,$

and at the surfaces of conductors

$$\nabla\phi_1 = -4\pi\rho_1\mathbf{n}, \quad \nabla\phi_2 = -4\pi\sigma_2\mathbf{n}$$

Substitution in (1), we get

$$\int_V [\phi_1(-4\pi\rho_2) - \phi_2(-4\pi\rho_1)\, dv$$

$$= \int_S [\phi_1(-4\pi\sigma_2\mathbf{n}) - \phi_2(-4\pi\sigma_1\, n)]\,.\, d\mathbf{a}$$

or $\qquad \int_V (\phi_1\rho_2 - \phi_2\rho_1)\, dv = -\int_S (\phi_1\sigma_2 - \sigma_1\phi_2)\, dS,$

negative sign occurring in the R.H.S. as **n**.d**a** is negative; **n** being directed outwards and d**a** being directed inwards

$$\therefore \int_V \phi_2\rho_2\, dv + \int_S \phi_1\sigma_2\, dS = \int_V \phi_2\rho_1\, dv + \int_S \phi_2\sigma_1\, dS \qquad ...(10)$$

which is Green's Reciprocal Theorem. It states that the potential ϕ_1 of a given distribution when multiplied by densities ρ_2, σ_2 of another distribution and summed over all space is equal to the potential of ϕ_2 the second distribution multiplied by densities ρ_1, σ_1 of the first distribution and summed over all space.

In particular let $\rho_1 = 0$, $\rho_2 = 0$. Since potentials are constant over the surfaces of conductors, (10) becomes

$$\phi_1 \int \sigma_2\, dS = \phi_2 \int \sigma_1\, dS,$$

i.e., if in one distribution the charges on the conductors are $e_1, e_2, ..., e_n$ and potentials are $\phi_1\ \phi_2, ... \phi_n$ and in the second distribution, the charges are e'_1,

$e'_2, ..., e'_n$ and potentials are $\phi_1', \phi_2', ..., \phi_n'$, then we have

$$\sum_{i=1}^{n} e_i\phi_i' = \sum_{i=1}^{n} e_i'\phi_i \qquad ...(11)$$

9.5.6. Uniqueness of the solution of an Electrostatic problem

Let ϕ_1 and ϕ_2 be the two possible potential distributions due to given charge distribution ρ and surface distribution σ on the surfaces of given conductors. Then in space, we have

$$\nabla^2\phi_1 = \nabla^2\phi_2 = -4\pi\rho,$$

and on the surfaces of conductors, we have

$$\nabla\phi_1 = \nabla\phi_2 = -4\pi\sigma\mathbf{n}$$

and that if $\quad \theta = \phi_1 - \phi_2,$

then $\quad \nabla^{2\theta} = 0$ in all space

and $\quad \nabla^{\theta} = \mathbf{0}$ on the surface of all conductors.

Now from Divergence theorem

$$\int_S (\theta\nabla\theta) \,.\, d\mathbf{a} = \int_V \text{div}\,(\theta\nabla\theta)\, dv$$

$$= \int_V [\theta\nabla^2\theta + (\nabla\theta)^2\, dv \qquad ...(12)$$

But $\nabla^2\theta = 0$ in all space and $\nabla\theta = \mathbf{0}$ on the surface of the conductors. Also from considerations of magnitude, the surface integral vanishes over the sphere at infinity.

$$\therefore \quad \int_V (\nabla\theta)^2\, dv = 0.$$

which implies that $\quad \nabla\theta = \mathbf{0}$ in all space.

or $\quad \theta =$ **constant**.

or $\quad \phi_1 - \phi_2 =$ constant

But $\quad \phi_1 = \phi_2$ over the surfaces of conductors.

$\therefore \quad \phi_1 = \phi_2$ in all space.

Thus the potential distribution must be unique.

9.5.7. Green's equivalent stratum

Let S be a closed surface enclosing a charge distribution and let P be a point outside S. In (1), let $\phi_1 = \phi$, the potential due to this distribution and let $\phi_2 = \dfrac{1}{r}$, then

$$\int_S \left[\phi\nabla\left(\frac{1}{r}\right) - \frac{1}{r}\nabla\phi\right] .\, d\mathbf{a}$$

$$= \int_V \left[\phi\nabla^2\left(\frac{1}{r}\right) - \frac{1}{r}\nabla^2\phi\right] dv. \quad ...(13)$$

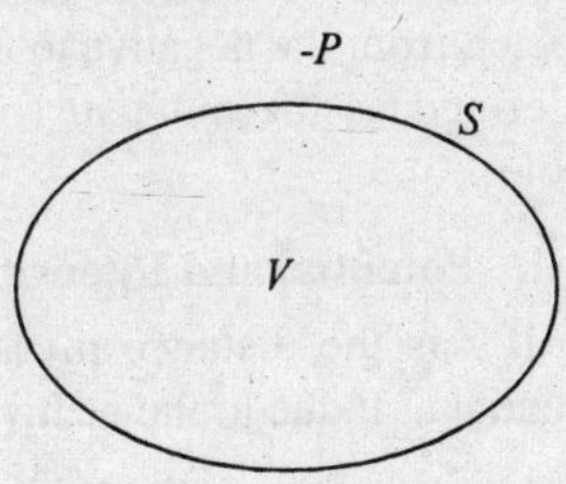

Fig. 9.12

Now let S be a closed equipotential surface so that ϕ is constant over it. Then

$$\int_S \phi\nabla\left(\frac{1}{r}\right) . d\mathbf{a} = \phi\int_S \nabla\left(\frac{1}{r}\right) . d\mathbf{a}$$

But $\frac{1}{r}$ is the potential of a unit charge placed at P and, therefore, by Gauss's theorem, since the charge is outside S, the surface integral vanishes.

Also throughout V,

$$\nabla^2\phi = -4\pi\rho,\ \nabla^2\left(\frac{1}{r}\right) = 0$$

$\therefore$ (13) becomes

$$\int_S \frac{1}{r}\nabla\phi . d\mathbf{a} = \int_V \frac{1}{r}\nabla^2\phi\, dv$$

or

$$\int_S \frac{1}{r}\nabla\phi . \mathbf{n}dS = -4\pi\int_V \frac{\rho\, dv}{r}$$

But $\int_V \frac{\rho}{r} dv$ is the potential ϕ(P) at P, due to charges inside S.

$$\therefore \qquad \phi(P) = -\frac{1}{4\pi}\int_S \frac{\nabla\phi . \mathbf{n}}{r} dS \qquad ...(14)$$

Thus the *potential at a point outside an equipotential of the field is the same as if a distribution of charge were distributed over this surface with density* $\left(-\frac{1}{4\pi}\nabla\phi.\mathbf{n}\right)$.

If the equipotential is replaced by a conductor, this will be the density on its outer surface. This distribution is called *Green's equivalent* layer or stratum.

9.6. DIPOLES

An electric dipole (or doublet) may be regarded as the limiting case of two equal and opposite charges e and $-e$ placed at a small distance l apart, as $e \to \infty$, $l \to 0$ in such a way that their product approaches a finite quantity m. The vector $\mathbf{M} = m\mathbf{a}$ where $\mathbf{a}$ is the *unit* vector directed from the negative to the positive charge, is defined as the *moment* (or strength) of the dipole.

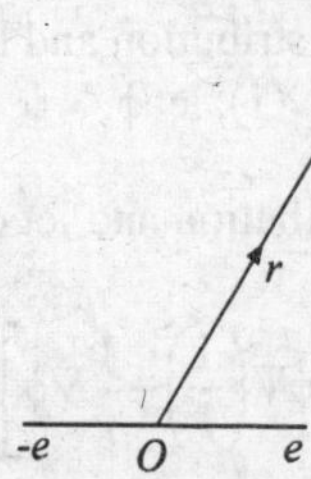

Fig. 9.13

9.6.1. Potential and Intensity

If r is the distance measured from O, the potential at P due to the neative charge is

$$\frac{-e}{r},$$

and that due to the positive charge is

$$e\left[\frac{1}{r} + l\mathbf{a}.\nabla\left(\frac{1}{r}\right)\right].$$

Therefore, the potential due to the doublet at P is

$$\phi = el\mathbf{a}\,.\,\nabla\left(\frac{1}{r}\right)$$

$$= \mathbf{M}\,.\,\nabla_0\left(\frac{1}{r}\right)$$

$$= -\mathbf{M}\,.\,\nabla^{P}\left(\frac{1}{r}\right) = -\,\mathrm{M}\,.\left(\frac{1}{r^3}\,\mathbf{r}\right)$$

or $$\phi = \frac{\mathbf{M.r}}{r^3} \qquad ...(1)$$

For a number of dipoles, the potential at P is given by

$$\phi = \sum_{i=1}^{n} \frac{\mathbf{M}_i\,.\,\mathbf{r}_i}{r_i^3} \qquad ...(2)$$

where $\mathbf{r}_i$ is the position vector of P relative to the *i*th doublet.

The electrostatic intensity **E** due to a single doublet is given by

$$\mathbf{E} = -\,\mathrm{grad}\,\phi = -\,\nabla\left[\frac{\mathbf{M.r}}{r^3}\right] = \frac{3(\mathbf{r.M})}{r^3}\,\mathbf{r} - \frac{1}{r^3}\,\mathbf{M}. \qquad ...(3)$$

From (1)

$$\phi = \frac{m\cos\theta}{r^2},$$

and, therefore, the resolved parts of the intensity along and perpendicular to OP are respectively.

$$-\frac{\partial\phi}{\partial r} = \frac{2m\cos\theta}{r^3} \qquad ...(3)$$

$$-\frac{1}{r} = \frac{\partial\phi}{\partial\theta}\,\frac{m\sin\theta}{r^3}.$$

9.6.2. Potential Energy

The potential energy W of the dipole in a field of intensity **E** and potential function ϕ, is given by

$$\mathrm{W} = -\,e\phi + e(\phi + l\mathbf{a}\,.\,\nabla\phi) = el\mathbf{a}\,.\,\nabla\phi$$

or $$\mathrm{W} = -\,\mathbf{M.E.} \qquad ...(5)$$

9.6.3. Force and couple on the dipole

The force on the negative charge is – *e*E and that on the positive charge is $e[\mathbf{E} + l(\mathbf{a}\,.\,\nabla)\,\mathbf{E}]$. The resultant force on the doublet is, therefore,

$$= el\,(\mathbf{a}\,.\,\nabla)\mathbf{E} = (\mathbf{M}\,.\,\nabla)\mathbf{E} \quad \text{...(5)}$$

and the resultant couple is

$$= l\mathbf{a} \times e\mathbf{E} = \mathbf{M} \times \mathbf{E} \quad \text{...(6)}$$

9.6.4. Mutual Potential Energy of Two Dipoles

The potential energy of a dipole of strength $\mathbf{M}_2$ in a field due to a dipole of strength $\mathbf{M}_1$ is

$$\begin{aligned} \mathbf{W} &= -\mathbf{M}_2\,.\,\mathbf{E} \\ &= -\mathbf{M}_2\,.\,(-\text{ grad }\phi) \\ &= \mathbf{M}_2\,.\,\nabla\left(\frac{\mathbf{M}_1\mathbf{r}}{r^3}\right) \\ &= \mathbf{M}_2\,.\left(\frac{\mathbf{M}_1}{r^3} - \frac{3(\mathbf{r}\,.\,\mathbf{M}_1)}{r^5}\,\mathbf{r}\right) \\ &= \frac{\mathbf{M}_1\,.\,\mathbf{M}_2}{r^3} - \frac{3(\mathbf{M}_1\,.\,\mathbf{r})\,(\mathbf{M}_2\,.\,\mathbf{r})}{r^5} \quad \text{...(7)} \end{aligned}$$

or

$$\mathbf{W} = \frac{m_1 m_2}{r^3}\,(\cos\varepsilon - 3\cos\theta_1\cos\theta_2) \quad \text{...(8)}$$

where ε is the angle between the axes of the doublets and θ_1, θ_2 are the angles which the line joining the doublets makes with the axes.

For coplanar dipoles, $\varepsilon = \theta_2 - \theta_1$ and (8) gives

$$\mathrm{W} = m_1 m_2\,(\sin\theta_1\sin\theta_2 - 2\cos\theta_1\cos\theta_2)\,r^{-3}. \quad \text{...(9)}$$

9.6.5. Couple exerted by one Dipole on another

The couple exerted the $\mathbf{M}_2$ on $\mathbf{M}_1$ is by (6) and (3)

$$\begin{aligned} \mathrm{G}_{21} &= \mathbf{M}_1 \times \left[\frac{3(\mathbf{r}\,.\,\mathbf{M}_2)}{r^5}\,\mathbf{r} - \frac{1}{r^3}\,\mathbf{M}_2\right] \\ &= 3(\mathbf{M}_1 \times \mathbf{r})\,(\mathbf{M}_2\,.\,\mathbf{r})\,r^{-5} - (\mathbf{M}_1 \times \mathbf{M}_2)\,r^{-3}. \quad \text{...(10)} \end{aligned}$$

Similarly

$$\mathrm{G}_{12} = 3(\mathbf{M}_2 \times \mathbf{r})\,(\mathbf{M}_1\,.\,\mathbf{r})\,r^{-5} - (\mathbf{M}_2 \times \mathbf{M}_1)\,r^{-3}. \quad \text{...(11)}$$

9.6.6. Force exerted by one Dipole on another

Let $\mathbf{F}_{21}$ denote the force exerted by $\mathbf{M}_2$ on $\mathbf{M}_1$ tending to move the latter without rotation. If $\mathbf{M}_1$ undergoes a displacement $\delta\mathbf{r}$ in which the dipole axis remains parallel to itself, the work done by the force is $\mathbf{F}_{21}\,.\,\delta\mathbf{r}$. The mutual potential energy changes from W to W + grad W. $\delta\mathbf{r}$.

$$\therefore \quad \mathbf{F}_{21}\,.\,\delta\mathbf{r} = -\text{ grad W}\,.\,\delta\mathbf{r}.$$

Since $\delta\mathbf{r}$ is arbitrary.

$$\begin{aligned} \mathbf{F}_{21} &= -\text{ grad W} \\ &= -\nabla[(\mathbf{M}_1\,.\,\mathbf{M}_2)r^{-3} - 3(\mathbf{M}_1\,.\,\mathbf{r})\,(\mathbf{M}_2\,.\,\mathbf{r})r^{-5}] \\ &= 3\,(\mathbf{M}_1\,.\,\mathbf{M}_2)r^{-5}\,\mathbf{r} - 15(\mathbf{M}_1\,.\,\mathbf{r})\,(\mathbf{M}_2\,.\,\mathbf{r})r^{-7}\,\mathbf{r} \end{aligned}$$

$$+ 3r^{-5} (\mathbf{M}_1 . \mathbf{r}) \nabla (\mathbf{M}_2 . \mathbf{r}) + 3r^{-5} (\mathbf{M}_2 . \mathbf{r}) \nabla (\mathbf{M}_1 . \mathbf{r}).$$

But $\nabla(\mathbf{M}_2 . \mathbf{r}) = \mathbf{M}_2,$

and $\nabla(\mathbf{M}_1 . \mathbf{r}) = \mathbf{M}_1.$

$$\therefore \quad \mathbf{F}_{21} = (\alpha\mathbf{r} + \beta\mathbf{M}_1 + \gamma\mathbf{M}_2)r^{-5}, \quad ...(12)$$

where
$$\left.\begin{aligned} \alpha &= 3\{(\mathbf{M}_1 . \mathbf{M}_2) - 5(\mathbf{M}_1 . \mathbf{r})(\mathbf{M}_2 . \mathbf{r})r^{-2}, \\ \beta &= 3(\mathbf{M}_2 . \mathbf{r}), \\ \gamma &= 3(\mathbf{M}_1 . \mathbf{r}). \end{aligned}\right\} \quad ...(13)$$

9.7. MULTIPLE EXPANSION OF ELECTRIC FIELD

Let us consider Value V′ occupied by the charge distribution. Let dV′ be an element of volume at position **r′** from an arbitrary point O with in the charge distribution and $\rho(\mathbf{r}')$ the charge density at point **r′**. Then the potential at point **r** due to the volume element dV′ is

$$d\phi(\mathbf{r}) = \frac{1}{4\pi\varepsilon_0} \frac{f(\mathbf{r}')\, d\mathrm{V}'}{|\mathbf{r} - \mathbf{r}'|}$$

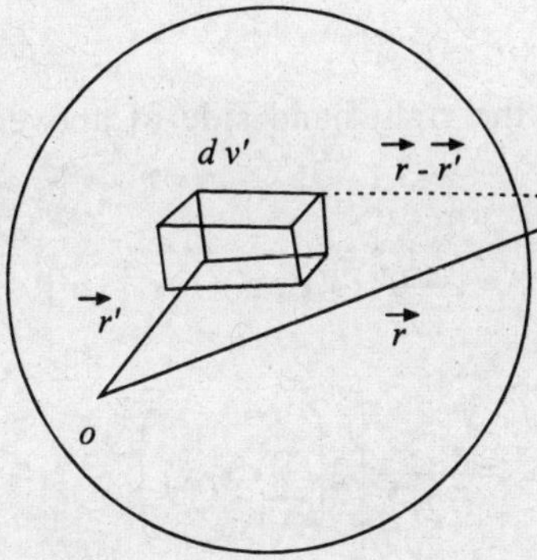

Fig. 9.14

Hence the potential at **r** due to entire charge distribution is given by

$$\phi(\mathbf{r}) = \frac{1}{4\pi\varepsilon_0} \int \frac{\rho(\mathbf{r}')}{|\mathbf{r} - \mathbf{r}'|}\, d\mathrm{V}' \quad ...(1)$$

If the point of observation D is remote from the origin, then $r' << r$ and then

$$\begin{aligned} |\mathbf{r} - \mathbf{r}'|^{-1} &= |(\mathbf{r} - \mathbf{r}'), (\mathbf{r} - \mathbf{r}')^{-1/2} \\ &= (r^2 - 2\mathbf{r} . \mathbf{r}' + r'^2)^{-1/2} \\ &= \frac{1}{r}\left[1 + \left\{-\frac{2\mathbf{r}.\mathbf{r}'}{r^2} - \frac{\mathbf{r}'^2}{r^2}\right\}\right]^{-1/2} \end{aligned}$$

$$= \frac{1}{r}\left[1 - \frac{1}{2}\left\{-\frac{2\mathbf{r}.\mathbf{r}'}{r^2} + \frac{\mathbf{r}'^2}{r^2}\right\}\right.$$

$$\left. + \frac{1}{2}\cdot\frac{3}{2}\left\{-\frac{2\mathbf{r}.\mathbf{r}'}{r^2} + \frac{r'^2}{r^2}\right\}^2 \;.....\right]$$

As $r' << r$, hence $\left(\frac{r'}{r}\right)$ may be neglected as compared to $\frac{2\mathbf{r}.\mathbf{r}'}{r^2}$, hence we may omit the terms involving the cube and higher powers of $\left(\frac{r'}{r}\right)$ so that

$$|\mathbf{r} - \mathbf{r}'|^{-1} = \frac{1}{r} + \frac{\mathbf{r}.\mathbf{r}'}{r^3} + \frac{1}{2}\left\{\frac{3(\mathbf{r}.\mathbf{r}')^2}{r^5} - \frac{r'^2}{r^3}\right\} + ...\Big] \quad ...(2)$$

Then (1) yields

$$\phi(\mathbf{r}) = \frac{1}{4\pi\varepsilon_0}\int_{V'}\left[\frac{1}{r} + \frac{\mathbf{r}.\mathbf{r}'}{r^3} + \frac{1}{2}\left\{\frac{3(\mathbf{r}.\mathbf{r}')^2}{r^5} - \frac{r'^2}{r^3}\right\} + ...\right]$$

$$\rho(\mathbf{r}')\,dV' \quad ...(3)$$

The third term on the right hand side of above equation may be expressed as

$$\frac{1}{2}\left\{\frac{3(\mathbf{r}.\mathbf{r}')}{r^5} - \frac{(\mathbf{r}')^2}{r^3}\right\} = \frac{1}{2r^5}\,[3(\mathbf{r}\,.\,\mathbf{r}')^2 - \mathbf{r}'^2\,.\,\mathbf{r}^2]$$

$$= \frac{1}{2r^5}\left[3\left(\sum_{i=1}^{3} x_i\, x'_j\right)^2 - (\mathbf{r}')^2\left(\sum_{j=1}^{3} x_j^2\right)\right]$$

$$= \frac{1}{2r^5}\sum_{i=1}^{3}\sum_{j=1}^{3} x_i\, x_j\,[3x'_j\, xy' - \delta_{ij}\,(\mathbf{r}')^2]$$

Substituting this in (3) and keeping in mind that the variable of interaction is r' only, therefore r can be taken outside the integral sign, we get

$$\phi(\mathbf{r}) = \frac{1}{4\pi\varepsilon_0}\left[\frac{1}{r}\int_{V'}\rho(\mathbf{r}')\,d\,V'\right.$$

$$+\frac{r}{r^3}\int_{V'}\mathbf{r}'\rho(\mathbf{r}')\,d\,V' + \sum_{i=1}^{3}\sum_{j=1}^{3}\frac{1}{2}\frac{x_i x_j}{r^5}$$

$$\left.\int_{V'}(3x'_j\, x'_j - \delta_{ij}\, r'^2)\rho(\mathbf{r}')\,d\,V' + ...\right] \quad ...(4)$$

where x_i, x_j are the cartesian components of $\mathbf{r}$ x'_i, x'_j are those of $\mathbf{r}'$ and δ_{ij} is Kronecker delta symbol defined as

$$\delta_{ij} = \begin{cases} 0 \text{ for } i \neq j \\ 1 \text{ for } i = j \end{cases}$$

It is obvious from eqn. (4) that the potential due to a charge distribution may be expressed as a summation of an infinite number of terms, viz.,

$$\phi(\mathbf{r}) = \phi_1(\mathbf{r}) + \phi_2(\mathbf{r}) + \phi_3(\mathbf{r}) + \ldots.. \qquad \ldots(5)$$

This equation may be interpreted as follows:

The first term, $\phi_1(\mathbf{r})$ represents the potential at $\mathbf{r}$ if the total charge were concentrated at the origin and is called the potential due to **monopole moment of the charge distribution.**

The second term $\phi_2(\mathbf{r})$ represents the potential which would result if a point dipole having dipole moment equal to that if the charge distribution were located at origin of coordinates and is called the potential due to **dipole moment of the charge distribution.**

It is obvious that $\phi_3(\mathbf{r})$ contains nine components corresponding to $i, j = 1, 2, 3$, out of which six are equal in pairs, thus leaving six distinct components. This set of quantities form the **quadrupole moment tensor.** The term $\phi_3(\mathbf{r})$ is called the potential due to **quadrupole moment of charge distribution**. The higher order terms $\phi_4(\mathbf{r})$, $\phi_5(\mathbf{r})$ etc., are called **multipole potentials.**

9.8. POLARISATION

When an insulator like glass or sulphur is placed in an electric field, each of its molecules, containing equal and opposite electric charges becomes a small dipole (the electrons, unlike those of the conductors, are not free to leave the atoms) whose strength will, for isotropic insulators or dielectrics, be proportional to the field strength $\mathbf{E}$. Each molecule becomes a dipole of strength $\alpha\mathbf{E}$ and if there are N molecules per unit volume, the total vector moment induced by the electric field per unit volume is $N\alpha\mathbf{E}$. The dielectric is said to be polarised and the vector moment per unit volume is called the *Polarisation Vector* or simply *Polarisation*. If $\mathbf{P}$ denotes this vector

$$\mathbf{P} = N\alpha\mathbf{E} = k\mathbf{E} \qquad \ldots(1)$$

where k is called *Dielectric Susceptibility.* With anistropic materials like crystals, there is also polarisation, but it need not be related to $\mathbf{E}$ in the simple manner of (1).

Thus in a volume element dv, the dipole strength is $\mathbf{P}dv$. Hence the potential at a point due to the polarisation of the dielectric is

$$\phi = -\int_V \frac{\mathbf{P}dv}{r^3} \cdot \mathbf{r} = \int_V \mathbf{P} \cdot \text{grad}\left(\frac{1}{r}\right) dv, \qquad \ldots(2)$$

where integration extends over the entire volume V of the dielectric and $\mathbf{r}$ is the position vector of the element dv relative to the point. We take negative sign, as unlike in (1) of § 9.6 the position vector $\mathbf{r}$ is here directed *towards* the dipole.

Now

$$\text{div}\left(\frac{\mathbf{P}}{r}\right) = \frac{1}{r}\text{ div }\mathbf{P} + \mathbf{P}\,.\left(\text{grad }\frac{1}{r}\right),$$

$$\therefore \quad \phi = \int_V \mathbf{P}\,.\text{ grad}\left(\frac{1}{r}\right) dv$$

$$= \int_V \text{div}\left(\frac{\mathbf{P}}{r}\right) dv - \int_V \frac{1}{r}\text{ div }\rho\, dv$$

$$= \int_S \frac{\mathbf{P}}{r}\,.\, d\mathbf{a} - \int_V \frac{1}{r}\text{ div }\mathbf{P}\, dv, \qquad \text{...(3)}$$

by using the Divergence theorem. Here the surface integral extends over the entire surface S of the dielectric.

Comparing (3) with equation (7) of § 9.5, we see that the potential due to polarisaiton **P** in a dielectric is the same as that due to

(*i*) a volume distribution of charge – div **P** throughout the dielectric and

(*ii*) a surface distribution **P . n** or P_n on the boundary of the dielectric.

We should take into account these two distributions in addition to any volume distribution ρ or surface distribution σ that may be there already in the field.

We now examine the effects of these two distributions on the laws of electrostatic field in vacuum established earlier.

9.8.1. Gauss's unrestricted theorem

(1) of § 9.4 becomes

$$\text{div }\mathbf{E} = 4\pi(\rho - \text{div }\mathbf{P})$$

or

$$\text{div }(\mathbf{E} + 4\pi\mathbf{P}) = 4\pi\rho \qquad \text{...(4)}$$

Define a new vector

$$\mathbf{D} = \mathbf{E} + 4\pi\mathbf{P}, \qquad \text{...(5)}$$

so that

$$\text{div }\mathbf{D} = 4\pi\rho$$

(1) of § 9.3 becomes

$$\int_S \mathbf{D}\,.\,d\mathbf{a} = 4\pi Q, \qquad \text{...(6)}$$

so that the flux of **D** out of any closed surface is equal to 4π times the total charge enclosed. For an isotropic medium, from (1) and (4)

$$\mathbf{D} = \mathbf{E} + 4\pi K\mathbf{E} = \mathbf{E}(1 + 4\pi K) = \kappa\mathbf{E}. \qquad \text{...(7)}$$

where

$$\kappa = 1 + 4\pi K, \qquad \text{...(8)}$$

κ is called the *Dielectric constant* or the *specific inductive capacity* of the medium.

For vacuum

$$\mathbf{P} = 0, K = 0, \kappa = 1, \mathbf{D} = \mathbf{E}.$$

D is an important vector and, in the presence of dielectrics, plays the same fundamental role as **E** in vacuum. In fact **D** is more fundamental, as it reduces to **E** in vacuum. **D** is called the *electric displacement* or the *displacement vector* of the field.

Henceforth we consider only isotropic media.

9.8.2. Coloumb's Law

Proceeding as for (5) of § 9.4, we get

$$\mathbf{D} = 4\pi\sigma\mathbf{n}. \qquad ...(9)$$

or

$$\mathbf{E} = \frac{4\pi\sigma}{\kappa}\mathbf{n}, \qquad ...(10)$$

9.8.3. Mechanical force on a conductor

Proceeding as for (7) of § 9.8, we get

$$\mathbf{D}_1 + \mathbf{D}_2 = 4\pi\sigma\,\mathbf{n},$$

$$\mathbf{D}_1 - \mathbf{D}_2 = \mathbf{0},$$

so that

$$\mathbf{D}_2 = 2\pi\sigma\mathbf{n},$$

or

$$\mathbf{E}_2 = \frac{2\pi\sigma}{\kappa}\mathbf{n}. \qquad ...(11)$$

$\therefore$ mechanical force on a charged conductor is

$$\frac{2\pi\sigma^2}{\kappa}$$

per unit area.

9.8.4. Force between point charges

Applying Gauss's theorem to a sphere of radius r, at the centre of which is placed a charge e, we have by spherical symmetry

$$4\pi r^2\,\mathbf{D} = 4\pi e\mathbf{n},$$

or

$$\mathbf{D} = \frac{e}{r^2}\mathbf{n},$$

or

$$\left.\begin{aligned} \mathbf{E} &= \frac{e}{\kappa r^2}\mathbf{n} \\ \phi &= \frac{e}{\kappa r} \end{aligned}\right\}. \qquad ...(12)$$

If follows that the force between two charges e_1, e_2 placed at a distance r apart in a medium of *s.i.c.* k is

$$\frac{e_1 e_2}{\kappa r^2}.$$

9.8.5. Capacity of a condenser

Consider two similar condensers, one containing a dielectric of constant κ and the other vacuum. If we charge both with the same surface density σ, it follows from the above discussion, that the field and, therefore, the potential difference between the plates of the first condenser is

$$\frac{1}{\kappa}$$

times that of the second. But capacity is merely charge divided by potential difference. So the capacity of the dielectric condenser is κ times the capacity of the vacuum condenser. This was the way in which Faraday defined the dielectric constant.

9.8.6. Poisson's equation

From (5) and (7)

$$\text{div } (\kappa \mathbf{E}) = 4\pi\rho,$$

or

$$\text{div } (\kappa \text{ grad } \phi) = -4\pi\rho, \quad ...(13)$$

which is Poisson's equation for a dielectric medium. If κ is constant, we get

$$\text{div (grad } \phi) = \nabla^2\phi = -\frac{4\pi\rho}{\kappa},$$

and if further $\rho = 0$, we get Laplace's equation.

9.8.7. Discontinuities of D and E at the boundary of two dielectrics

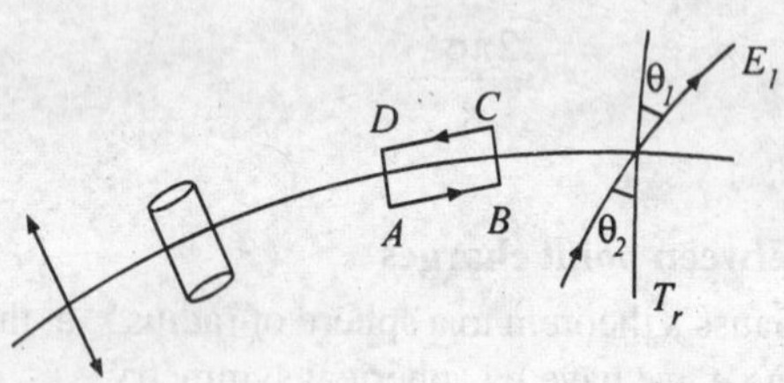

Fig. 9.15

Applying Gauss's theorem to the flat pill box with its faces in the two dielectrics, we get

$$\mathbf{D}_1.\mathbf{n}_1 + \mathbf{D}_2.\mathbf{n}_2 = 4\pi\rho\mathbf{n}_1$$

or

$$D_{N_1} - D_{N_2} = 4\pi\sigma \quad ...(14)$$

where D_{N_1} D_{N_2}

denote the normal components of the Displacement Vector in two media.

If $\sigma = 0$, we get

$$D_{N_2} = D_{N_2} \text{ or } \kappa_1 E_{N_1} = \kappa_2 E_{N_2} \quad ...(15)$$

Now consider a closed curve T with its sides parallel to the surface of separation and edges of negligible size. Since the field is conservative,

$$\oint_T \mathbf{E} \,.\, d\mathbf{r} = 0.$$

$$\therefore \quad E_{T_1} = E_{T_2} \quad ...(16)$$

i.e., the tangential component is continuous along a surface of discontinuity.

From (15) and (16),

$$\kappa_1 \frac{E_{N_1}}{E_{T_1}} = \kappa_2 \frac{E_{N_2}}{E_{T_2}}$$

or $$\frac{\tan\theta_1}{\tan\theta_2} = \frac{\kappa_1}{\kappa_2} \qquad \text{...(17)}$$

which is the law of refraction of the lines of force at an uncharged interface.

9.9. ENERGY OF THE ELECTROSTATIC FIELD

Let the densities of the volume and surface distributions in the dielectric medium be ρ and σ respectively. Let ϕ be the potential at any point. Suppose the charges are being steadily brought from infinity to build up the final field. At any stage suppose a fraction x of the final charges has been brought, where $0 < x < 1$. Since the potential depends linearly on the charges, the potential at any point would be $x\phi$ where ϕ is the final potential. The work done by an external agency in increasing the charges to a fraction $x + dx$ of the final charges is

$$\int_V (x\phi)(\rho dx)\, dv + \sum_i \int_{S_i} (x\phi)(\sigma dx)\, dS. \qquad \text{...(1)}$$

The total work done against the field forces is, therefore,

$$\left[\int_V \rho\phi\, dv \sum_i \int_{S_i} \phi\sigma\, dS\right] \int_0^1 x dx,$$

and, therefore, the electrostatic energy of the field is

$$W = \frac{1}{2}\left[\int_V \rho\phi\, dv + \sum_i \int_{S_i} \phi\sigma\, dS\right], \qquad \text{...(2)}$$

the surface integral being taken over all the surface distributions and volume integral over the volume distribution.

If the field consists of only a number of charged conductors. $\rho = 0$ and ϕ = constant over the surface of each conductor (2) then reduces to

$$W = \frac{1}{2}\sum_i e_i\phi_i \qquad \text{...(3)}$$

where e_i is the total charge and ϕ_i is the potential of the ith conductor.

Now $$\nabla.\mathbf{D} = 4\pi\rho, \qquad \text{...(4)}$$

and $$\nabla.(\phi\mathbf{D}) = (\nabla.\mathbf{D})\phi + (\nabla\phi).\mathbf{D}$$

∴ (2) becomes

$$W = \frac{1}{8\pi}\int_V \nabla.(\phi\mathbf{D})dv - \frac{1}{8\pi}\int_V \nabla\phi.\mathbf{D}dv + \sigma\phi dS$$

Since there is no charge inside conductors, V is the space bounded by the sphere at infinity and the surfaces of the conductors S_i. Applying Divergence theorem to the first term, replacing $\nabla\phi$ by $-\mathbf{E}$ in the second term and using $\mathbf{D} = 4\pi\sigma\mathbf{n}$ in the third, we get

$$W = \frac{1}{8\pi}\int_S \phi\mathbf{D}.d\mathbf{a} - \frac{1}{8\pi}\sum_i \int_{S_i} \phi\mathbf{D}.d\mathbf{a}$$

$$+ \frac{1}{8\pi}\int_V \mathbf{E}.\mathbf{D}dv + \frac{1}{8\pi}\sum_i \int_{S_i} \phi\mathbf{D}.d\mathbf{a}$$

or $$W = \frac{1}{8\pi}\int_S \phi\mathbf{D}.d\mathbf{a} + \frac{1}{8\pi}\int_V \mathbf{E}.\mathbf{D}dv$$

Now for large r, ϕ is of order $\frac{1}{r}$, $\mathbf{D}$ is of order $\frac{1}{r^2}$, $d\mathbf{a}$ is of order r^2; and we may take the volume of integration as large as we please, provided we assume that $\rho = 0$ outside some sphere of finite radius. The integral over the sphere at infinity vanishes and we get

$$W = \frac{1}{8\pi}\int_V \mathbf{E}.\mathbf{D}dv, \qquad ...(5)$$

where integration is carried over the whole of space except inside conductors. But there $\mathbf{E} = \mathbf{0}$ and, therefore, the integration can be extended over the whole of space.

(5) shows that energy may be regarded as distributed throughout space with density $\mathbf{E}.\mathbf{D}/8\pi$. In an isotropic medium, the energy density is

$$\frac{1}{8\pi\kappa}\mathbf{D}^2 \text{ or } \frac{1}{8\pi}\kappa\mathbf{E}^2$$

In vacuum, it reduces to $\frac{1}{8\pi}.\mathbf{E}^2$

EXAMPLES

1. *Charges $e_1, e_2, ..., e_n$ are placed at points $A_1, A_2, ..., A_n$. Find the intensity* **E** *at any point of the field.*

Let $\mathbf{a}_i$ be the position vector of A_i and let $\mathbf{r}$ be the position vector of P. The force on a unit-charge placed at P due to the charge e_i is $\frac{e_i}{A_i P^2}$ in the direction $\overrightarrow{A_iP}$.

$$\therefore \quad \mathbf{E} = \sum_{i=1}^{n}\left[\frac{e_i}{A_iP^2}\cdot\frac{1}{A_iP}\overrightarrow{A_iP}\right]$$

$$= \sum_{i=1}^{n} \frac{e_i}{A_iP^3} \overrightarrow{A_iP}$$

$$= \sum_{i=1}^{n} \lambda_i [\mathbf{r} - \mathbf{a}_i]$$

$$= \mathbf{r} \, (\sum_{i=1}^{n} \lambda_i) - \sum_{i=1}^{n} (\lambda_i \, \mathbf{a}_i)$$

$$= (\sum_{i=1}^{n} \lambda_i) \left(\mathbf{r} - \frac{\sum_{i=1}^{n} (\lambda_i \mathbf{a}_i)}{\sum_{i=1}^{n} \lambda_i} \right),$$

where $\lambda_i = \dfrac{e_i}{A_i P^3}$.

But $\dfrac{\sum_{i=1}^{n} \lambda_i \mathbf{a}_i}{\sum_{i=1}^{n} \lambda_i}$

gives the position-vector $\mathbf{r}_0$ of the centroid G of masses $\lambda_1, \lambda_2, ..., \lambda_n$ placed at $A_1, A_2, ..., A_n$ respectively.

$$\therefore \quad \mathbf{E} = (\sum_{i=1}^{n} \lambda_i) \, (\mathbf{r} - \mathbf{r}_0) = (\sum_{i=1}^{n} \lambda_i) \, \overrightarrow{GP}.$$

Hence the direction of the line of force at P is obtained by joining the centroid G to P and the magnitude of the resultant force on the unit charge at P is $(\sum_{i=1}^{n} \lambda_i)$ GP.

2. *Write down the interaction potential energy of two short electric dipoles seperated by a distance. If one of the dipoles is inclined at an angle θ_1 to the radius vector joining them, show that in the state of equilibrium, the other dipole would make an angle θ_2 with it given by*

$$\tan \theta_2 = -\frac{1}{2} \tan \theta_1.$$

The interaction potential energy of two short electric dipoles is given by

$$U = \frac{\mathbf{M}_1 . \mathbf{M}_2}{r^3} - \frac{3}{r^5} (\mathbf{M}_1 . \mathbf{r}) (\mathbf{M}_2 . \mathbf{r}) \qquad ...(1)$$

If one of the dipoles of dipole moment $\mathbf{M}_1$ is inclined at an angle θ_1 to the radius vector $\mathbf{r}$ and other dipole of moment $\mathbf{M}_2$ is inclined at an angle θ_2 with the same radius vector $\mathbf{r}$, then the angle between the two dipoles would be $\theta_1 - \theta_2$. So equ. (1) takes the form

$$U = \left[\frac{M_1 M_2 \cos(\theta_1 - \theta_2)}{r^3} - \frac{3}{r^5}(M_1 r \cos\theta_1)(M_2 r \cos\theta_2)\right]$$

$$= \frac{M_1 M_2}{r^3}[\cos(\theta_1 - \theta_2) - 3\cos\theta_1 \cos\theta_2] \qquad \text{...(2)}$$

In the state of equilibrium of dipoles, the force must be zero and potential energy must be extremum. To apply the condition of extremum potential energy

$$\frac{\partial U}{\partial \theta_2} = \frac{M_1 M_2}{r^3}$$

$$[\sin(\theta_1 - \theta_2) + 3\cos\theta_1 \sin\theta_2] = 0$$

$$\Rightarrow \quad \sin(\theta_1 - \theta_2) + 3\cos\theta_1 \sin\theta_2 = 0$$

$$\Rightarrow \quad \sin\theta_1 \cos\theta_2 + 2\cos\theta_1 \sin\theta_2 = 0$$

$$\Rightarrow \quad \tan\theta_2 = -\frac{1}{2}\tan\theta_1$$

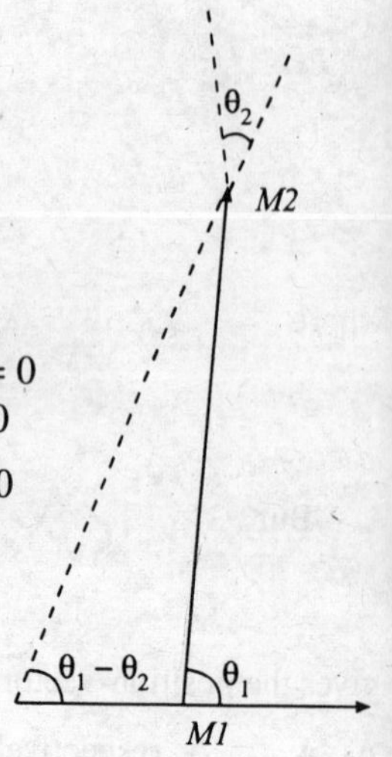

Fig. 9.16

3. *Find the electric field outside an earthed conducting sphere, due to an external point charge and the charge which it induces on the sphere; also determine the surface density of the induced charge.*

Let O be the centre of the conducting sphere of radius *a* and let charge *e* be placed at point A, where OA = *f*.

We shall first show that we can find a charge *e'* at inverse point A′ such that sphere becomes an equipotential with $\phi = 0$ due to charges *e* at A and *e'* at A′. This would be so if for every point P on the sphere

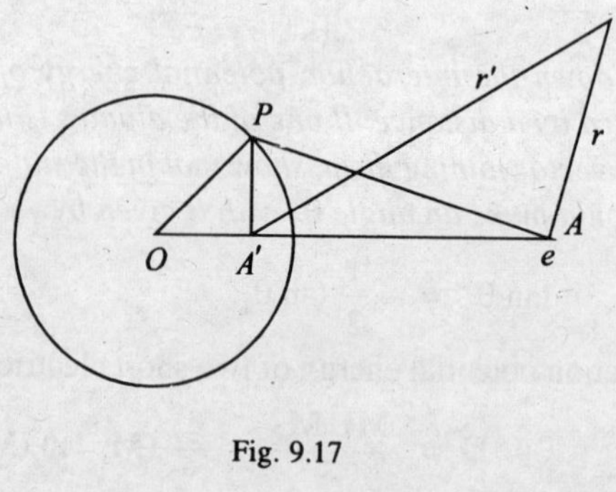

Fig. 9.17

$$\frac{e}{\text{AP}} + \frac{e'}{\text{A'P}} = 0$$

i.e., if
$$e' = -e\frac{\text{A'P}}{\text{AP}} = -e\frac{\text{AP}}{\text{OP}}$$

from similar triangles OA′P and OPA, since OA.OA′ = a^2.

$$\therefore \qquad e' = -e\frac{a}{f}.$$

By Green's equivalent layer theorem, the field outside the sphere due to e' is the same as that due to the equilibrium layer of the induced charge on it. Thus

$$\phi = \frac{e}{f} - \frac{ea/f}{r'}$$

will give a solution; but by the uniqueness theorem this is the solution of our problem.

To find the surface density σ induced, we use Coloum's Law

$$4\pi\sigma\mathbf{n} = \mathbf{E} = \frac{e}{r^3}\mathbf{r} - \frac{ea}{f}\frac{\mathbf{r}'}{r'^3},$$

where **r** and **r′** are the vectors $\overrightarrow{\text{AP}}, \overrightarrow{\text{A'P}}$, respectively.

Now
$$\mathbf{r} = \overrightarrow{\text{AO}}, \overrightarrow{\text{OP}} = \overrightarrow{\text{AO}} + a\mathbf{n} = -f\mathbf{m} + a\mathbf{n}.$$

$$\mathbf{r}' = \overrightarrow{\text{A'O}} + \overrightarrow{\text{O'P}} = \overrightarrow{\text{A'O}} + a\mathbf{n} = -\frac{a^2}{f}\mathbf{m} + a\mathbf{n},$$

where **m** is the unit vector along $\overrightarrow{\text{OA}}$.

$$\therefore \quad 4\pi\sigma\mathbf{n} = \frac{e}{\text{AP}^3}(-f\mathbf{m} + a\mathbf{n}) - \frac{ea}{f}\frac{1}{\text{A'P}^3}\left(-\frac{a^3}{f}\mathbf{m} + a\mathbf{n}\right).$$

$$= e\left(-\frac{f}{\text{AP}^3} + \frac{a^3}{f^2}\cdot\frac{1}{\text{A'P}^3}\right)\mathbf{m} + ea\left(\frac{f}{\text{AP}^3} - \frac{a}{f}\frac{1}{\text{A'P}^3}\right)\mathbf{n}.$$

But
$$\frac{\text{AP}}{\text{A'P}} = \frac{f}{a}.$$

$$\therefore \quad 4\pi\sigma\mathbf{n} = \frac{e}{\text{AP}^3}\left(-f + \frac{a^3}{f^2}\cdot\frac{f^3}{a^3}\right)\mathbf{m} + \frac{ea}{\text{AP}^3}\left(1 - \frac{a}{f}\frac{f^3}{a^3}\right)\mathbf{n}.$$

$$\therefore \quad 4\pi\sigma = \frac{ea}{\text{AP}^3}\left(1 - \frac{f^2}{a^2}\right)$$

or $$\sigma = -\frac{1}{4\pi a}\frac{e(f^2-a^2)}{AP^3}.$$

Note : — An *electrical image* is defined as a set of one of more point charges which produce on the other side of the surface the same electrical field as would the charges induced on that surface by the inducing system situated there.

Thus in the above example, the charge $-\frac{ea}{f}$ placed at the inverse point is the image of charge e in the earthed conducting sphere. If the sphere were however at potential V instead of at potential zero, the image system would consist of charge $-\frac{ea}{f}$ at A′ and a charge Va at centre. This would give $Va - \frac{ea}{f}$ as the induced charge on the sphere. If the sphere is insulated and uncharged, the image system would, therefore, consist of charge $-\frac{ea}{f}$ at A′ and charge $\frac{ea}{f}$ at the centre.

4. *Show that the introduction of a new conductor lessens the energy of an electrostatic field, the new conductor being insulated and uncharged.*

Let V′ be the volume finally occupied by the conductor and let the medium be air (or vacuo). Let accents refer to the field variables after the introduction of the conductor. Then the decrease of energy of the field is

$$\begin{aligned} W - W' &= \frac{1}{8\pi}\int_V \mathbf{E}^2 dv - \frac{1}{8\pi}\int_{V-V'} \mathbf{E}'^2\, dv \\ &= \frac{1}{8\pi}\int_V \mathbf{E}^2 dv + \frac{1}{8\pi}\int_{V-V'} (\mathbf{E}^2 - \mathbf{E}'^2)\, dv \\ &= \frac{1}{8\pi}\int_V \mathbf{E}^2 dv + \frac{1}{8\pi}\int_{V-V'} \{(\mathbf{E}-\mathbf{E}')^2 + 2(\mathbf{E}-\mathbf{E}')\,.\,\mathbf{E}'\}dv. \end{aligned}$$

But

$$\mathbf{E} = -\nabla\phi;\ \mathbf{E}' = -\nabla\phi'$$

$$\begin{aligned} \therefore \quad & \int_{V-V'} (\mathbf{E}-\mathbf{E}')\,.\,\mathbf{E}'dv \\ &= \int_{V-V'} (\nabla\phi')\,.\,[\nabla(\phi-\phi')]\, dv. \\ &= \int_{V-V'} \{\nabla\,.\,[\phi\nabla(\phi-\phi')] - [\nabla\,.\,\nabla(\phi-\phi')]\phi'\}\, dv \\ &= \int_{V-V'} [div\,[\phi'\,\nabla(\phi-\phi')]dv \end{aligned}$$

Since $\nabla^2(\phi-\phi') = 0$

Using Divergence theorem, we obtain

$$\int_{V|V'} (\mathbf{E}-\mathbf{E}')\,.\,\mathbf{E}'dv = \int_S [\phi'\nabla(\phi-\phi')]\,.\,d\mathbf{a};$$

the surface integral extends over the surfaces of the conductors and the sphere at infinity. Over each conductor ϕ' is constant and $\int \nabla\phi \, . \, d\mathbf{a} = \int \nabla\phi' \, . \, d\mathbf{a}$, since the total charge does not vary over each conductor. Also the surface integral over the sphere at infinity vanishes.

$$\therefore \quad W - W' = \frac{1}{8\pi}\int_{V'} \mathbf{E}^2 dv + \frac{1}{8\pi}\int_{V-V'} (\mathbf{E} - \mathbf{E}')^2 \, dv < 0$$

and hence $\quad W' < W.$

Exercises

1. Show that the dimensions of the electrostatic charge potential, capacity and intensity are respectively.

$$M^{\frac{1}{2}} L^{\frac{3}{2}} T^{-1}, \quad M^{\frac{1}{2}} L^{\frac{1}{2}} T^{-1}, \quad L, \; M^{\frac{1}{2}} L^{-\frac{1}{2}} T^{-1}.$$

2. A field of force is due to three charges e_1, e_2, e_3 at A, B, C. Show that the direction of the line of force at a point P is given by joining P to the centre of gravity of masses

$$\frac{e_1}{AP^2}, \frac{e_2}{BP^3}, \frac{e_3}{CP^3}$$

at A, B, C, respectively, and find the magnitude of the force at P.

3. (*i*) Show that under the attraction of a nucleus of charge Ze, an electron of mass M and charge $-e$ can describe a circle of any radius r with velocity v if

$$mv^2 = \frac{Ze^2}{r}.$$

(*ii*) If there are two charges $\pm e$ at $(\pm a, 0, 0)$ then a third charge $-e$ can rotate about x-axis with velocity v in a circle of radius r if

$$mv^2 = \frac{2e^2 r^2}{(r^2 + a^2)^{\frac{3}{2}}}.$$

4. Draw lines of force and equipotential surfaces for the following system of charges

(*i*) two point charges e_1, e_2 of same sign,

(*ii*) two point charges e and $-e$,

(*iii*) two point charges $4e$ and $-e$.

5. Prove that

(*i*) no line of force can begin and end on the same conductor,

(*ii*) in a field containing many charges the conductor of highest potential must be positively charged and that of the lowest potential must be negatively charged,

(*iii*) if a closed equipotential surface contains no charge; the potential is constant throughout the interior.

6. Apply Gauss's theorem and the symmetry of the problem to show that

(*i*) For an infinite line-wire carrying charge e per unit length, the intensity outside is $\frac{2e}{r}$ perpendicular to the wire.

(*ii*) For a charged spherical conductor $\mathbf{E} = 0$ at a point inside and $\mathbf{E} = \frac{e}{r^2}\mathbf{r}$ at a point outside.

(*iii*) At all distances from an infinite plane, charged on each side with uniform surface density σ, the electric field is of constant magnitude $4\pi\sigma$.

7. Use of fact that at a point, algebraic sum of the resolved parts of the forces along the normal to the line of the force must vanish, to obtain the equation of the lines of force

(*i*) in the form $\Sigma e_i \cos\theta_i$ = constant for a system of collinear point charges.

(*ii*) in the form $\sum_i e_i\theta_i$ = constant for a system of infinite line charges all perpendicular to a given plane where e_i denotes the charge per unit length on the ith wire and θ_i is the angle which the line joining the point of intersection of the wire with the plane, and the point on the line of force, makes with a line fixed in the plane.

8. Obtain mechanical force on a conductor by regarding the surface density to be distributed in a layer of small thickness t.

9. Prove that the capacity of a condensor

(*i*) formed by two infinite coaxial circular cylinders of radii a and b, the inner being charged and the outer earthed is

$$\frac{1}{2\log e(ba^{-1})}$$

per unit length.

(*ii*) formed by two concentric spherical conduction of radii a and b, the inner being earthed and outer insulated is

$$\frac{b^2}{b-a}.$$

(*iii*) formed by two infinite parallel plates distant d apart with a slab of dielectric of s.i.c. κ and thickness t between them is

$$\frac{1}{4\pi\left(d-t+\frac{t}{\kappa}\right)}$$ per unit area.

(*iv*) formed by two concentric spherical conductors of radii, *a, d,* modified by insertion of concentric spherical shell of dielectric of s, i, c, κ, radii

$b, c\ (d > c > b > a)$ is

$$\frac{\kappa abcd}{[(\kappa cd(b-a)+\kappa ab(d-c)+ad(c-b)]}.$$

(*v*) a spherical condenser radii *a, b*; space on one side of a diametral plane being filled with a dielectric of constant κ is

$$\frac{1}{2}(1+\kappa)\frac{ab}{b-a}.$$

10. Prove that the work done in bringing charges e_1, e_2, ..., e_n from infinity is

$$\sum_i \sum_j \frac{e_i e_j}{r_{ij}} = \frac{1}{2}\sum_i e_i \phi_i$$

where r_{ij} is the distance between the final positions of the *i*th and *j*th charges and ϕ_i is the potential at the position of the *i*th charge due to the rest of the charges.

11. Show that the electrostatic energy of a system of conductors can be expressed as $\frac{1}{2}\sum_i e_i \phi_i$.

12. Verify the formula $\frac{1}{8\pi}\int \mathbf{E}^2 dv$ for electrostatic energy for

(*i*) a spherical conductor,

(*ii*) a spherical condenser,

(*iii*) a parallel plate condenser.

13. A point charge *e* is placed at a point A in front of an infinite earthed plane conductor at a distance *a* from it. Show that its electrical image is a charge – *e* at the image point and the surface density induced at a point P in the conductor is

$$\sigma = \frac{-ea}{2\pi\, AP^3}.$$

14. A point charge *e* is placed at a point A outside a conducting sphere of radius *a* at a distance *f* from the centre. Show that the surface density induced at a point P on the sphere is

(*i*) $\sigma = -\dfrac{e(f^2-a^2)}{4\pi a\, AP^3}$ if the sphere is earthed, (*ii*) $\sigma = -\dfrac{e(f^2-a^2)}{4\pi a\,\mathrm{AP}^3} + \dfrac{e}{4\pi af}$ if the sphere is insulated and without charge

(*iii*) $\sigma = -\dfrac{e(f^2-a^2)}{4\pi a\,\mathrm{AP}^3} + \dfrac{e}{4\pi af} + \dfrac{Q}{4\pi a^2}$ where the sphere is insulated and has a total charge Q.

(*iv*) $\sigma = -\dfrac{e(f^2-a^2)}{4\pi a\,\mathrm{AP}^3} + \dfrac{\phi}{4\pi a}$, when the sphere is maintained at a given potential ϕ.

15. Two charged conducting spheres with centres A, B and radii a, b intersect at right angles. The plane of the circle in which they intersect meets AB in C. Show that the external field is equivalent to the field of certain point charges at A, B, C.

Prove also that the capacity of the conductors formed by the spheres is

$$a + b - ab\,(a^2 + b^2)^{-1/2}$$

and find the ratio in which the charge is divided between the spheres.

16. In § 9.6, prove that

(*i*) $\mathbf{F}_{12} = -\mathbf{F}_{21}$,

(*ii*) $\mathbf{G}_{12} = -\mathbf{G}_{21} - \mathbf{r} \times \mathbf{F}_{21}$,

(*iii*) For coplanar magnets,

$$G_{21} = m_1 m_2 r^{-3}\,(2 \sin\theta_1 \cos\theta_2 + \sin\theta_2 \cos\theta_1).$$

17. Write down the conditions satisfied by displacement **D** and the electric intensity **E** in the field due to n closed conductors with surfaces S_1, S_2, ..., S_n and charges $e_1, e_2, \ldots e_n$, in an unbounded uncharged medium in which the dielectric co-efficient κ is a function of position. Show that the energy of the field is

$$\frac{1}{8\pi}\int_V \mathbf{E}\,.\,\mathbf{D}\,dv,$$

where V is the volume outside the dielectrics. Vectors **D′**, **E′** are such

$$\text{div}\,\mathbf{D}' = 0, \quad \mathbf{D}' = \kappa\mathbf{E}'$$

that at all points of V and such that

$$\int_{S_i} \mathbf{D}'\,.\,d\mathbf{a} = 4\pi e_i \; (i = 1, 2, \ldots, n), \text{ prove that}$$

$$\int_V \mathbf{E}'\,.\,\mathbf{D}'\,dv \geq \int_V \mathbf{E}\,.\,\mathbf{D}\,dv$$

if **D′**, **E′** are of order $\dfrac{1}{r^2}$ at a great distance r from the conductors.

9.10. MAGNETOSTATICS

The mathematical theory of stationary magnets can be derived in almost the same manner as that of electrostatics, assuming that

(*i*) the field of a magnet is the same as that due to two equal and opposite magnetic charges (poles) placed at its extremities; and

(*ii*) like poles repel and unlike poles attract according to the inverse square law.

The main difference is that a magnetic pole, unlike an electrostatic charge cannot be isolated and in any volume, we cannot have an excess of positive over negative poles or *vice versa*. Hence ρ is always zero. Also the magnetic susceptibility κ can be positive or negative. The magnetic substance is called *paramagnetic* of $\kappa > 0$ and *diamagnetic* if $\kappa < 0$. The parallelism is illustrated in the following table :—

Electrostatics		*Magnetostatics*
Electrostatic field	$\mathbf{E} \leftarrow\rightarrow \mathbf{H}$,	the magnetic field
Electrostatic charge	$e \leftarrow\rightarrow m$,	the magnetic pole strength
Electrostatic potential	$\phi \leftarrow\rightarrow \Omega$,	the magnetic potential
Electric polarisation	$\mathbf{P} \leftarrow\rightarrow \mathbf{I}$,	the intensity of magnetisation
Electric susceptibility	$K \leftarrow\rightarrow \kappa$	the magnetic susceptibility
Dielectric constant	$\kappa \leftarrow\rightarrow \mu$	the magnetic permeability
Electric Displacement	$\mathbf{D} \leftarrow\rightarrow \mathbf{B}$,	the magnetic induction

Electrostatic energy density $\frac{1}{8\pi}\mathbf{E.D} \leftarrow\rightarrow \frac{1}{8\pi}\mathbf{H.B}$ the magnetostatic energy density.

$$\mathbf{P} = \kappa\mathbf{E} \leftarrow\rightarrow \mathbf{I} = \kappa\mathbf{H}$$

$$\mathbf{D} = \kappa\mathbf{E} \leftarrow\rightarrow \mathbf{B} = \mu\mathbf{H}$$

$$\kappa = 1 + 4\pi\kappa \leftarrow\rightarrow \mu = 1 + 4\pi\kappa$$

$$\mathbf{D} = \mathbf{E} + 4\pi\mathbf{P} \leftarrow\rightarrow \mathbf{B} = \mathbf{H} + 4\pi\mathbf{I}$$

Gauss's divergence theorem

$$\nabla.\mathbf{D} = 4\pi\rho \leftarrow\rightarrow \nabla.\mathbf{B} = 0$$

[flux of magnetic induction out of any closed surface is zero].

At a surface of discontinuity except when there is a surface charge. At a surface of discontinuity except when there is a surface current.

$$D_{N_1} = D_{N_2} \leftarrow\rightarrow B_{N_1} = B_{N_2}$$

$$E_{T_1} = E_{T_2} \leftarrow\rightarrow H_{T_1} = H_{T_2}$$

9.10. MAGNETIC SHELL

A surface distribution of dipoles whose direction is everywhere normal to the surface is called a magnetic shell. If τ is the dipole strength per unit area, the potential at P due to small dipole of strength τ *d***a** is

$$\frac{\tau d\mathbf{a}.\mathbf{r}}{r^3} = \kappa d\omega. \qquad ...(1)$$

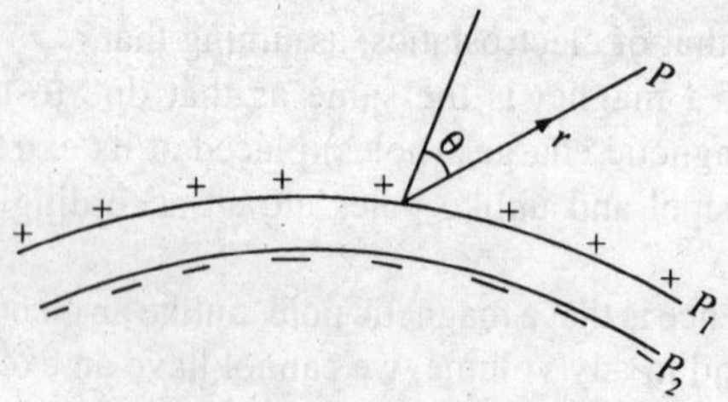

Fig. 9.18

The shell is said to be of *uniform strength* if its strength τ is the same at all points. In this case, the potential at P due to the whole shell is $\tau\omega$ where ω is the solid angle subtended at P by the whole shell. The formula holds when the lines drawn from P meet the shell first on the positive side and the sign will have to be changed if the lines first meet the shell on the negative side.

If the shell is closed, the potential at all external points is zero and at all internal points is $-4\tau\pi$.

If $\quad \phi(P_1) = \tau\omega,\ \phi(P_2) = -\tau(4\pi - \omega) = \tau\omega - 4\pi\tau,$ Fig. 9.19

we have $\quad \phi(P_1) - \phi(P_2) = 4\pi\tau,$ ...(2)

9.12. MOVING CHARGES (OR CURRENTS)

Let a charge of a density ρ be flowing with mean velocity **V**. Then the charge crossing the element $d\mathbf{a}$ per unit time is

$$\rho\mathbf{V} \cdot d\mathbf{a} = \mathbf{J}.d\mathbf{a} \qquad ...(1)$$

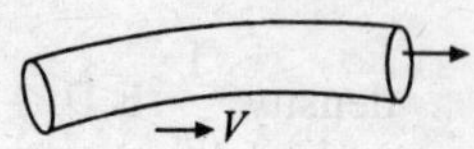

Fig. 9.19

where **J** is defined as the *conduction current vector* or *current density vector*. The current across any surface drawn in the medium is

$$i = \int_S \mathbf{J}.d\mathbf{a} \qquad ...(2)$$

The total charge inside a closed surface is

$$Q = \int_V \rho dv,$$

and if there are no sources or sinks inside the surface, the rate at which this is decreasing is

$$-\int_V \frac{\partial\rho}{\partial t}\, dv$$

As it is due to the outward flow of charge,

$$-\int_V \frac{\partial\rho}{\partial t} = \int_S \mathbf{J}.d\mathbf{a}$$

Applying Divergence theorem,

$$\int_V \left(\frac{\partial\rho}{\partial t} + \text{div}\,\mathbf{J}\right) dv = 0$$

Since this result holds for any arbitrary volume, the integrand must vanish identically and we have the *equation of continuity* or the *equation of conservation of charge,*

$$\text{div } \mathbf{J} = -\frac{\partial \rho}{\partial t} \qquad ...(3)$$

For steady state, ρ is independent of time, and we have

$$\text{div } \mathbf{J} = 0 \qquad ...(4)$$

It has been found experimentally that if **E** is the field causing the current,

$$\mathbf{J} = \sigma \mathbf{E} \qquad ...(5)$$

which is known as *Ohm's Law.* Here σ is defined as the *conductivity* of the material. It is large for conductors and small for insulators. Its reciprocal $\frac{1}{\sigma}$ is called the *specific resistance* of the medium.

9.13. MAGNETIC EFFECTS OF CURRENTS

Oersted (1832) discovered that electric currents produce magnetic effects and Ampere found that two small coils carrying currents exert on each other a force of precisely the same type as exists between dipoles. We associate, therefore, with each loop, a vector **m**, the magnetic moment of the loop. The direction of the moment is perpendicular to the plane of the loop and is such that the right-handed rotation about it takes us in the same direction as the current i in the loop and to area dS of the loop but is independent of the shape of the loop. Thus

$$\mathbf{m} = k\, i\, d\mathbf{a}$$

We define a C.G.S. electromagnetic unit (e.m.u.) as that current which would give **m** equal to id**a** so that $k = 1$ and

$$\mathbf{m} = i\, d\mathbf{a} \qquad ...(1)$$

We shall, for the present use these units and assume also that the medium is vacuum.

Let a current i flow in a large coil s. Draw a surface S spanning it and on it a network of lines dividing it into small loops. If we imagine the current i flowing in each network, currents along the lines of the network cancel out and only a current i along s remains. The magnetic effect of the coil is the same, therefore, as that of a set of magnetic dipoles of moment id**a** covering the surface. This is of

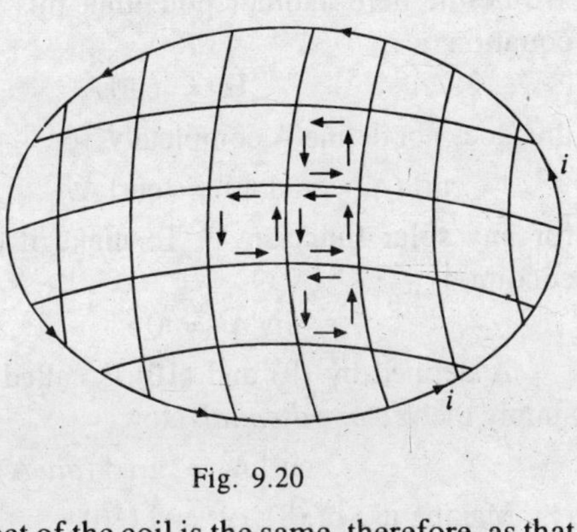

Fig. 9.20

course, the same, as our magnetic shell of strength i. Consequently as in (2) of § 9.11, if we carry a unit positive pole round a closed path embracing a current i, a work equal to $4\pi i$ is done. This gives

$$\int \mathbf{H}.d\mathbf{S} = 4\pi i \quad \text{...(2)}$$

If no current is embraced

$$\int \mathbf{H}.d\mathbf{S} = 4 \quad \text{...(3)}$$

which gives at once

$$\text{div } \mathbf{H} = 0 \quad \text{...(4)}$$

Also by using Stokes theorem, we have for any surface, closed or open

$$\int \text{curl } \mathbf{H}.d\mathbf{a} = 0$$

As this is true for any arbitrary surface

$$\text{curl } \mathbf{H} = 0 \quad \text{...(5)}$$

H may, therefore, be expressed as the gradient of a scalar quantity. This is the magnetostatic potential Ω, so that

$$\mathbf{H} = -\text{grad } \Omega \quad \text{...(6)}$$

Combining (4) and (6), we have

$$\text{div (grad } \Omega) = 0$$

or

$$\nabla^2 \Omega = 0 \quad \text{...(7)}$$

This holds only in regions not occupied by currents and in vacuum. When currents are present, from (2) of § 9.12 and (2) above

$$\int \mathbf{H}.d\mathbf{S} = 4\pi \int \mathbf{J}.d\mathbf{a}$$

Transforming by Stokes Theorem

$$\int (\text{curl } \mathbf{H} - 4\pi \mathbf{J}) \, . \, d\mathbf{a} = 0$$

Since this is true for all arbitrary surfaces,

$$\text{curl } \mathbf{H} = 4\pi \mathbf{J} \quad \text{...(8)}$$

In this case **H** cannot be expressed as the gradient of a scalar potential. We define here another potential, this time a vector potential **A** by the equation

$$\mathbf{H} = \text{curl } \mathbf{A}, \quad \text{...(9)}$$

this does not define **A** completely, for

$$\text{curl } (\mathbf{A} + \text{grad } \psi) = \text{curl } \mathbf{A},$$

for any solar function ψ. To make it determinate, we impose another condition

$$\text{div } \mathbf{A} = 0 \quad \text{...(10)}$$

A defined by (9) and (10) is called the *magnetic vector potential* or simply the *vector potential.* Now

$$\text{curl } \mathbf{A} = \text{grad } (\textit{div } \mathbf{A}) - \nabla^2 \mathbf{A}$$

Making use of (8), (9) and (10),

$$\nabla^2 \mathbf{A} = -4\pi \mathbf{J}. \quad \text{...(11)}$$

This is a vector equation, each component of which resembles Poisson's equation. If $\mathbf{A} \to \mathbf{0}$ at infinity, we have from (4) of § 9.5,

$$\mathbf{A} = i\int \frac{\mathbf{J}dv}{r}. \qquad ...(12)$$

This takes a simple form, if currents flow in a thin wire; for then $\mathbf{J}dv$ may be replaced by $i\mathbf{ds}$, so that

$$\mathbf{A} = \int \frac{\mathbf{ds}}{r}. \qquad ...(13)$$

Each element of the wire may be said to make a contribution $i\dfrac{\mathbf{ds}}{r}$ to $\mathbf{A}$. Now

$$\text{curl } \frac{i\mathbf{ds}}{r} = \frac{i}{r}\text{ curl }\mathbf{ds} + \left(\text{grad } \frac{i}{r}\right) \times \mathbf{ds}.$$

Differentiation here is with respect to the co-ordinates of the point P at which the field is being determined, whose co-ordinates do not appear in $\mathbf{ds}$, so that curl $\mathbf{ds} = \mathbf{0}$. Also if $\mathbf{r}$ is measured from P to $\mathbf{ds}$,

$$\text{grad } \frac{i}{r} = \frac{i}{r^3}\mathbf{r},$$

so that

$$\text{curl } \frac{i\mathbf{ds}}{r} = \frac{i}{r^3}\mathbf{r} \times \mathbf{ds} \qquad ...(14)$$

From (9), (13) and (14)

$$\mathbf{H} = i\int \frac{\mathbf{r} \times \mathbf{ds}}{r^3} \qquad ...(15)$$

This is known as *Biot-Savart's* Law. They had established it for a straight line current only. In that case

$$\text{H} = i\int \frac{\sin\theta}{r^2}\,dx, \qquad ...(16)$$

and for an infinite straight wire

$$\text{H} = i\int_{-\infty}^{\infty} \frac{\text{R}}{(\text{R}^2 + x^2)^{3/2}}\,dx$$

$$= \frac{2i}{\text{R}}. \qquad ...(17)$$

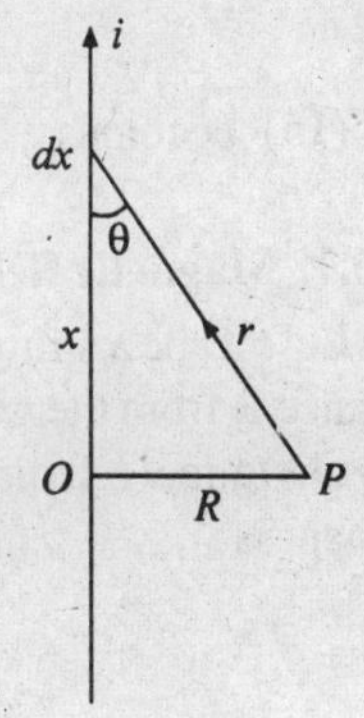

Fig. 9.21

An alternative approach is to take (15) as fundamental and to build up the theory on the Bio-Savart's Law as basis.

From (15)

$$d\mathbf{H} = i\,\frac{\mathbf{r} \times \mathbf{ds}}{r^3}.$$

If the current is flowing in a closed path

$$\mathbf{H} = i \oint \frac{\mathbf{r} \times d\mathbf{s}}{r^3}$$

$$= i \oint \nabla\left(\frac{1}{r}\right) \times d\mathbf{s}$$

$$= \oint \nabla \times \left(\frac{1}{r} d\mathbf{s}\right)$$

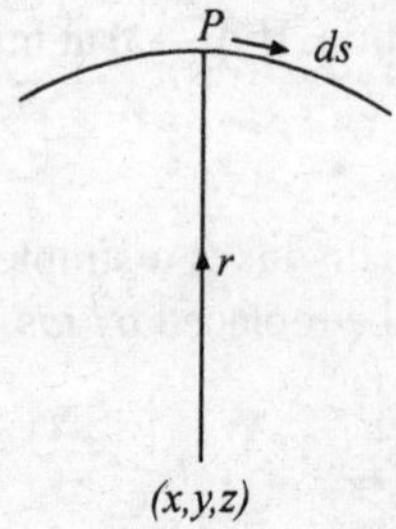

Fig. 9.22

as ∇ does not operate on $d\mathbf{s}$ and i is constant,

$$\therefore \quad \mathbf{H} = \nabla \times \mathbf{A} \quad ...(9)$$

where $\quad \mathbf{A} = \int \frac{i}{r} d\mathbf{s} = \oint \frac{\mathbf{J} dv}{r}, \quad ...(12, 13)$

integration in the last case being taken over all space containing currents.

$$\nabla . \mathbf{A} = \int \nabla . \left(\frac{\mathbf{J}}{r}\right) dv = \int \frac{\mathbf{J}}{r} . d\mathbf{a} \quad ...(18)$$

If all currents lie within a finite sphere, then as in (18), pushing the boundary of the region to infinity, where $\mathbf{J} = \mathbf{0}$, we get

$$\nabla . \mathbf{A} = 0. \quad ...(10)$$

Also $\quad \text{curl } \mathbf{H} = \nabla \times (\nabla \times \mathbf{A}) = \nabla(\nabla . \mathbf{A}) - \nabla^2 \mathbf{A} = -\nabla^2 \mathbf{A}.$

$$\therefore \quad \text{curl } \mathbf{H} = -\nabla^2 \left[\int \frac{\mathbf{J} dv}{r}\right] = 4\pi \mathbf{J} \quad ...(8)$$

For *steady currents in magnetic material,* we use the magnetic induction **B** instead of **H** and define vector potential **A** by

$$\mathbf{B} = \text{curl } \mathbf{A}, \; div \; \mathbf{A} = 0. \quad ...(19)$$

(11) then becomes $\quad \nabla^2 \mathbf{A} = 4\pi\mu \mathbf{J} \quad ...(20)$

(12) becomes $\quad \mathbf{A} = \mu \int \frac{\mathbf{J} dv}{r} \quad ...(21)$

(13) becomes $\quad \mathbf{A} = \mu i \int \frac{d\mathbf{s}}{r} \quad ...(22)$

(15) becomes $\quad \mathbf{B} = \mu i \int \frac{d\mathbf{s} \times \mathbf{r}}{r^3}. \quad ...(23)$

9.13.1. Magnetic field due to the current flowing in a straight wire

Let CD be a straight wire carrying a current I ampere and P the point at a distance R from the wire. Consider a small element *ab* of length $\delta \mathbf{l}$ of the wire at a distance l from M, the magnetic field induction $\delta \mathbf{B}$ due to this element is

$$\delta \mathbf{B} = \frac{\mu I}{4\pi} \frac{\delta \mathbf{l} \times \mathbf{r}}{r^3} \quad ...(1)$$

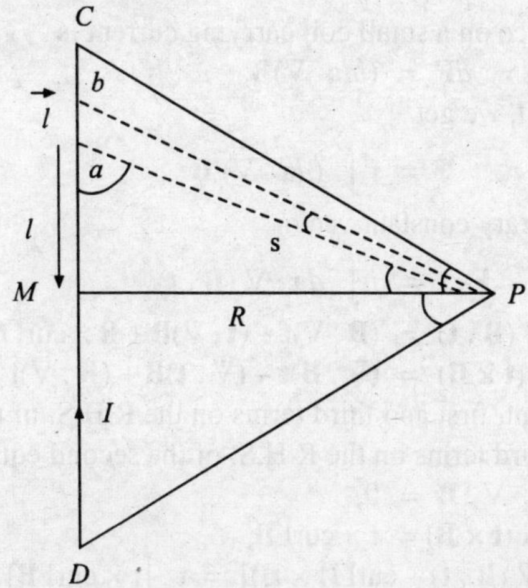

Fig. 9.23

Here **r** is the position vector of point P relative to the element *ab*.

The direction of $\delta\mathbf{B}$ at point P is perpendicular to the plane of the paper nd is directed into the paper.

$$\delta\mathbf{B} = |\delta\mathbf{B}| = \frac{\mu I}{4\pi}\frac{\delta l \sin\theta}{r^2}$$

ve have

$$l = R \tan\phi \Rightarrow \delta l = R \sec^2\phi\, \delta\phi,$$

$$r = \sec\phi \quad \text{and} \quad \sin\theta = \sin(90 - \phi) = \cos\phi$$

$$\therefore \quad \delta B = \frac{\mu I}{4\pi}\frac{(R \sec^2\phi\, \delta\phi)\cos\phi}{(R\sec\phi)^2}$$

$$= \frac{\mu I}{4\pi R}\cos\phi\, \delta\phi$$

Therefore, the total magnetic induction B at point P due to the whole vire is

$$B = \frac{\mu I}{4\pi R}\int_{-\beta}^{\alpha}\cos\phi\, d\phi = \frac{\mu I}{4\pi R}(\sin\alpha + \sin\beta) \qquad ...(2)$$

If the wire is infinitely long, then $\alpha = \pi/2$ and $B = \pi/2$ and

$$B = \frac{\mu I}{4\pi R}.$$

.14. FORCE ON CURRENT IN A MAGNETIC FIELD. AMPERES RULE

From (5) of § 9.6 the force on a magnetic particle of moment **M** in an nhomogeneous field is

$$(\mathbf{M} \cdot \nabla)\, \mathbf{B},$$

and, therefore, force on a small coil carrying current is

$$d\mathbf{F} = (id\mathbf{a} \,.\, \nabla)\mathbf{B},$$

and for a large coil, we get

$$\mathbf{F} = i \int (d\mathbf{a} \,.\, \nabla)\, \mathbf{B}.$$

It **t** is an arbitrary constant vector

$$\mathbf{F} \,.\, \mathbf{t} = i \int d\mathbf{a} \,.\, \nabla\, (\mathbf{B} \,.\, \mathbf{t}) \qquad ...(1)$$

Now $\nabla\, (\mathbf{B} \,.\, \mathbf{t}) = (\mathbf{B} \,.\, \nabla)\mathbf{t} + (\mathbf{t} \,.\, \nabla)\mathbf{B} + \mathbf{B} \times \text{curl}\, \mathbf{t} + \mathbf{t} \times \text{curl}\, \mathbf{B}$ and

$$\text{curl}\, (\mathbf{t} \times \mathbf{B}) = (\nabla \,.\, \mathbf{B})\mathbf{t} - (\nabla \,.\, \mathbf{t})\mathbf{B} + (\mathbf{B} \,.\, \nabla)\mathbf{t} - (\mathbf{t} \,.\, \nabla)\, \mathbf{B}.$$

As **t** is constant, first and third terms on the R.H.S. of the first equation and second and third terms on the R.H.S. of the second equation vanish.

Also, $\nabla \,.\, \mathbf{B} = 0,$

$\nabla(\mathbf{B} \,.\, \mathbf{t}) - \text{curl}\, (\mathbf{t} \times \mathbf{B}) = \mathbf{t} \times \text{curl}\, \mathbf{B},$

$\therefore \quad \mathbf{t} \,.\, [\nabla\, (\mathbf{B} \,.\, \mathbf{t}) - \text{curl}\, (\mathbf{t} \times \mathbf{B})] = \mathbf{t} \,.\, [\mathbf{t} \times \text{curl}\, \mathbf{B}] = 0.$

But **t** is any arbitrary vector.

$$\therefore \quad \nabla\, (\mathbf{B} \,.\, \mathbf{t}) = \text{curl}\, (\mathbf{t} \times \mathbf{B}).$$

$\therefore$ (1) becomes

$$\mathbf{F} \,.\, \mathbf{t} = i \int d\mathbf{a} \,.\, [\text{curl}\, (\mathbf{B} \times \mathbf{t})],$$

$$= i \int d\mathbf{s} \,.\, (\mathbf{B} \times \mathbf{t}), \qquad \text{[By Stokes' Theorem]}$$

$$= i \int (d\mathbf{s} \times \mathbf{B}) \,.\, \mathbf{t}$$

$$\therefore \quad \mathbf{F} = i \int d\mathbf{s} \times \mathbf{B} \qquad ...(2)$$

This shows that the total force on the circuit is the same as if each element $d\mathbf{s}$ experiences a force $id\mathbf{s} \times \mathbf{B}$, which is perpendicular to both **B** and $d\mathbf{s}$. This is known as *Ampere's Rule*. It states that the total force is equivalent to a force $i\mathbf{B} \sin \theta$ per unit length of the wire in the sense and direction given by (2).

If we replace $id\mathbf{s}$ by $\mathbf{J}dv$, we get

$$\mathbf{F} = \int \mathbf{J} \times \mathbf{B} dv, \qquad ...(3)$$

so that for volume currents, the mechanical force is equivalent to a force $\mathbf{J} \times \mathbf{B}$ per unit volume. If we put $\mathbf{J} = \rho\mathbf{V}$ as in (1) of § 111, the force becomes $\rho\mathbf{V} \times \mathbf{B}$ per unit volume in e.m.u., and $\dfrac{\rho}{c}\, (\mathbf{V} \times \mathbf{B})$ in mixed units.

Thus force on a charge e moving with velocity **V** in an electric field **E** and magnetic induction **B** is given by

$$\mathbf{F} = e\left[\mathbf{E} + \frac{1}{c}\, \mathbf{V} \times \mathbf{B}\right]. \qquad ...(4)$$

9.15. CO-EFFICIENT OF MUTUAL INDUCTION

Let us find the mutual potential energy of two coils carrying currents i_1

and i_2. The first can be replaced by a magnetic shell of strength i_1. Let $\mathbf{B}_2$ be the field due to the second, their mutual potential energy is [(4) of § 9.6].

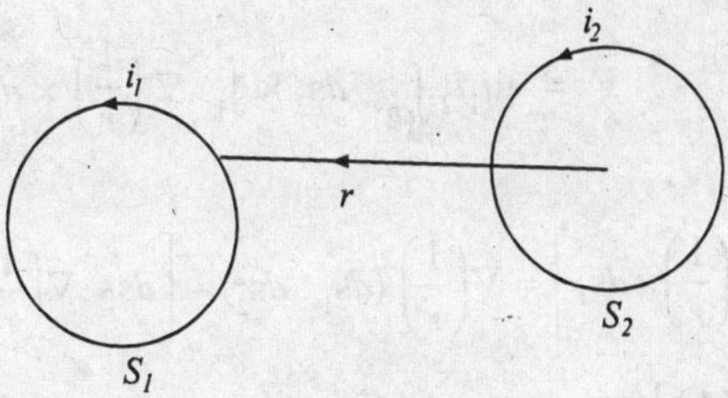

Fig. 9.24

$$\begin{aligned} W &= i_1 \int_{S_1} \mathbf{B}_2 . d\mathbf{a} \\ &= i_1 \int_{S_1} \text{curl } \mathbf{A}_2 . d\mathbf{a} \\ &= -i_1 \int \mathbf{A}_2 . d\mathbf{s}_1 \quad \text{[by using Stokes' theorem]} \\ &= -i_1 i_2 \mu \int_{(2)} \int_{(1)} \frac{d\mathbf{s}_1 . d\mathbf{s}_1}{r} \quad \text{using [(22), § 9.13]} \\ &= -M_{12} i_1 i_2, \qquad ...(1) \end{aligned}$$

where $$M_{12} = \mu \int_{(2)} \int_{(1)} \frac{d\mathbf{s}_1 . d\mathbf{s}_1}{r}. \qquad ...(2)$$

(2) given *Neumann's formula* for M_{12}, called the *co-efficient of mutual induction*.

We have proved,

$$\int_{S_1} \mathbf{B}_2 . d\mathbf{a} = M_{12} i_2,$$

if $$i_2 = 1, M_{12} = \int_{S_1} \mathbf{B}_2 . d\mathbf{a} \qquad ...(3)$$

Therefore, M_{12} can be defined as the flux of **B** through circuit (1) when unit current flows through (2).

The symmetry of (2) shows that $M_{12} = M_{21}$. If (1) and (2) coincide, we get **L**, the co-efficient of self-induction of a circuit which may be defined as the flux of **B** through the circuit when unit current flows through it.

9.16. FORCE BETWEEN TWO COILS

The force on the element $i_2 ds_2$ is

$$d\mathbf{F} = i_2 d\mathbf{s}_2 \times \mathbf{B}_1.$$

But $$\mathbf{B}_1 = \mu i_1 \int_1 \frac{d\mathbf{s}_1 \times \mathbf{r}}{r^3} = -i_1 \mu \int_1 d\mathbf{s}_1 \times \nabla \left(\frac{1}{r}\right)$$

$$\therefore \quad d\mathbf{F} = \mu i_2 i_1 \int_{(1)} \left[d\mathbf{s}_1 \times \nabla\left(\frac{1}{r}\right)\right] \times d\mathbf{s}_2.$$

Integrating over the second circuit.

$$\mathbf{F} = \mu i_1 i_2 \int_{(2)} d\mathbf{s}_2 \times \int_1 \nabla\left(\frac{1}{r}\right) \times d\mathbf{s}_1.$$

Now

$$d\mathbf{s}_2 \times \left[\nabla\left(\frac{1}{r}\right) \times d\mathbf{s}_1\right] = \nabla\left(\frac{1}{r}\right)(d\mathbf{s}_1 \,.\, d\mathbf{s}_2) - \left[d\mathbf{s}_2 \,.\, \nabla\left(\frac{1}{r}\right)\right] d\mathbf{s}_1,$$

and $\int_{(2)} \left[d\mathbf{s}_2 \,.\, \nabla\left(\frac{1}{r}\right)\right] d\mathbf{s}_1 = \int_{(2)} d\left(\frac{1}{r}\right) d\mathbf{s}_1 = 0,$

as $d\mathbf{s}_1$ is constant for integration round (2) and $\frac{1}{r}$ is single valued.

$$\tau = \mu i_1 i_2 \int_{(2)} \int_{(1)} \nabla\left(\frac{1}{r}\right) d\mathbf{s}_1 \,.\, d\mathbf{s}_2 \qquad \text{...(1)}$$

This is the force of loop (1) on loop (2). It is equal and opposite to the force of loop (2) on loop (1) as

$$\nabla\left(\frac{1}{r_{12}}\right) = -\nabla\left(\frac{1}{r_{21}}\right).$$

The force may be looked upon as due to a force of attraction $\mu i_1 i_2 \cos \varepsilon \; ds_1 ds_2 / r^2$ between two elements of the circuits of lengths ds_1 and ds_2 and mutual inclination ε.

In mixed units, we have to divide the R.H.S. of (1) by c.

As an example, consider the force between long thin parallel wires at a distance R apart in vacuum. This is

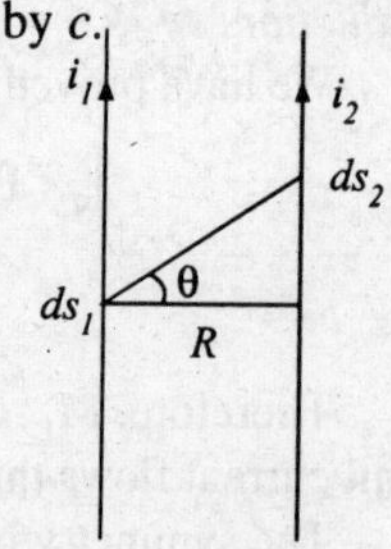

Fig. 9.25

$$= i_1 i_2 ds_1 \int_{-\frac{\pi}{2}}^{\frac{\pi}{2}} \frac{d(\mathrm{R} \tan \theta)}{\mathrm{R}^2 \sec^2 \theta} \cos \theta$$

$$= \frac{2 i_1 i_2 \, ds_1}{\mathrm{R}}$$

i.e., force per unit length is $\frac{2 i_1 i_2}{\mathrm{R}}$. The force is of attraction or repulsion according as the currents are in the same or in opposite directions.

9.17. ELECTROMAGNETIC INDUCTION

Faraday observed that a change in the magnetic field produces an electromotive force in a circuit. If **B** is magnetic induction, the flux through a surface S bounded by a curve Γ is

$$\int \mathbf{B}.d\mathbf{a}$$

If **E** is the electric field at any point, the law of electromagnetic induction states that

$$\frac{\partial}{\partial t}\int \mathbf{B}.d\mathbf{a} = -\int \mathbf{E}.d\mathbf{s}$$

By Stokes' theorems $\int \left(\text{curl}\,\mathbf{E} + \frac{\partial \mathbf{B}}{\partial t}\right).\, d\mathbf{a} = 0.$

This is possible if

$$\text{curl}\,\mathbf{E} = -\frac{\partial \mathbf{B}}{\partial t} \qquad ...(1)$$

This is a generalisation of

$$\nabla \times \mathbf{E} = 0,$$

which is true

(*i*) in an electrostatic field where $\mathbf{B} = \mathbf{0}$.

(*ii*) in steady state when $\frac{\partial \mathbf{B}}{\partial t} = \mathbf{0}$.

9.18. A CONDUCTING ROD MOVING THROUGH A UNIFORM MAGNETIC FIELD

Let *ab* be a thin conducting rod of length *l* moving with velocity **v** parallel to the direction of Y-axis in a uniform field of magnetic induction **B**. The force acting on any charge *q* of the conducting rod is given by

$$\mathbf{F} = q\,(\mathbf{v} \times \mathbf{B}) \qquad ...(1)$$

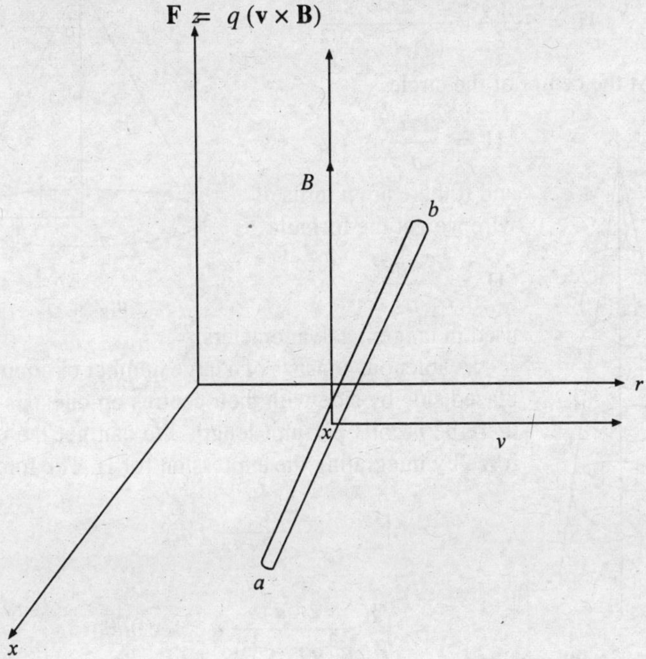

Fig. 9.26

The direction of this force **F** is perpendicular to both the directions of velocity **v** and magnetic induction **B**. If the charge q is positive, the direction of the force is along positive X-axis and if the charge q is negative, the direction of the force is along negative X-axis.

In the **steady state** when the conductor moves with a constant velocity **v**, there must be developed an equal and opposite force at every point inside the rod to balance the magnetic force given by (1). The magnetic force **F** pushes the negatively charged electrons towards one end of the rod and the other end becomes positively charged. This process of accumulation of charges establishes an electric field **E** such that

$$q\mathbf{E} + q(v \times \mathbf{B}) = 0$$

$$\Rightarrow \qquad \mathbf{E} = -\mathbf{v} \times \mathbf{B}$$

9.19. SOLENOIDS

The potential due to a circular current at a point on its axis

$$= i\omega$$

$$= i2\pi(1 - \cos a)$$

$$= i2\pi\left(1 - \frac{z}{\sqrt{a^2 + z^2}}\right).$$

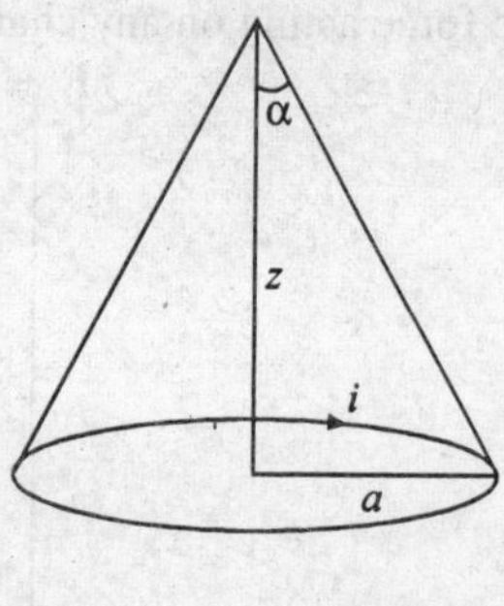

Fig. 9.27

Magnetic force

$$H = -\frac{\partial \phi}{\partial z} = \frac{2\pi i a^2}{(a^2 + z^2)^{3/2}}$$

At the centre of the circle

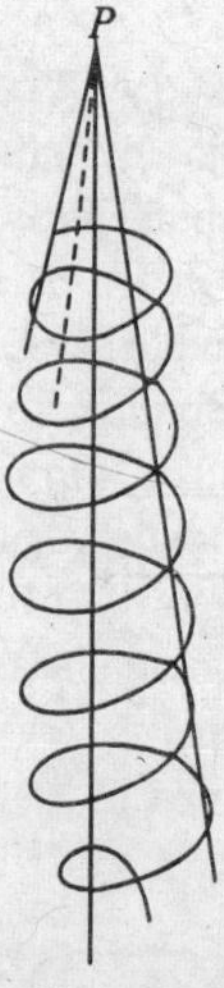

Fig. 9.28

$$H = \frac{2\pi i}{a},$$

and if there are n turns of wire we get the formula

$$H = \frac{2\pi i n}{a},$$

used in tangent galvanometers.

A solenoid consists of a large number of circular coils placed side by side with their centres on one axis OP. Let there be n coils per unit length. We can get the resultant force by integrating the expression for H. The force

$$= \int_{z_1}^{z_2} \frac{2\pi i a^2}{(a^2 + z^2)^{3/2}}\, n dz,$$

$$= \int_{\alpha}^{\beta} \frac{-2\pi i a^2 n}{a^3 \operatorname{cosec}^3 \theta}\, a \operatorname{cosec} \theta d\theta,$$

where $z = a \cot \theta$.

$\therefore$ $H = 2\pi in(\cos \beta - \cos \alpha)$.

If the solenoid is very long, we may put

$$x = \pi, \qquad \beta = 0,$$

so that $H = 4\pi in$,

gives the field strength at a point inside a long solenoid and its direction is parallel to the axis.

EXAMPLES

1. *Use Biot Savart law to obtain an expression for the magnetic field at the centre of a coil in the form of a square of side 2d* carrying a current *I.*

(Bombay 1996, Rohilkhed 1998)

P be the centre of a square ABCD carrying a current I. Then by eqn. (2) § 9.13.1, the magnetic field at the centre P due to the side CD

$$= \frac{\mu I}{4\pi d}(\sin 45° + \sin 45°) = \frac{\sqrt{2}\,\mu I}{4\pi d}$$

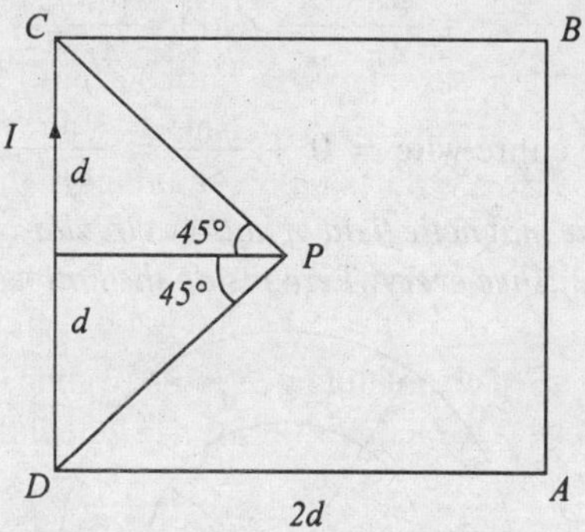

Fig. 9.29

By symmetry the magnetic field at P due to each side will be the same

$$\therefore \qquad B = 4 \times \frac{\sqrt{2}\,\mu I}{4\pi d} = \frac{\sqrt{2}\,\mu I}{\pi d}.$$

2. *A long wire having a semi-circular loop of radius r earries a current i. Find the magnetic induction at the centre O due to entire wire.*

Magnetic induction at a point due to a current carrying element $i\,\delta\mathbf{l}$ is

$$\delta\mathbf{B} = \frac{\mu_0 i}{4\pi} = \frac{\delta\mathbf{l} \times \mathbf{r}}{r^3}$$

$$= \frac{\mu_0 i}{4\pi} = \frac{\delta l \sin\theta}{r^2}.$$

Field due to lnear portion:

If we consider any element of length $\delta\mathbf{l}$ of either linear portion *ab* or *de*, the angle between $\delta\mathbf{l}$ and $\mathbf{r}$ is 0 or π.

$\therefore$ Magnetic induction due to both linear portions of wire = 0.

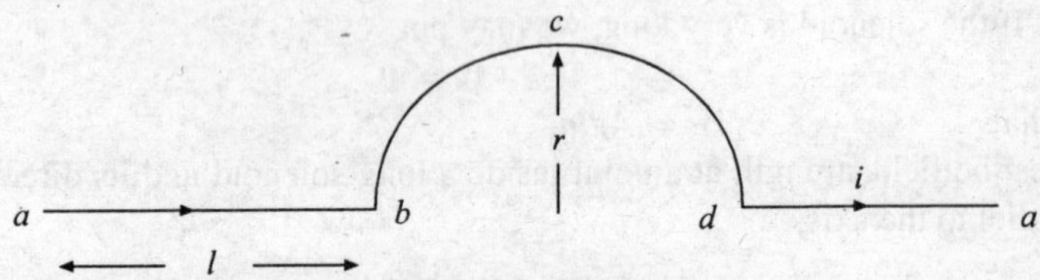

Fig. 9.30

Field due to semi-circular arc:

Magnitude of **B** at O, due to small element

$$\delta B = \frac{\mu_0}{4\pi}\frac{i\,\delta l \sin \pi/2}{r^2} = \frac{\mu_0}{4\pi}\frac{i\delta l}{r^2}$$

$$\therefore \quad \mathbf{B} = \sum \delta B = \sum \frac{\mu_0}{4\pi}\frac{i\delta l}{r^2} = \frac{\mu_0}{4\pi}, \frac{i}{r^2}\sum \delta l$$

$$= \frac{\mu_0}{4\pi}\cdot\frac{i}{r^2}\cdot(\pi r) = \frac{\mu_0 i}{4r}.$$

$$\therefore \text{ Field due to entire wire } = 0 + \frac{\mu_0 i}{4r} = \frac{\mu_0 i}{4r}$$

3. *Show that the magnetic field of hollow circular cylindrical conductor carrying current vanishes everywhere inside the hollow region.*

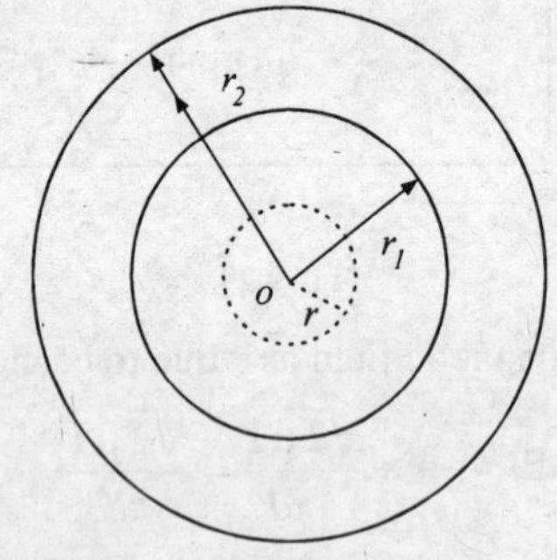

Fig. 9.31

The axis of the cylindrical conductor passes through the point O. Let the internal and external radii of the cylindrical conductor be r_1 and r_2 respectively. The magnetic field produced due to the current flowing in the conductor at any point will be tangential and equal in magnitude everywhere.

$$\therefore \quad \oint \mathbf{B}\,.\,d\mathbf{l} = \text{B}\,.\,2\pi r$$

But $\oint \mathbf{B}\,.\,d\mathbf{l} = \mu \times$ current enclosed by the circular path of radius r)

$$= \mu \times 0$$

$$\Rightarrow \quad B = 0.$$

Exercises

1. Prove that at a change of medium, B_n and H_s are continuous, but if a current i per unit width flows along the boundary, the tangential component in a direction perpendicular to the current satisfies the equation

$$H_{s_1} - H_{s_2} = 4\pi i.$$

2. Show that if the line of **B** are refracted at a change of medium and if the angles with the normal are θ_1, θ_2, then

$$\mu_1 \cot \theta_1 = \mu_2 \cot \theta_3.$$

3. Show that the effect of magnetisation **I** in a certain volume is the same as that of a volume distribution of pole strength $\rho = -\, div\, \mathbf{I}$ per unit volume and a surface distribution of pole strength $\sigma = \mathbf{n} \,.\, \mathbf{I}$ per unit area (Poisson's imaginary matter).

4. Show that the potential at a point P due to a uniformly magnetised body is

$$\Omega = \mathbf{I} . \int \frac{\mathbf{I}}{r}\, ds = \int \text{div}\, \frac{\mathbf{I}}{r}\, dv = \mathbf{I} . \int \nabla\left(\frac{1}{r}\right) dv,$$

where r is the distance from P to element dv.

Hence prove that the potential due to a uniformly magnetised solid sphere of radius a is the same as that due to a short magnet at its centre, whose moment is $\frac{4}{3}\pi a^3 \mathbf{I}$.

5. Deduce from Question (3) that for a straight bar magnet uniformly magnetised in the direction of its length, $\rho = 0$, $\sigma = 0$ over the sides and $\sigma = \mathbf{I}$ at flat ends.

6. Show that the action of a non-homogeneous field **H** on a body magnetised to intensity **I** is equivalent to a force

$$\mathbf{F} = . \int \mathbf{I} . \nabla \mathbf{H} dv$$

through the origin and a couple

$$\mathbf{L} = \int [\mathbf{r} \times (\mathbf{I} . \nabla \mathbf{H}) + \mathbf{I} \times \mathbf{H}]\, dv.$$

7. Show that in vacuum the vector potential **A** has the following value

(*i*) for a small magnetic dipole of moment **M**,

$$\mathbf{A} = \frac{\mathbf{M} \times \mathbf{r}}{r^2}$$

where **r** is measured from dipole to P,

(*ii*) In the field due to a long straight current i parallel to unit vector **a**,

$$\mathbf{A} = -2i \log r\, \mathbf{a}$$

where r is the perpendicular distance from the wire.

(*iii*) In the field due to a current a flowing in a circular wire $r = a, z = 0$ in cylindrical co-ordinates, the vector potential at r, θ, z is given by

$$\mathbf{A}_r = \mathbf{A}_z = \theta$$

$$\mathbf{A}_\theta = 2ia \int_0^\pi \frac{\cos\theta \, d\theta}{(a^2 + r^2 + z^2 - 2ar\cos\theta)^{\frac{1}{2}}}.$$

If the loop is small, show that this is approximately

$$\frac{\pi i a^2 r}{(r^2 + z^2)^{3/2}}.$$

(*iv*) For a uniform field of intensity $\mathrm{H}_\mathbf{k}$

$$\mathbf{A} = \frac{1}{2} \mathrm{H}(x\mathbf{j} - y\mathbf{i}).$$

(*v*) In the field due to a body whose intensity of magastisation is **I**,

$$\mathbf{A} = \int \nabla\left(\frac{\mathbf{I}}{r}\right) \times \mathbf{I} dv$$

8. A current i flows round a circular wire of radius a and a current i' flows in a long straight wire in the same plane as the circle and at a distance c from the centre. Show that co-efficient of mutual induction,

$$\mathrm{M}_{12} = 4 \int_{-a}^{a} \frac{\sqrt{a^2 - x^2}}{c + x} dx = 4n\left[c - \sqrt{c^2 - a^2}\right]$$

and that the force between the two circuits is

$$4\pi i i' \left\{\frac{c}{\sqrt{c^2 - a^2}} - 1\right\}.$$

9. Two circular circuits are planes at right angles to the line joining their centres. Show that the co-efficient of induction is

$$2\pi(a^2 - c^2) \int_0^{\frac{1}{2}\pi} \frac{\cos 2\theta \, d\theta}{\sqrt{a^2 \sin^2\theta + c^2 \cos^2\theta}}$$

where a, c are the longest and shortest lines which can be drawn from one circuit to other. Find the force between the circuits.

Deduce that if the radii of the two circuits are a each and the distance f between them is large,

$$\mathrm{M}_{12} = \frac{2\pi^2 a^4}{f^3}.$$

Show that if unit currents flow in the same direction round the coils, they attract each other with a force $\frac{6\pi^2 a^4}{f^4}$.

10. Show that the magnetic field at the centre of a square coil of side $2a$ carrying current i is $\frac{4i\sqrt{2}}{a}$.

11. A pair of circular coils of radii, a, b and of m, and n turns respectively carry the same current in the same sense and are placed parallel on a common axis at a distance $\frac{1}{2}(a+b)$ apart. Find the magnetic force they produce at a point on the axis at a distance $\frac{1}{2}a$, $\frac{1}{2}b$ from the coil and show that on the axis near the point, the field is so uniform that its first two differential co-efficients vanish at the point if $\frac{m}{a^2}=\frac{n}{b^2}$ and the third in addition if $m = n$, $a = b$.

12. Show that the force between two circuits can be expressed as

$$\mathbf{F} = \int d\mathbf{s}_2 \times (\nabla \times \mathbf{B}_1),$$

where integration extends over any surface spinning the second circuit.

13. Verify that $\frac{e}{\sqrt{\kappa\mu}}$ has dimensions of velocity.

14. An electron of mass m and charge e is projected in any direction in a uniform magnetic field $\mathbf{H}$ parallel to z-axis. Show that the equations of motion are

$$m\ddot{x} = e\mathrm{H}\dot{y},\ m\ddot{y} = -e\mathrm{H}\dot{x},\ m\ddot{x} = 0.$$

Deduce that in general the path of the electron is a helix described at constant speed.

If the electron moves between the plates of a parallel plate condenser producing a constant field $\mathbf{H}$, show how the equations are modified and integrate them, when

$$x = y = \dot{x} = \dot{y} = 0 \text{ at time } t = 0.$$

15. Explain why equation curl $\mathbf{H} = 4\pi\mathbf{J}$ is not affected by the presence of magnetisable matter.

9.20. MAXWELL'S EQUATIONS

For electrostatic field and steady currents, we have already established the equations

$$\nabla \times \mathbf{E} = \mathbf{0}, \quad \text{...(1)}$$

$$\nabla . \mathbf{D} = 4\pi\rho, \quad \text{...(2)}$$

$$\nabla . \mathbf{B} = 0, \quad \text{...(3)}$$

$$\nabla \times \mathbf{H} = 4\pi\mathbf{J}, \quad \text{...(4)}$$

$$\nabla . \mathbf{J} = 0. \quad \text{...(5)}$$

For the non-steady state, (1) has already been generalised in § 9.16 to

$$\nabla \times \mathbf{E} = -\frac{\partial \mathbf{B}}{\partial t}, \qquad ...(6)$$

and from (3) of § 9.12, (5) is generalised to

$$\nabla . \mathbf{J} = -\frac{\partial \rho}{\partial t}, \qquad ...(7)$$

(4) and (7) are inconsistent; for (4) implies

$$\nabla . \mathbf{J} = \frac{1}{4\pi} \nabla . (\nabla \times \mathbf{H}) = 0.$$

We retain (7) and try to modify (4). From (2) and (7)

$$\nabla \mathbf{J} = -\frac{\partial}{\partial t}\left(\frac{1}{4\pi} \nabla . \mathbf{D}\right) = -\nabla . \left(\frac{\partial}{\partial t}\frac{\mathbf{D}}{4\pi}\right),$$

or $$\nabla . \left(\mathbf{J} + \frac{1}{4\pi}\frac{\partial \mathbf{D}}{\partial t}\right) = 0 \qquad ...(8)$$

We generalise (4) to

$$\nabla \times \mathbf{H} = 4\pi \left(\mathbf{J} + \frac{1}{4\pi}\frac{\partial \mathbf{D}}{\partial t}\right), \qquad ...(9)$$

which reduces to (4) in the steady state. $\frac{1}{4\pi}\frac{\partial \mathbf{D}}{\partial t}$ is called the *Displacement Current.* Using mixed or Gaussian units in which we measure **D, E,** ρ in e.s.u. and **B, H, J** in e.m.u., we can write the fundamental equations of electromagnetic theory, known as Maxwell's equations, as,

$$\text{div } \mathbf{D} = 4\pi\rho, \qquad ...(2)$$

$$\text{div } \mathbf{B} = 0, \qquad ...(3)$$

$$\text{curl } \mathbf{H} = 4\pi \mathbf{J} + \frac{1}{c}\frac{\partial \mathbf{D}}{\partial t}, \qquad ...(9')$$

$$\text{curl } \mathbf{E} = -\frac{1}{c}\frac{\partial \mathbf{B}}{\partial t}. \qquad ...(6')$$

We supplement these by the constitutive relations

$$\mathbf{D} = \kappa \mathbf{E}, \qquad ...(10)$$

$$\mathbf{B} = \mu \mathbf{H}, \qquad ...(11)$$

$$\mathbf{J} = \sigma \mathbf{E}.$$

We assume, hereafter that κ, μ, σ are constants.

Since $\mathbf{B} = \text{curl } \mathbf{A}$

(6′) gives $$\text{curl } \mathbf{E} = -\frac{1}{c}\frac{\partial}{\partial t}(\text{curl A})$$

or $$\text{curl}\left(\mathbf{E} + \frac{1}{c}\frac{\partial \mathbf{A}}{\partial t}\right) = 0$$

or $$\mathrm{E}+\frac{1}{c}\frac{\partial \mathbf{A}}{\partial t} = -\text{grad}\ \phi. \qquad ...(13)$$

In steady and in electrostatic case, it reduces to

$$\mathbf{E} = -\text{grad}\ \phi$$

and thus ϕ is the analogue of the electrostatic potential. From (9′), using (3), (10) and (12)

$$\text{curl}\left(\frac{\mathbf{B}}{\mu}\right) = 4\pi\mathbf{J}+\frac{\kappa}{c}\frac{\partial}{\partial t}$$

$$= 4\pi\mathbf{J}+\frac{\kappa}{c}\left(-\text{grad}\ \frac{\partial \phi}{\partial t}-\frac{1}{c}\frac{\partial^2 \mathbf{A}}{\partial t^2}\right)$$

or $$\frac{1}{\mu}\nabla\times(\nabla\times\mathbf{A}) = 4\pi\mathbf{J}-\frac{\kappa}{c}\left(\text{grad}\ \frac{\partial \phi}{\partial t}+\frac{1}{c}\frac{\partial^2 \mathbf{A}}{\partial t^2}\right)$$

or $$\nabla(\nabla\ .\ \mathbf{A})-\nabla^2\mathbf{A} = 4\pi\mathbf{J}-\frac{\mu\kappa}{c}\left(\text{grad}\ \frac{\partial \phi}{\partial t}+\frac{1}{c}\frac{\partial^2 \mathbf{A}}{\partial t^2}\right)$$

or $$\nabla^2\mathbf{A} = \frac{\mu\kappa}{c^2}\frac{\partial^2 \mathbf{A}}{\partial t^2}-4\pi\mu\mathbf{J}$$

$$+\ \text{grad}\left(\text{div}\,\mathbf{A}+\frac{\mu\kappa}{c^2}\frac{\partial \phi}{\partial t}\right) \qquad ...(14)$$

We impose on **A** the condition

$$\text{div}\ \mathbf{A}+\frac{\mu\kappa}{c^2}\frac{\partial \phi}{\partial t} = 0. \qquad ...(15)$$

which reduces to (19) § 9.13 in the steady case. (14) now becomes

$$\nabla^2\mathbf{A} = \frac{\mu\kappa}{c^2}\frac{\partial^2 \mathbf{A}}{\partial t^2}-4\pi\mu\mathbf{J}. \qquad ...(16)$$

Taking divergence of both sides of (12)

$$\text{div (grad}\ \phi) = -\left[div\,\mathbf{E}+\frac{1}{e}\frac{\partial}{\partial t}(\text{div}\,\mathbf{A})\right].$$

or $$\nabla^2\phi = -\left[\frac{4\pi\rho}{\mathrm{K}}+\frac{1}{c}\frac{\partial}{\partial t}\left(-\frac{\mu\kappa}{c}\frac{\partial \phi}{\partial t}\right)\right],$$

or $$\nabla^2\phi = \frac{\mu\kappa}{c^2}\frac{\partial^2 \phi}{\partial t^2}-\frac{4\pi\rho}{\mathrm{K}} \qquad ...(17)$$

9.21. ELECTROMAGNETIC WAVES

In a non-conducting medium, where $\sigma = 0$, $\mathbf{J} = \mathbf{0}$ and in which there is no residual charge ($\rho = 0$) (16) and (17) of the last section give

$$\nabla^2 \mathbf{A} = \frac{\mu\kappa}{c^2}\frac{\partial^2 \mathbf{A}}{\partial t^2}, \qquad \text{...(1)}$$

$$\nabla^2 \phi = \frac{\mu\kappa}{c^2}\frac{\partial^2 \phi}{\partial t^2}, \qquad \text{...(2)}$$

we thus see that the magnetic potential **A** and the electric potential ϕ both satisfy the standard equation of wave motion. We can easily see that in this case ($\rho = 0$, $\mathbf{J} = 0$, $\sigma = 0$) both **H** and **E** also satisfy the wave equation, for Maxwells equations reduce to

$$\text{div } \mathbf{E} = 0, \text{ div } \mathbf{H} = 0, \text{ curl } \mathbf{E} = -\frac{\mu}{c}\frac{\partial \mathbf{H}}{\partial t}, \text{ curl } \mathbf{H} = \frac{\kappa}{c}\frac{\partial \mathbf{E}}{\partial t}. \qquad \text{...(3)}$$

Now
$$\nabla^2 \mathbf{H} = \text{grad (div } \mathbf{H}) - \text{curl curl } \mathbf{H}$$
$$= -\text{curl (curl } \mathbf{H})$$
$$= -\text{curl}\left(\frac{\kappa}{c}\frac{\partial \mathbf{E}}{\partial t}\right)$$
$$= -\frac{\kappa}{c}\frac{\partial}{\partial t}(\text{curl } \mathbf{E}) = \frac{\kappa}{c}\frac{\partial}{\partial t}\left(\frac{\mu}{c}\frac{\partial \mathbf{H}}{\partial t}\right),$$

or
$$\nabla^2 \mathbf{H} = \frac{\kappa\mu}{c^2}\frac{\partial^2 \mathbf{H}}{\partial t^2}, \qquad \text{...(4)}$$

Similarly
$$\nabla^2 \mathbf{E} = \text{grad } (div\ \mathbf{E}) - \text{curl curl } \mathbf{E}$$
$$= -\text{curl}\left(-\frac{\mu}{c}\frac{\partial \mathbf{H}}{\partial t}\right) = \frac{\mu}{c}\frac{\partial}{\partial t}\text{ curl } \mathbf{H}.$$

or
$$\nabla \mathbf{E}^2 = \frac{\mu\kappa}{c^2}\frac{\partial^2 \mathbf{E}}{\partial t^2}. \qquad \text{...(5)}$$

From (1), (2), (4), (5) we deduce that waves can be propagated in a medium where μ, κ are constants and ρ, σ, **J** vanish, with velocity $\frac{c}{\sqrt{\mu\kappa}}$. In free space where $\kappa = 1$, $\mu = 1$ the velocity is just c. But we have defined c as the ratio of the two sets of units and its experimental value is approximately 2.998×10^{10} cms, per second. But this is precisely the velocity of light in vacuum. This suggests that light waves are electromagnetic in nature. This has subsequently received complete verification. Infrared, ultra violet, X-rays, γ-rays, wireless waves are of this nature and differ only in their wave lengths. For visible light in a non-concluding medium, we may put $\mu = 1$ and then velocity of light is $\frac{e}{\sqrt{K}}$. It is known that this velocity $= \frac{e}{n}$ where n is refractive index of the medium. We deduce Maxwell's relation

$$\kappa = n^2.$$

Exercises

1. Prove curl $\mathbf{H} = 4\pi\rho\mathbf{E} = \frac{\kappa}{c}\frac{\partial \mathbf{E}}{\partial t}$.

Deduce $\frac{\kappa}{c}\frac{\partial \rho}{\partial t} + 4\pi\sigma\rho = 0$, $\rho = \rho_0\, e^{-1}\, \frac{4\pi\sigma c}{\kappa}$.

2. When $\sigma \neq 0$, prove the equation of telegraphy

$$\nabla^2\mathbf{E} = \frac{\kappa\mu}{c^2}\frac{\partial^2\mathbf{E}}{\partial t^2} + \frac{4\pi\sigma\mu}{c}\frac{\partial \mathbf{E}}{\partial t}$$

and establish a similar equation for **H**.

3. If $\mathbf{B} = \text{curl}\,\mathbf{A}'$, $\mathbf{E} = -\frac{1}{c}\frac{\partial \mathbf{A}'}{\partial t}$ grad ϕ'.

$$\text{div}\,\mathbf{A}' = 4\pi\sigma' - \frac{\kappa\mu}{c}\frac{\partial \phi'}{\partial t}$$

show that if $\rho = 0$, $\mathbf{A}'$and ϕ satisfy the equation

$$\nabla^2 = \frac{4\pi\sigma\mu}{c}\frac{\partial}{\partial t} + \frac{\kappa\mu}{c^2}\frac{\partial^2}{\partial t^2}.$$

4. (*i*) *Prove div* $\left(\frac{\partial \mathbf{B}}{\partial t}\right) = 0$.

(*ii*) In § 111 deduce (7) from (2) and (9′).

5. *Verify that*

$$\mathbf{E} = a\cos p\left(t - \frac{x}{V}\right)\mathbf{i},\ \mathbf{H} = \sqrt{\frac{K}{\mu}}\ a\cos p\left(t - \frac{z}{V}\right)\mathbf{j}$$

in which $V = \frac{c}{\sqrt{K\mu}}$ and a and p are constants satisfy (6) § 118 and (3), (4), (5) § 119.

Interpret the result as a plane wave of freuency $\frac{p}{2\pi}$ travelling in the z-direction with velocity V. Verify also that the electromagnetic waves are transverse by showing that the vectors **E, H** and the direction of propagation form a right handed set. Also show that the protential **A** and ϕ which describe the plane wave are

$$\phi = -ax\cos p\left(t - \frac{z}{V}\right),\ \mathbf{A} = -\frac{acx}{V}\cos p\left(t - \frac{z}{V}\kappa\right).$$

6. Let

$$\mathbf{H} = H_1\mathbf{i} + H_2\mathbf{J} + H_3\mathbf{K},\ \mathbf{E} = E_1\mathbf{i} + E_2\mathbf{j} + E_3\mathbf{K}$$

Let H_i, E_i be functions of u, where $u = x - t\text{V}$, then, in the case when $\sigma = 0$, the equations

$$\text{curl } \mathbf{H} = \frac{\kappa}{c}\frac{\partial \mathbf{E}}{\partial t}, \text{curl } \mathbf{E} = -\frac{\mu}{c}\frac{\partial \mathbf{H}}{\partial t},$$

become

$$-\frac{\partial H_2}{\partial u}\mathbf{j} + \frac{\partial H_2}{\partial u}\mathbf{K} = \frac{\kappa}{c}\left(\frac{\partial E_1}{\partial u}\mathbf{i} + \frac{\partial E_2}{\partial u}\mathbf{j} + \frac{\partial E_3}{\partial u}\mathbf{K}\right)(-\text{V})$$

$$-\frac{\partial E_3}{\partial u}\mathbf{j} + \frac{\partial E_2}{\partial u}\mathbf{K} = -\frac{\kappa}{c}\left(\frac{\partial H_1}{\partial u}\mathbf{i} + \frac{\partial H_2}{\partial u}\mathbf{j} + \frac{\partial H_3}{\partial u}\mathbf{K}\right)(-\text{V})$$

Deduce $\mathbf{E}_1 = \mathbf{0}$, $\mathbf{H}_1 = \mathbf{0}$, $\text{V} = \frac{c}{\sqrt{\mu\kappa}}$. ...(5)

Show also that,

$$\mathbf{i} \times \mathbf{H} = -\frac{\text{V}_k}{c}\mathbf{E}$$

$$\mathbf{i} \times \mathbf{E} = \frac{\mu \text{V}}{c}\mathbf{H}$$

$$\mathbf{i}\,.\,\mathbf{H} = 0, \ \mathbf{i.E} = 0, \ \mathbf{E.H} = 0.$$

Deduce that **E, H, n** form a right handed system and the waves are transverse.

7. Write down some particular solutions of **E** and **E** in the last question. Verify that they satisfy the corresponding equations and find the corresponding values of ϕ and **A**.

OBJECTIVE QUESTIONS

For each of the following questions, four alternatives are given for the answer. Only one of them is correct. Choose the correct alternative.

1. For a single point charge, the equipotential surfaces are

(*a*) Parallel planes (*b*) Concentric spheres

(*c*) Confocal Ellipsoids (*d*) None of these

2. The flux of **E** out of the surface of a sphere of radius *a*, at the centre of which is placed a charge *e*, is

(*a*) E*e* (*b*) E*ea* (*c*) 4πE (*d*) 4π*e*

3. The total solid angle subtended by a closed surface at a point on the surface is

(*a*) 4π (*b*) 2π (*c*) zero (*d*) π

4. The solid angle subtended by a line on a sphere of dihedral angle α at its centre is

(*a*) 2α (*b*) 4α (*c*) 6α (*d*) 8α

5. The strength of a tube of force along its path

(*a*) is zero (*b*) is infinite

(*c*) remains constant (*d*) None of these

6. A charged conductor experiences a mechanical force $2\pi\sigma^2$ per unit area along

(*a*) the tangent plane to the surface
(*b*) the outward normal to the surface
(*c*) the inward normal to the surface
(*d*) any direction

7. The potential energy of the dipole of strength **M** in a field intensity **E** is

(*a*) **M . E** (*b*) ME (*c*) – **M . E** (*d*) |**M** × **E**|

8. The electric field strength at any point outside a spherical charge distribution as if the whole charge were

(*a*) Distributed at the surface
(*b*) Distributed uniformly inside the sphere
(*c*) Concentrated at the centre
(*d*) Distributed round the centre

9. For polar atom the potential is mainly due to

(*a*) monopole moment of the charge distribution
(*b*) dipole moment of the charge distribution
(*c*) quadrupole moment of charge distribution
(*d*) multipole moments of charge distribution

10. The induced field in the dielectric is

(*a*) opposite in direction to the external applied field
(*b*) in the direction of the external applied field
(*c*) perpendicular to the external applied field
(*d*) not concerned with the external applied field

11. Electric displacement is defined in such away that it

(*a*) depends on distribution and magnitude of charges which produce electric field and the nature of the medium
(*b*) depends on the distribution and magnitude of charges which produce electric field but is independent of the nature of the medium
(*c*) is independent of the distribution and magnitude of charges which produce electric field but depends on the nature of the medium
(*d*) is independent of the distribution and magnitude of charges which produce electric field and of the nature of the medium

12. If the current in parallel wires are in the same direction, the force between them

(*a*) is attractive (*b*) is repulsive
(*c*) does not exist (*d*) Neither attractive nor repulsive

13. The divergence of the magnetic induction **B** is

(*a*) constant (*b*) depends on **H**
(*c*) always zero (*d*) None of these

14. The net outward flux of electric displacement vector through the surface enclosing a volume is equal to

(*a*) zero

(*b*) net charge contained within that volume

(*c*) negative rate of change of magnetic flux linked with the path

(*d*) conduction current plus displacement current through any surface bounded by the path

15. Electromagnetic waves

(*a*) are longitude is nature

(*b*) travel with the speed of light

(*c*) field vectors **E** and **H** are in opposite phases

(*d*) the direction of flow of electromagnetic energy is perpendicular to the direction of wave propagation

ANSWERS

1. (*b*)	**2.** (*d*)	**3.** (*b*)	**4.** (*a*)	**5.** (*c*)
6. (*b*)	**7.** (*c*)	**8.** (*c*)	**9.** (*b*)	**10.** (*a*)
11. (*b*)	**12.** (*a*)	**13.** (*c*)	**14.** (*b*)	**15.** (*b*)